Teacher Development for Climate Action:

Climate Change Education

Dumsani Wilfred Mncube
University of South Africa, South Africa

Sikhulile Bonginkosi Msezane
University of South Africa, South Africa

Anriët Van Deventer
University of South Africa, South Africa

IGI Global
Scientific Publishing
Publishing Tomorrow's Research Today

Vice President of Editorial	Melissa Wagner
Director of Acquisitions	Mikaela Felty
Director of Book Development	Jocelynn Hessler
Production Manager	Mike Brehm
Cover Design	Jose Rosado

Published in the United States of America by
IGI Global Scientific Publishing
701 East Chocolate Avenue
Hershey, PA, 17033, USA
Tel: 717-533-8845 | Fax: 717-533-7115
Website: https://www.igi-global.com E-mail: cust@igi-global.com

Library of Congress Cataloging-in-Publication Data

LCCN: 2026011055 (CIP Data Pending)
ISBN13: 979-8-3373-6975-4
Isbn13Softcover: 979-8-3373-6976-1
EISBN13: 979-8-3373-6977-8
British Cataloguing in Publication Data
A Cataloguing in Publication record for this book is available from the British Library.

Editorial Advisory Board

Table of Contents

Detailed Table of Contents

and pre-service teachers during the initial implementation phase, 2022-23, and highlighting some of the challenges and benefits of the framework. In doing this it emphasises the potential for the theoretical framework underpinning the initial teacher education framework as a theory of change model for embedding climate and sustainability education throughout teacher education. This chapter also offers insight into further work the university are engaged in in relation to nature education and carbon dioxide monitoring in schools. Further scrutiny of these approaches in relation the framework alignment with UK Government 'Curriculum and Assessment Review' (DfE, 2025) for England highlights the importance of aligning policy, pedagogy and practice.

Climate change education presents unique challenges that demand interdisciplinary approaches and innovative teaching strategies. This chapter explores how South African teachers can be supported and empowered to integrate climate change into their teaching across multiple subjects, including languages, science, geography, and mathematics. Drawing on national curriculum policies and teacher development programmes, it highlights the importance of connecting classroom learning to real-world environmental issues to foster critical thinking, collaboration, and problem-solving among learners. The chapter argues that empowering teachers requires more than enhancing subject knowledge; it involves building confidence, encouraging teamwork, and equipping educators with practical tools to design meaningful lessons.

In 2015, UN member countries also adopted the 2030 Agenda for Sustainable Development a comprehensive global plan of action for 'people, planet and prosperity' comprising 17 SDGs and 169 targets to be achieved by 2030, including SDG 4 Quality Education and SDG13 on climate action (Nerini et al 2019). This chapter is guided by three research questions: First, What theoretical bases explain the role of teacher education in promoting climate action and societal resilience? Second: Identify the gaps related to teacher education in the context of climate change education through previous research studies. And Third: Creating a model

teaching in this way, my own practice as a teacher educator at a distance university in South Africa.

Climate change is a pressing global challenge, with Africa facing heightened vulnerability due to its dependence on climate-sensitive sectors such as agriculture, water resources, and energy production. Climate Change Education (CCE) plays a critical role in fostering environmental literacy, ethical awareness, and action-oriented learning. Teachers are central to effective CCE implementation; however, evidence suggests that many African educators, despite understanding climate change issues, face limitations in subject knowledge, pedagogy, instructional resources, and continuous professional development. Reflective practice is increasingly recognised as a key mechanism for strengthening teachers' capacities, supporting adaptive teaching, and empowering learners to engage with climate challenges. This chapter highlights the need to integrate experiential, inquiry-based, and action-focused approaches into teacher professional development to promote transformative climate education and sustainable practices in African contexts.

This literature review synthesises the philosophical and conceptual foundations, international practices, and contextual relevance of promoting sustainability through STEM education in Ghana and Africa. It argues that addressing complex socio-environmental challenges such as climate change, resource scarcity, and social inequality requires integrating sustainability principles into STEM education. The review highlights constructivism, inquiry-based learning, design thinking, and systems thinking as key pedagogical approaches that support active, contextualised, and interdisciplinary problem-solving. Evidence from international case studies demonstrates the effectiveness of Phenomenon-Based Learning, Project-Based Learning, and Maker Education in developing sustainability competencies. In Africa, sustainability demands are intensified by limited resources, inadequate teacher capacity, and gender imbalances in STEM, requiring coordinated policy action, curriculum reform, sustained investment, and continuous teacher professional development

Despite Namibia's dependence on the resources harvested off the Benguela Current ecosystems, climate change education in Namibia primarily focuses on land with minimal integration of marine content. This chapter examines barriers and opportunities for integrating climate change education in marine science content among secondary school teacher training programmes in Namibia. The research employs a qualitative, interpretivist approach involving document analysis, interviews, and case studies of existing climate education initiatives. The findings reveal significant gaps in teachers' ocean climate literacy, a lack of resources and limited understanding of the ocean's role in climate regulation. There is an opportunity to introduce context-specific marine learning activities into the secondary school curriculum due to policy frameworks and institutional structures supporting climate change education in Namibia. There is a strong policy and institutional support and growing interest in marine and environmental studies. The Namibia's Education for Sustainable Development (ESD) strategy for 2030 is actively working on a template to integrate climate awareness into formal and non-formal education, including teacher education. The chapter suggests teacher development strategies that can enhance climate teaching and promote marine ecological connections following transformative learning pedagogies. The recommendations focus on curriculum reform, improving teacher competence, and utilising Namibia's unique maritime and coastal settings as starting points for teaching about climate change.

This chapter explores Environmental Change Education (ECE) in the Foundation Phase, framed within the context of escalating global ecological crises and the urgent need for transformative educational responses. It positions environmental degradation—manifested in climate change, biodiversity loss, pollution, and resource scarcity—as both a planetary emergency and a profound educational challenge. It furthermore argues that these crises are largely rooted in unsustainable human practices characteristic of the Anthropocene, and therefore require a fundamental shift in human behaviour that must begin in early childhood education. The chapter conceptualises environmental change not only as an ecological issue but also as a matter of children's rights, highlighting the disproportionate vulnerability of young people to environmental risks. ECE is a multidimensional process that not only transmits scientific knowledge but also fosters environmental awareness and ethical responsibility, critical thinking, and active citizenship. The chapter aligns with the Curriculum and Assessment Policy Statement (CAPS), particularly in the Life Skills curriculum. Although CAPS promotes holistic development and integrated, interdisciplinary learning, teaching is knowledge-driven rather than transformative. The chapter argues that early, integrated, and creative pedagogical approaches are essential for fostering meaningful environmental understanding and long-term behavioural change. These approaches include interdisciplinary teaching and Arts-integrated teaching with the Creative and Performing Arts positioned as a powerful pedagogical vehicle. Arts-based approaches—encompassing music, drama, visual arts, and movement—enable experiential, affective, and imaginative engagement with environmental concepts, making abstract ideas accessible and memorable for young learners. As far as methodology is concerned, the researcher used a qualitative research design within a pragmatic paradigm, which prioritises practical, contextually relevant knowledge. Document analysis and literature review are used to examine curriculum policy, international practices, and pedagogical strategies related to ECE and Arts integration. To support the study's arguments, integrated lesson ideas are provided that show how environmental themes can be infused through Arts-based approaches, which help narrow the gap between theory and practice. The Integrated Curriculum Model (ICM) serves as a theoretical framework, providing a structured lens for organising content, understanding, and thematic integration. When aligned with the Foundation Phase, the ICM emphasises conceptual depth and the development of higher-order thinking while using overarching themes such as environmental sustainability. Along with ICM, other theoretical perspectives—including constructivism, socio-cultural theory, experiential learning, and critical pedagogy—also support the use of Arts integration for fostering active, collaborative, and reflective learning. Various models of curriculum integration are suggested, but particularly the synergy model, which can be meaningfully implemented while maintaining disciplinary integrity. Despite a supportive curriculum, impeding challenges include limited teacher preparation, insufficient resources, and a lack of methodological guidance for inquiry-based and integrated teaching. In countries such as Finland and Singapore, the effectiveness of inquiry-based and interdisciplinary approaches is a good example for the South African context. The critical role of teacher education in addressing challenges, such as the development of disciplinary, pedagogical, and integrative knowledge among pre-service teachers, is emphasised in the chapter. Arts-integrated pedagogy can address many of the identified challenges by enhancing learner engagement, supporting diverse learning styles, and fostering emotional and ethical connections to environmental issues. Indigenous Knowledge enriches ECE by promoting cultural

Foreword

Climate change stands as one of the most urgent and complex challenges of our time, calling for responses that are not only scientific and political, but deeply educational. Within this landscape, teacher education holds a pivotal role. Teachers do more than transmit knowledge; they shape how future generations understand the world, relate to it, and act within it. Preparing educators to engage meaningfully with climate change is therefore both a professional responsibility and an ethical commitment. *Teacher Development for Climate Action: Climate Change Education* enters this conversation at a critical moment, offering a thoughtful, research-informed, and contextually grounded contribution to the field.

What gives this volume its strength is its ability to bring together diverse perspectives while maintaining a clear focus on practice. It moves with ease between theory and application, offering insights that are both conceptually robust and grounded in lived educational realities. Early chapters situate the discussion within higher education and teacher preparation, exploring climate pedagogy (Chapter 1), policy and practice in international contexts (Chapter 2), and the need to empower teachers within the South African context (Chapter 3). Together, these contributions remind us that climate change education is not a uniform endeavour; it is shaped by context, culture and the conditions within which teachers work.

A recurring thread throughout the book is the recognition that teacher preparation remains uneven, in many cases, insufficient for the demands of climate-responsive education. Chapters addressing climate-ready teaching and resilient learning environments (Chapter 4), alongside teacher training and capacity-building (Chapter 5), speak directly to these gaps. Reflections on Creative Arts education (Chapter 6), professional development (Chapter 7), and integrated STEM approaches in African contexts (Chapter 8) extend this conversation, demonstrating that meaningful change requires both systemic support and pedagogical imagination.

The volume also highlights how climate change education cuts across disciplines in ways that challenge traditional curriculum boundaries. Contributions focusing on specialised and integrated approaches (Chapters 9 and 10) emphasise the need for holistic thinking, while the inclusion of arts-based pedagogies (Chapter 11) offers a compelling reminder that climate understanding is not only cognitive, but also emotional, cultural and creative.

At the same time, the book engages with emerging developments that are reshaping education globally. The integration of digital technologies and artificial intelligence (Chapters 12 and 13) reflects new possibilities, but also raises important questions about access, readiness, and ethics. The exploration of big data in climate science (Chapter 14) and the emphasis on Indigenous Knowledge Systems (Chapter 15) further expand the conversation, pointing towards more inclusive, plural, and context-sensitive ways of knowing and teaching about climate change.

What ultimately distinguishes this volume is its ability to hold complexity without losing clarity. It does not offer simple solutions, but rather invites thoughtful engagement, critical reflection, and purposeful action. By placing teacher development at the centre of climate change education, it affirms the vital role of educators as agents of change.

This is a timely and necessary contribution one that speaks to educators, researchers, and policymakers alike, and that encourages continued collaboration in shaping more responsive, equitable, and sustainable educational futures.

Philip Kwashi Atiso Ahiaku
University of Zululand, South Africa

Preface

Climate change has become one of the defining challenges of our time. Its consequences are not confined to environmental degradation alone but extend to social, economic, and political dimensions that influence human well-being and sustainable development. In response to these challenges, education systems across the world are increasingly recognising the importance of climate change education to equip learners with the knowledge, skills, and values required to engage with environmental issues and contribute to sustainable futures.

The idea for this edited volume emerged from ongoing academic conversations among scholars, teacher educators, and researchers who share a common interest in advancing climate change education. These discussions highlighted the growing need for scholarly work that explores innovative approaches to teaching climate change while addressing the practical challenges educators face across diverse educational contexts. While climate change is a global phenomenon, its impacts and educational responses vary significantly across regions. As a result, there is a need for platforms that bring together diverse perspectives from different countries and disciplines.

This book, therefore, brings together contributions from scholars across several countries, including South Africa, Ghana, Zambia, Namibia, and other international contexts. The chapters reflect a wide range of perspectives on climate change education, including curriculum development, teacher professional development, interdisciplinary teaching approaches, indigenous knowledge systems, digital technologies, and climate justice. By presenting both theoretical insights and empirical research, the contributors offer valuable perspectives on how education systems can respond to the urgent challenges posed by climate change.

A key strength of this volume lies in its interdisciplinary orientation. Climate change cannot be addressed effectively within the confines of a single discipline. Instead, it requires collaboration across fields such as environmental science, education, technology, and social sciences. The chapters in this book illustrate how interdisciplinary approaches can enrich climate education and enable learners to develop a more comprehensive understanding of environmental challenges and sustainability.

The book is organised into three thematic sections that reflect key areas of focus in contemporary climate change education research. The first section examines policy frameworks, curriculum development, and teacher education initiatives to strengthen educators' capacity to teach climate change effectively. The second section focuses on pedagogical approaches and interdisciplinary learning strategies that promote climate literacy and encourage learners to engage critically with environmental issues. The final section explores emerging trends in climate education, including the use of digital technologies, artificial intelligence, indigenous knowledge systems, and climate justice perspectives.

As editors, we would like to extend our sincere gratitude to all the contributing authors who generously shared their research and expertise. Their commitment to advancing climate change education made this volume possible. I also wish to acknowledge the reviewers and colleagues who provided valuable feedback during the development of the manuscript. Their insights significantly improved the quality and coherence of the chapters.

Finally, I hope that this book will serve as a useful resource for researchers, teacher educators, policymakers, and practitioners interested in climate change education. By sharing diverse perspectives and experiences, this volume seeks to contribute to ongoing efforts to strengthen climate education and support educators in preparing learners for the environmental challenges of the future.

CHAPTER OVERVIEW

Chapter 1: Climate Pedagogy in Higher Education

This chapter examines climate pedagogy in higher education as an interdisciplinary, transformative framework that equips students with knowledge, skills, and ethical perspectives to address climate change. It broadens traditional climate science education by integrating critical thinking, climate justice, indigenous knowledge, and experiential learning. It highlights challenges such as resource limitations and cultural barriers, explores theoretical foundations rooted in critical pedagogy and transformative learning, and discusses pedagogical strategies including collaborative, problem-based, and place-based approaches. Emphasis is placed on developing climate literacy across cognitive, affective, and behavioral dimensions, fostering systems thinking, and promoting ethical responsibility. Case studies illustrate best practices, while assessment frameworks focus on authentic, multidisciplinary methods aligned with sustainability competencies. Emerging trends and policy recommendations underscores higher education's vital role in advancing climate action.

Chapter 2: Climate Change Education Policy and Pedagogy for Training Teachers' Practices in England

This chapter focuses on climate education policy, pedagogy and practice and its implementation in one Higher Education Institution in England. By exploring the Climate Change Education Framework developed for pre-service teaching in England, United Kingdom in 2022 this chapter will draw on insights from teacher educators and pre-service teachers during the initial implementation phase, 2022-23, and highlighting some of the challenges and benefits of the framework. In doing this it emphasises the potential for the theoretical framework underpinning the initial teacher education framework as a theory of change model for embedding climate and sustainability education throughout teacher education. This chapter also offers insight into further work the university are engaged in in relation to nature education and carbon dioxide monitoring in schools. Further scrutiny of these approaches in relation the framework alignment with UK Government 'Curriculum and Assessment Review' (DfE, 2025) for England highlights the importance of aligning policy, pedagogy and practice.

Chapter 3: Empowering South African Teachers for Interdisciplinary Climate Change Education

Climate change education presents unique challenges that demand interdisciplinary approaches and innovative teaching strategies. This chapter explores how South African teachers can be supported and empowered to integrate climate change into their teaching across multiple subjects, including languages, science, geography, and mathematics. Drawing on national curriculum policies and teacher development programmes, it highlights the importance of connecting classroom learning to real-world environmental issues to foster critical thinking, collaboration, and problem-solving among learners. The chapter argues that empowering teachers requires more than enhancing subject knowledge; it involves building confidence, encouraging teamwork, and equipping educators with practical tools to design meaningful lessons.

Chapter 4: Climate-Ready Teachers, Resilient Future: Linking Teacher Education, Climate Action, and Societal Resilience

In 2015, UN member countries also adopted the 2030 Agenda for Sustainable Development a comprehensive global plan of action for 'people, planet and prosperity' comprising 17 SDGs and 169 targets to be achieved by 2030, including SDG 4 Quality Education and SDG13 on climate action [Nerini et al 2019]. This chapter is guided by three research questions: First, What theoretical bases explain the role of teacher education in promoting climate action and societal resilience? Second: Identify the gaps related to teacher education in the context of climate change education through previous research studies. And Third: Creating a model for a climate-ready learning environment through teacher preparedness? The paper follows a qualitative approach secondary data sources and references managed with Zotero.To address the first RQ, the theoretical base was discussed and for the second RQ identifies gaps in teacher education related to SDG 13, as described in previous research and for third RQ, the chapter proposed a 6-step model to prepare climate-ready teachers to develop a learning environment.

Chapter 5: Teacher Training Needs and Capacity Building for Interdisciplinary Climate Change Curriculum Implementation

The growing urgency of the climate crisis calls for an education system that transcends traditional subject boundaries to promote holistic understanding and sustainable actions. Teachers stand at the centre of this transformation. However, many teachers remain inadequately trained to teach climate change through interdisciplinary frameworks. This raises a critical question: do teachers possess the requisite capacity and confidence to prepare learners for the realities of a climate-impacted world? To address this question, Change Theory in Education theory developed by Fullan in 1982 was used to underpin the study. In this light, conceptualization approach was adopted. This approach involves a systematic process of defining, clarifying, and organizing main concepts to deepen understanding of a research issue.

Chapter 6: A Reflection on Preparing New Creative Arts Teachers: Climate Change and Curriculum Redevelopment

Currently in South African primary schools, Creative Arts forms part of the Life Skills curriculum (on the current CAPS curriculum). This chapter explores how climate change educational learning opportunities may occur in this curriculum, and the forms these opportunities take or might take. The chapter is also a lecturer's reflection on her practice in teaching student teachers about how visual arts in the

Creative Arts education has value in educating learners on climate change in the following ways: through teaching about local artists whose work reflects climate change ideas, through materiality: in use of materials, through stewardship of materials and the environment, and through designing projects which directly address climate change, on the curriculum. These three areas are key to how future teachers might be prepared to teach this curriculum and empowered to enact climate change in future classrooms. This chapter is a reflection about incorporating this knowledge and teaching in this way, my own practice as a teacher educator at a distance university in South Africa.

Chapter 7: Reflective Professional Development for Climate Change Education: A Framework for Strengthening Teacher Competence and Instructional Practice

Climate change is a pressing global challenge, with Africa facing heightened vulnerability due to its dependence on climate-sensitive sectors such as agriculture, water resources, and energy production. Climate Change Education (CCE) plays a critical role in fostering environmental literacy, ethical awareness, and action-oriented learning. Teachers are central to effective CCE implementation; however, evidence suggests that many African educators, despite understanding climate change issues, face limitations in subject knowledge, pedagogy, instructional resources, and continuous professional development. Reflective practice is increasingly recognised as a key mechanism for strengthening teachers' capacities, supporting adaptive teaching, and empowering learners to engage with climate challenges. This chapter highlights the need to integrate experiential, inquiry-based, and action-focused approaches into teacher professional development to promote transformative climate education and sustainable practices in African contexts.

Chapter 8: Fostering Sustainability Through Integrated STEM Education: Policy, Pedagogical, and Systemic Perspectives From Ghana and Africa

This literature review synthesises the philosophical and conceptual foundations, international practices, and contextual relevance of promoting sustainability through STEM education in Ghana and Africa. It argues that addressing complex socio-environmental challenges such as climate change, resource scarcity, and social inequality requires integrating sustainability principles into STEM education. The review highlights constructivism, inquiry-based learning, design thinking, and systems thinking as key pedagogical approaches that support active, contextualised, and interdisciplinary problem-solving. Evidence from international case studies demonstrates the effectiveness of Phenomenon-Based Learning, Project-Based Learning, and Maker Education in developing sustainability competencies. In Africa, sustainability demands are intensified by limited resources, inadequate teacher capacity, and gender imbalances in STEM, requiring coordinated policy action, curriculum reform, sustained investment, and continuous teacher professional development

Chapter 9: Assessing Barriers and Opportunities for Integrating Climate Change Education in Marine Science Content Among Secondary School Teacher Training Programs

Despite Namibia's dependence on the Benguela Current ecosystems, climate change education in Namibia primarily focuses on land with minimal integration of marine content. This chapter examines barriers and opportunities for integrating climate change education in marine science content among secondary school teacher training programmes in Namibia. The research employs a qualitative, interpretivist approach involving document analysis, interviews, and case studies of existing climate education initiatives. The findings reveal significant gaps in teachers' ocean climate literacy, a lack of resources and limited understanding of the ocean's role in climate regulation. There is an opportunity to introduce context-specific marine learning activities into the secondary school curriculum due to policy frameworks and institutional structures supporting climate change education in Namibia. There is a strong policy and institutional support and growing interest in marine and environmental studies.

Chapter 10: Climate Change Education: An Integrated Approach

These words were spoken by environmental activist Greta Thunberg, who compared the world's environmental crisis to a house on fire. The urgency captured in this statement reflects the broader global recognition that the world is facing an environmental crisis of unprecedented scale. Climate change, biodiversity loss, water scarcity, pollution, and extreme weather events signal that the "house" of our planetary home is indeed under threat. While it may not be possible to extinguish the fire entirely, there is an urgent need to contain and manage it responsibly. Prideaux et al. (2020) propose a "flattening of the curve" approach to global warming, advocating for sustainable mitigation strategies rather than abrupt, unrealistic reversals.

Chapter 11: Climate Change Education Through Creative Arts: Global Perspectives and the ICA1501 Module Case Study

This chapter examines the integration of climate change education into higher education curricula through a case study of ICA1501, Introduction to Creative Arts, at a South African university. Situated in teacher education, the module advances a transformative pedagogy that integrates eco-pedagogy and arts-based learning to foster climate change literacy in Visual Arts and Drama. Informed by Freirean critical pedagogy, it frames climate change education as a global, holistic, justice-oriented approach encompassing social, ethical, cultural, and environmental dimensions. Arts-based practices, including visual production, performance, and reflective inquiry, render scientific knowledge affective and contextually meaningful. Climate change functions as an integrating curricular lens through eco-art, drama, and critical reflection. Positioned alongside regional models across five continents, ICA1501 demonstrates how climate change pedagogy can be adapted across contexts and disciplines to cultivate socially responsive, imaginative leaders for sustainability.

Chapter 12: Teachers' Use of Information and Communication Technologies and Artificial Intelligence in Climate Change Lessons

Climate change is widely recognised as one of the most demanding global challenges of the 21st century, affecting environmental sustainability, social equity, economic growth and human wellbeing. Teaching can play a pivotal role by supporting learners to develop the knowledge, skills, values and attitudes needed to understand climate change and respond to its impacts. However, teaching it is challenging, as the topic is complex, interdisciplinary, value-laden and emotionally demanding. This chapter examines how information and communication technologies (ICTs), including artificial intelligence (AI) tools, can enhance climate change teaching. It highlights how ICTs and AI can improve lessons while recognising pedagogical, ethical and practical challenges. Their effectiveness depends on pedagogically informed, context-sensitive and carefully planned implementation. The chapter also discusses ongoing professional development, ethics and alignment between curriculum, policy and ICT integration.

Chapter 13: Teachers and Teacher Educators' Readiness for Integrating AI in the Teaching of Climate Change Education

The rapid advancement of artificial intelligence (AI) is reshaping teaching and learning across education systems worldwide. This study examined teachers, teacher educators, and student teachers readiness to integrate AI into climate change and sustainability education in Zambia. Drawing on a descriptive survey of 2,634 participants across ten provinces, the research reveals that while awareness and positive attitudes toward AI are widespread, pedagogical integration remains limited. AI use is largely exploratory and concentrated on tasks such as lesson preparation rather than transformative teaching practices. A significant gap exists in climate change training, with the majority of educators lacking recent professional development. The findings highlight that educator readiness is shaped by knowledge, pedagogical competence, ethical awareness, and institutional support. Structural challenges, including uneven infrastructure and limited policy guidance, further constrain progress.

Chapter 14: Decoding Climate Complexity: The Big Data Revolution in Climatology

Big data analytics allows researchers study satellite images, climate models, and weather station data. Climate division. Big data analytics and climateology need good data. Weather stations, satellites, and climate models collect data. Errors, biases, and arguments may occur. Researchers streamlined data merging security. Researchers use ML to modify climate data. Updated protocols facilitate data sharing and use. Climate modeling is hard with nonlinear dynamics. Analysing and modeling data is tricky. Climate data rarely change linearly. To solve these difficulties, scientists are creating new climate data management and interpretation technologies. Climate data is simplified by T-SNE. Nonlinear, non-stationary machine learning is studied. Climate scientists use big data analysis to identify complicated patterns and trends in vast databases. Big data analytics may benefit climate researchers despite its downsides. Big data analytics and new innovation can help scientists fight climate change. Future sustainability would improve.

Chapter 15: Decolonising Climate Change Education through Indigenous Knowledge Systems

The chapter focuses on India and South Africa for their rich cultural biodiversity and colonial histories. Both these countries share rich ecological diversity and histories of colonialism. In South India, the ritual water tanks, sacred temple groves, community-led conservation tradition, and many showcase how biodiversity is supported by local knowledge and traditional resource management practices. In South Africa, philosophies like Ubuntu and indigenous practices around medicinal plants (Muthi) showcase similar community-oriented approaches to foster environmental protection and climate resilience. Therefore, the chapter interconnects the varied knowledge boards and showcases how the IKS can be imparted into climate change education as a significant step. To make this integration into an effective practice, the following three-dimensional methodology is proposed: 1. Curriculum Reform 2. Pedagogical Change for teacher trainees 3. Experiential Learning that actively imparts local community knowledge.

WRITTEN JUSTIFICATION OF SCHOLARLY CONTRIBUTION

This edited volume, *Developing Teachers' Futures for Climate Action: Climate Change Education*, makes a significant contribution to scholarship by advancing original, interdisciplinary research in climate change education, with a particular focus on teacher development and educational transformation. The book draws on diverse methodologies, including qualitative, quantitative, and mixed-methods research, as well as systematic reviews, conceptual analyses, and context-based case studies. The chapters present original empirical findings from school, higher education, and community contexts, with strong representation from African and Global South perspectives, thereby addressing gaps in existing literature. The volume contributes to knowledge production by integrating key thematic areas, including curriculum development, teacher education, interdisciplinary pedagogies, and sustainability education, and by offering both theoretical insights and practical frameworks for implementation.

A distinctive contribution of this book lies in its integration of emerging and critical dimensions of climate change education, including artificial intelligence and digital technologies, indigenous knowledge systems, and climate justice perspectives. Furthermore, all chapters underwent a rigorous double-masked peer-review process to ensure scholarly quality and originality. Each submission was initially screened by the editors, followed by evaluation by at least two independent expert reviewers based on criteria such as methodological rigour, originality, and relevance. Authors revised their chapters in response to reviewer feedback before final acceptance. This process ensured that the volume disseminates credible, original research and reflects new developments within the discipline, making it a valuable resource for advancing both scholarship and practice in climate change education.

DECLARATION OF ORIGINALITY AND TARGET AUDIENCE

I hereby declare that this book, *Developing Teachers' Futures for Climate Action: Climate Change Education*, is an original scholarly work. No part of this publication has been plagiarised or previously published elsewhere in its current form. All sources have been appropriately acknowledged, and the chapters included represent original research contributions by the respective authors. All chapters underwent a rigorous double-masked peer-review process conducted by independent experts to ensure academic integrity and quality.

The primary target audience for this book includes researchers, academics, and postgraduate students in education, environmental studies, and sustainability science. It is also intended for teacher educators, curriculum developers, and policymakers involved in climate change education and education for sustainable development. In addition, the book will benefit practitioners and educators seeking to integrate climate change education into diverse teaching and learning contexts.

ACKNOWLEDGMENT

The editors would like to express sincere gratitude to all the contributing authors for their valuable scholarly work and commitment to advancing climate change education. Your dedication, insightful research, and willingness to engage with this collaborative project have made this volume possible.

Appreciation is also extended to the publishers for their guidance, professionalism, and support throughout the publication process. Special thanks go to the editorial team for their careful coordination and attention to detail in ensuring the quality and coherence of the manuscript.

The editors further acknowledge the reviewers for their critical insights and constructive feedback, which greatly strengthened the quality and academic rigour of the chapters. Your contributions have been instrumental in maintaining the scholarly standard of this book.

Finally, heartfelt thanks are extended to all who contributed, directly and indirectly, to the successful completion of this work.

CONCLUSION

The chapters presented in this volume collectively highlight the growing importance of climate change education in preparing societies to address environmental challenges and promote sustainable development. As climate-related risks continue to intensify globally, education systems must play a central role in equipping learners with the competencies needed to understand climate processes, evaluate environmental impacts, and participate actively in climate action. This book has brought together diverse scholarly perspectives that demonstrate how climate change education can be strengthened through curriculum reform, innovative pedagogical practices, technological integration, and the inclusion of indigenous knowledge systems.

One of the key insights emerging from this volume is the critical role of teacher education and professional development in advancing climate change education. Teachers are central agents in translating policy frameworks and curriculum objectives into meaningful classroom experiences. However, several chapters reveal that many educators still face challenges in teaching climate-related topics, including limited training opportunities, insufficient teaching resources, and uncertainty about appropriate pedagogical approaches. Addressing these challenges requires sustained investment in teacher preparation programmes that integrate climate change education into both pre-service and in-service training. Higher education institutions, therefore, have a crucial responsibility to ensure that future teachers are equipped with the knowledge, pedagogical skills, and interdisciplinary understanding necessary to teach climate change effectively.

Another important theme emerging from this volume is the value of interdisciplinary and systems-based approaches to climate education. Climate change is a complex phenomenon that intersects with environmental, economic, technological, and social systems. Consequently, effective climate education requires teaching approaches that integrate knowledge from multiple disciplines. The chapters in this book demonstrate how interdisciplinary learning frameworks can foster climate literacy by encouraging learners to examine environmental issues from diverse perspectives. Such approaches also foster critical thinking and problem-solving skills, enabling learners to engage with real-world sustainability challenges.

The integration of digital technologies and emerging innovations also represents an important development in climate change education. Several contributions in this volume explore the potential of digital tools, artificial intelligence, and big data to enhance climate learning and research. These technologies can provide learners with access to climate data, simulation tools, and interactive learning platforms that support deeper engagement with environmental concepts. At the same time, the adoption of digital technologies in education requires careful consideration of teacher readiness, digital literacy, and equitable access to technological resources.

Equally significant is the recognition of indigenous knowledge systems and climate justice perspectives in climate education. Indigenous communities possess valuable ecological knowledge and sustainable practices that have developed over generations of interaction with the natural environment. Integrating these perspectives into formal education can enrich climate learning while also promoting cultural inclusivity and respect for diverse knowledge systems. Furthermore, climate justice perspectives remind educators and policymakers that the impacts of climate change are not distributed equally across societies. Education, therefore, has a role in fostering critical awareness of environmental inequalities and in promoting socially responsible responses to climate challenges.

Taken together, the contributions in this book illustrate that climate change education is a dynamic and evolving field that requires collaboration among educators, researchers, policymakers, and communities. Strengthening climate education involves not only revising curricula but also transforming teaching practices, expanding professional development opportunities, and embracing diverse knowledge systems. As climate challenges intensify, education systems must adapt to ensure learners are prepared to engage with environmental uncertainties and contribute to sustainable solutions.

Looking forward, several key priorities emerge for future research and practice in climate change education. First, there is a need for more empirical research examining the effectiveness of different pedagogical approaches in promoting climate literacy and behavioural change. Second, greater attention should be given to integrating climate change education across all levels of education, from primary schooling to higher education and teacher training. Third, collaborative partnerships among universities, schools, communities, and policymakers are essential for developing contextually relevant climate education initiatives.

In conclusion, climate change education represents a vital component of global efforts to achieve sustainable development and environmental resilience. By empowering educators and learners with the knowledge and skills needed to address climate challenges, education can play a transformative role in shaping a more sustainable and equitable future. The insights presented in this volume provide valuable guidance for strengthening climate education and supporting the collective pursuit of climate action in diverse educational contexts.

Dumsani Wilfred Mncube

Department of Curriculum and Instructional Studies, College of Education, School of Education, University of South Africa, South Africa

Sikhulile Bonginkosi Msezane

Department of Science and Technology Education, College of Education, School of Education, University of South Africa, South Africa

Anriët Van Deventer

Department of Language Education, Arts and Culture, College of Education, School of Education, University of South Africa, South Africa

Acknowledgment

The editorial team expresses its sincere appreciation and acknowledges all contributing authors for their meaningful contributions to the book, *Teacher Development for Climate Action: Climate Change Education.* Your chapter represents an important and insightful addition to this volume, enriching the scholarly discourse on climate change education and the preparation of teachers for climate-responsive pedagogy.

Your collective expertise, spanning curriculum studies, environmental and climate change education, and innovative teaching and learning approaches, has significantly strengthened the depth and rigour of this publication. The perspectives and research you shared not only advance academic understanding but also offer practical guidance for educators, policymakers, and institutions committed to addressing the global climate crisis through transformative education.

We are grateful for your dedication, time, and scholarly excellence throughout the development of this chapter. It has been a privilege to include your work in this book, and we trust that your contribution will make a meaningful impact on research and practice in climate change education.

Thank you once again for your outstanding contribution.

Chapter 1
Climate Pedagogy in Higher Education

Phyllis Bernice Kwarteng Donkor
http://orcid.org/0000-0003-3249-3025
University of Ghana, Ghana

John Sedofia
http://orcid.org/0000-0002-5409-516X
University of Ghana, Ghana

Joyce Senya Ama Anku
http://orcid.org/0000-0001-9565-1065
University of Ghana, Ghana

Richard Kyere Asomah
University of Ghana, Ghana

ABSTRACT

This chapter examines climate pedagogy in higher education as an interdisciplinary, transformative framework that equips students with knowledge, skills, and ethical perspectives to address climate change. It broadens traditional climate science education by integrating critical thinking, climate justice, indigenous knowledge, and experiential learning. It highlights challenges such as resource limitations and cultural barriers, explores theoretical foundations rooted in critical pedagogy and transformative learning, and discusses pedagogical strategies including collaborative, problem-based, and place-based approaches. Emphasis is placed on developing climate literacy across cognitive, affective, and behavioral dimensions, fostering systems thinking, and promoting ethical responsibility. Case studies illustrate best

DOI: 10.4018/979-8-3373-6975-4.ch001

practices, while assessment frameworks focus on authentic, multidisciplinary methods aligned with sustainability competencies. Emerging trends and policy recommendations underscores higher education's vital role in advancing climate action.

INTRODUCTION

This chapter discusses climate pedagogy in higher education as a transformative interdisciplinary framework. It aims to provide students with the knowledge, skills, and ethical attitudes necessary to tackle the complex challenges of climate change. Climate pedagogy encompasses the natural sciences, social sciences, humanities, and Indigenous knowledge systems. This approach promotes climate literacy, critical engagement, and transformative action based on sustainability and climate justice principles.

Definition and Scope of Climate Pedagogy

Climate pedagogy stands apart from traditional climate science education by broadening its focus. It includes critical thinking, ethical responsibility, and active engagement in climate advocacy, policy, and community involvement. This approach uses various teaching methods, such as hands-on learning, local context exploration, teamwork in problem-solving, and reflective activities. These strategies aim to engage students on cognitive, emotional, and behavioral levels. For example, experiential learning directly involves students in the phenomena they study; they learn through action and reflection. This process promotes critical thinking and awareness of social and ethical issues (Montrose, 2002).

Climate change education highlights critical thinking, not just as a skill for assessing data but also to critique social justice issues and deconstruct power structures. This approach helps students analyze evidence and imagine informed actions (Oberman & Sainz, 2021). Place-based learning connects theoretical knowledge about climate to specific local environments, enabling students to engage meaningfully with their community and surroundings, and fostering personal and collective commitments to climate action. Research shows that local ecological restoration projects can serve as a foundation for awareness and transformative learning, demonstrating the capacity of place-based teaching to promote sustainable, community-focused environmental practices (Schindel Dimick, 2016). Together, these teaching strategies cultivate critical thinking, ethical reasoning, and active civic engagement, which are necessary to address climate change effectively.

Recent studies reveal the transformative potential of climate pedagogy. They emphasize climate justice, interdisciplinary learning, hands-on experiences, and institutional support. Including climate justice in curricula highlights the ethical issues and social inequalities worsened by climate change. Critical climate education encourages students to assess mitigation options with a justice perspective rooted in political ecology and environmental justice (Svarstad, 2021). Youth activism in climate justice movements shows how education can be hands-on and action-focused, developing skills in political advocacy and social change that go beyond scientific knowledge (Trott, 2024). Interdisciplinary methods that combine scientific knowledge with social and behavioral sciences promote teamwork across traditional academic areas. These methods are supported by leadership models that encourage peer learning and institutional change (Davison et al., 2013).

Hands-on learning and community involvement are crucial for improving students' climate competencies. They help bring about cognitive, emotional, and behavioral changes necessary for sustainable futures. Internships and service projects engage students with real-world environmental issues and encourage reflection on the social and ethical implications of climate knowledge (Lugg, 2007; Montrose, 2002). Institutional commitment to incorporating climate education across all disciplines is vital. Higher education institutions are key players in teaching about climate change mitigation. They promote the integration of knowledge, leadership in sustainability, inclusivity, and ethical responsibility. This commitment helps align academic programs with global climate goals and societal change (Davison et al., 2013; Priatna & Khan, 2024).

Importance of Climate Education in Higher Education

Universities and colleges contribute to a sustainable future through teaching, research, and commitment to carbon neutrality, as evidenced by sustainability initiatives worldwide (Sen et al., 2021). These institutions help mitigate climate change and advance sustainability by reducing carbon footprints and preparing future leaders to address environmental issues (Priatna & Khan, 2024; Suklun & Bengü, 2024). However, climate change topics are not consistently integrated into curricula. Faculty members recognize the need for sustainability education but face challenges like limited time, resources, and institutional support, creating gaps in curriculum integration (Kang et al., 2024; Rushton & Walshe, 2025). These challenges present opportunities for reforms that could better meet climate action needs (Henderson et al., 2017).

When universities promote climate justice and community engagement, they strengthen their roles as anchor institutions for societal change (Kinol et al., 2023). Universities set examples through sustainability policies, with many having carbon neutrality goals and leading in energy, waste management, and emissions reduction (Sen et al., 2021). These efforts offer lessons for integrating climate education into academic settings (Elsayed et al., 2025). Justice-centered frameworks link climate education with social equity, encouraging institutions to assume transformative roles in climate action (Filho et al., 2023; Kaklauskaite & Streimikiene, 2024). Focus on UN Sustainable Development Goal 13 emphasizes educational institutions' role in promoting knowledge and advocacy for climate mitigation, equipping students with skills to address climate challenges (Elsayed et al., 2025; Filho et al., 2023).

THEORETICAL FOUNDATIONS OF CLIMATE PEDAGOGY

Climate teaching combines environmental science, education, and social justice. This blend helps people understand and engage with climate issues. It is not only about sharing facts about climate change; it also teaches skills such as critical thinking, ethical reasoning, and teamwork for problem-solving. These skills contribute to making real changes. Climate teaching is grounded in learning theories that aim to empower students. It demonstrates how ecological, economic, and social systems are interlinked, encouraging sustainable actions and advocacy. By situating climate education within a broader social and political context, it addresses issues such as climate justice, fairness, and resilience. The main goal is to prepare individuals and communities to effectively confront global environmental challenges.

These theoretical foundations offer necessary conceptual support. Their value lies in how they are applied in higher education systems. The next section examines the current state of climate pedagogy. It highlights how these ideas inform institutional practices and emerging trends.

Pedagogical Theories Relevant to Climate Education

Pedagogical theories in climate education provide a foundation for effective teaching, especially through critical pedagogy based on Paulo Freire's work. Critical pedagogy focuses on developing critical awareness and action, essential for helping learners analyze the complex social, scientific, and justice aspects of climate change. Freire's critical pedagogy critiques traditional "banking" education models and promotes dialogical, problem-posing education. This encourages students to question and reshape their realities rather than passively receive information (Nugraha et al., 2024; Shih, 2018). This approach fosters critical awareness by prompting students to engage with their surroundings and with forces that shape them, empowering them to act against environmental injustices.

Expanding on Freirean traditions, ecopedagogy provides a critical environmental addition that links social injustice with ecological harm. Originating from Latin American critical education movements, ecopedagogy reveals the political aspects of environmental issues and questions dominant development models, neoliberalism, and colonial legacies (Misiaszek, 2015). This approach encourages students to think critically about social and environmental impacts of globalization (Misiaszek, 2015). Integrating ecopedagogy improves climate education by stressing justice, decoloniality, and human-ecological systems relationships, leading to transformative learning.

Critical thinking serves as both an evaluative skill for examining evidence and a tool for critiquing social justice, revealing power dynamics affecting climate vulnerability and responses (Oberman & Sainz, 2021). Critical thinking pedagogy, informed by Freirean and Foucauldian ideas, advances these goals by helping learners question dominant narratives, recognize systemic oppression, and engage in discussions linking knowledge to social change (Moodley & Chetty, 2024; Oberman & Sainz, 2021). This perspective enables students to imagine and advocate for fair strategies.

Linking critical pedagogy with participatory approaches increases awareness of political and ecological issues, especially in urban or marginalized communities, where learning is rooted in local contexts (Bellino & Adams, 2017; Breunig, 2005). Through participatory methods, students learn about climate issues and gain ability to act within social movements, fostering practical efforts for social change (Oberman & Sainz, 2021; Tarlau, 2014).

Climate education grounded in critical pedagogy and critical thinking promotes understanding and action. It challenges learners to engage critically with scientific data, social structures, and ethical issues, empowering them to become agents of change for climate justice and sustainability (Nugraha et al., 2024; Oberman & Sainz, 2021).

Outdoor experiential learning also supports interdisciplinary engagement by placing learners in natural settings. This fosters ecological awareness and sensitivity to the environment beyond traditional classroom boundaries by providing firsthand experiences of ecological processes, promoting understanding of connections in nature essential for developing environmentally responsible behaviors and values (Kagawa & Selby, 2012). Furthermore, place-responsive teaching draws on environmental and cultural contexts of specific areas. When combined with Critical Indigenous Pedagogies, which focus on Indigenous knowledge and ways of knowing, climate education becomes enriched with culturally responsive approaches. These methods validate Indigenous perspectives and promote educational sovereignty while challenging colonial legacies and providing opportunities for Indigenous learners and frameworks for non-Indigenous learners to engage with local knowledge (Bleazby et al., 2022; Martin et al., 2015; Mccarty & Lee, 2014).

Integrating Information and Communication Technologies (ICT) with active teaching improves climate education by providing context-specific learning experiences in diverse settings. ICT tools, from mobile devices to augmented reality, extend climate education beyond physical boundaries. This fosters student engagement and digital literacy, crucial for navigating climate information. For instance, augmented reality applications effectively explain climate phenomena by immersing students in interactive experiences, deepening their understanding of climate impacts (Chen et al., 2024). These technology-enhanced methods support motivation and provide pathways for knowledge acquisition, vital for preparing students for climate mitigation efforts (Anastasopoulou et al., 2024; Bharti et al., 2024).

Moreover, ICT helps develop digital skills critical in modern education. It facilitates collaborative learning, access to information, and critical thinking skills needed to analyze climate data. When integrated into experiential and culturally relevant teaching methods, ICT becomes a powerful tool for promoting inclusivity and empowerment in climate education, reaching marginalized groups who might face educational challenges (Heeks, 2017; Muccione et al., 2025; Yusuf & Balogun, 2011).

Capability- and competence-based approaches continue to be discussed in sustainability education. Capability-based education emphasizes transformative learning and critical thinking over skill acquisition. This approach supports Education for Sustainable Development (ESD) by promoting adaptability and inquiry, vital for addressing complex climate challenges (Holdsworth & Thomas, 2020). Studies indicate that sustainability competencies include systems thinking, futures thinking, values reflection, collaboration, and acting. These competencies develop through interactive learning and dialogue among students and teachers (Vesterinen & Ratinen, 2023) and aligns with the capability approach, emphasizing broad, adaptable learning over narrowly defined skills.

Competence-based education emphasizes specific skills and knowledge needed to complete tasks effectively. Research shows that integrating sustainability competencies into academic disciplines underscores the importance of content knowledge and new teaching methods (Diepolder et al., 2021; Lozano et al., 2021). Professional development programs focus on building proficiency by improving content knowledge and teaching techniques. This enables educators to address scientific facts and social aspects of climate education (Fix et al., 2020). Evidence suggests that action-oriented teaching methods empower students to engage critically and take responsible actions, aligning with both competency and capability goals (Chen & Liu, 2020).

Capability-based education enhances critical reflection and adaptability for managing climate uncertainties. Competence-based education strengthens specific skills for sustainable practices. Climate education incorporates disaster risk reduction through interactive, participatory learning methods grounded in real experiences. These approaches link climate change actions with risk assessment and community preparedness. This helps learners gain practical skills to support climate resilience. Effective climate education integrates multiple subjects, adapts to local contexts, and uses technology. It focuses on knowledge, skills, values, empowerment, and systems thinking needed for sustainable futures (Anderson, 2012; Gero et al., 2011; Kagawa & Selby, 2012).

Interdisciplinary Nature of Climate Studies

Tackling the scientific, social, ethical, and cultural aspects of climate change requires a blend of natural sciences, social sciences, humanities, and Indigenous knowledge. The natural sciences provide fundamental knowledge of climate systems, along with skills such as data analysis and modeling. Social sciences and humanities offer valuable insights into governance, ethics, and social justice. Indigenous knowledge gives alternative views and sustainable practices that enhance Western ideas.

Effective climate education uses various teaching methods, including hands-on learning, critical pedagogy, and problem-based learning. These approaches encourage systems thinking, awareness, and empowerment. Collaborative curriculum development with diverse stakeholders, including students, faculty, and community members, ensures that multiple perspectives are represented. This makes education more relevant and inclusive. Integrating technology through ICT facilitates access to diverse data sources, interactive learning, and essential digital skills in modern climate education.

Recognizing climate change as a complex issue requires skills such as critical thinking, adaptability, and a systemic perspective. Education focused on capabilities emphasizes transformative learning and critical reflection rather than isolated skill acquisition. Including interdisciplinary climate education in university programs requires strong institutional support and alignment with global sustainability goals, such as UN SDG 13. Thus, universities become centers for interdisciplinary research, education, and advocacy, merging climate science with social justice and sustainability to prepare graduates for effective climate action.

Critical Pedagogy and Transformative Learning in Climate Context

Critical pedagogy and transformative learning in climate education emphasize critical awareness, social justice, hands-on engagement, teacher effectiveness, and connections across disciplines. Drawing on Paulo Freire's ideas, critical pedagogy helps analyze climate issues in a comprehensive manner. It combines debate skills with social justice critique to empower students to take meaningful climate action. Experiential and place-based teaching connects theory to local contexts, fostering environmental awareness and relevant skills. Teaching environmental justice builds self-confidence and collective action by addressing social inequalities and challenging fear-based climate narratives that may foster disengagement.

Teacher confidence and professional development are crucial for effective, transformative climate education. Training programs that improve teachers' subject knowledge and teaching methods positively affect student engagement and learning outcomes, particularly in climate justice. Including disaster risk reduction education through hands-on and participatory approaches helps students understand and manage climate risks. Community engagement and empowerment are essential for sustainable adaptation and mitigation.

Interdisciplinary and systems-thinking methods help students address climate change as a complex issue. These methods develop the skills necessary to navigate and influence complex socio-environmental systems related to climate change. Some recent studies also highlight the need to incorporate emotional and psychological aspects into climate education, recognizing eco-anxiety and climate grief as important factors that affect student engagement and mental health (Hickman et al., 2021). Addressing these emotional aspects through supportive teaching strategies can boost resilience and commitment to ongoing climate action.

Moreover, the growth of digital and hybrid learning, accelerated by the COVID-19 pandemic, has created new opportunities and challenges for climate education. The innovative use of virtual reality, simulation tools, and online collaborative platforms enables immersive, hands-on learning and global connection, thereby expanding the reach and inclusivity of climate education (Barth et al., 2023). However, ensuring equitable access to digital skills is essential to ensure that these technologies effectively meet the needs of a diverse student body.

Artificial intelligence (AI) is also being applied in climate education to personalize learning pathways and address educational disparities, particularly in regions with uneven access to quality education such as Pakistani higher education institutions. AI-powered platforms enhance climate literacy by tailoring content to students' backgrounds and learning needs through adaptive learning systems, intelligent tutoring, and AI-driven analytics that monitor engagement and comprehension in real time (Abgaryan et al., 2023; Cheong et al., 2022; Mahmoud & Sørensen, 2024).

These innovations leverage machine learning and natural language processing to create more customized and effective learning experiences, thereby improving student motivation, academic performance, and equity in education (Katiyar et al., 2024; Taşkın, 2025). In higher education, AI-driven intelligent assistants and adaptive learning platforms have been shown to reduce cognitive load and support personalized pathways, addressing individual student needs and facilitating effective knowledge acquisition (Sajja et al., 2025; Vashishth et al., 2024).

Furthermore, ethical integration of AI with STEAM (Science, Technology, Engineering, Arts, and Mathematics) pedagogies is a subject of active research. Scholars emphasize sustainable learning models that blend technical AI knowledge with creativity and critical thinking, positioning AI as a catalyst for societal transformation rather than just a delivery tool (Ayanwale et al., 2024; Zhai & Krajcik, 2024). Trust and acceptance among educators towards AI-based educational technologies are key factors shaping successful AI adoption, with continuous professional development programs recommended to foster confidence and ethical use (Ayanwale et al., 2024).

Thus, climate pedagogy in higher education is an evolving, interdisciplinary, and transformative educational framework rooted in critical pedagogy and hands-on learning. It integrates diverse knowledge systems and teaching strategies to develop climate literacy, critical skills, and ethical values necessary for meaningful climate action. Commitment from institutions, teacher readiness, and alignment with global sustainability goals are crucial for advancing climate education that equips students to tackle the complex challenges of climate change responsibly and effectively.

CURRENT LANDSCAPE OF CLIMATE EDUCATION IN HIGHER EDUCATION

Climate education in colleges and universities now encompasses diverse forms of knowledge and pedagogical approaches, which are all geared towards improving students' understanding of climate issues, developing critical thinking, and encouraging ethical behavior for climate action. Schools and teachers play a key role in ensuring climate education supports global sustainability goals. This approach helps students tackle the complex challenges of climate change in a responsible and effective way.

Despite these positive developments, the implementation of climate pedagogy varies and faces constraints. Understanding these limitations is crucial for moving from hopes to action. Therefore, the next section examines the key barriers shaping climate pedagogy in higher education.

Integration of Climate Topics Across Disciplines

The UNESCO Education for Sustainable Development (ESD) framework incorporates climate change education across subjects by promoting a holistic, interdisciplinary approach. This approach addresses the environmental, social, and economic aspects of sustainability. It encourages education systems worldwide to include climate change topics not only in the natural sciences but also in the social sciences, the humanities, and technical fields. This fosters a better understanding and inspires action.

UNESCO's ESD framework highlights twelve key issues arranged into four groups:

i. approach (scope, policy, cooperation)
ii. content (environmental, social, economic pillars)
iii. teaching (methodologies, transformative learning, capacity building), and
iv. organization (metrics, documents, institutions).

Climate change education aligns with these pillars to ensure its integration across subjects. This integration helps raise awareness, promote participation, and encourage cooperation needed for significant societal change (Glavič, 2020).

Recent UNESCO (2023) initiatives on greening education systems promote whole-institution approaches that integrate sustainability into curricula, teaching methods, campus operations, and partnerships with communities. These frameworks ask for embedding sustainability skills across disciplines, connecting teaching with climate action goals, and encouraging global citizenship and systems thinking. This change marks a shift from fragmented sustainability education to a systematic curricular transformation that matches ESD priorities.

In practice, various countries apply this integrative approach in higher education and teacher training. For instance, at the Universitat de València in Spain, the training for secondary education teachers adopts a holistic, interdisciplinary approach that aligns with UNESCO's ESD competencies and the UN 2030 Agenda. This encourages collaboration across subjects to prepare teachers to engage effectively with sustainability changes, including climate change (Aznar et al., 2018).

Globally, higher education institutions have transitioned from traditional environmental education (EE) to comprehensive ESD models. These models incorporate climate change into curricula in engineering, the social sciences, and the humanities. This shift helps students understand the complexities of climate-related challenges through integrated learning that combines technical knowledge with social dimensions such as governance and economics, thereby leading to more effective climate action (Acosta Castellanos & Queiruga-Dios, 2021).

Moreover, UNESCO Chairs in various Northern Hemisphere countries, such as Germany, Iceland, Portugal, and South Korea, illustrate interdisciplinary research and teaching that contribute to several Sustainable Development Goals (SDGs), including those related to climate change. These Chairs operate across academic fields to instill sustainability principles. They also highlight models of sustainability that inform climate-related education and policy integration (Bergman et al., 2018).

This integration helps countries equip learners with the knowledge, skills, and attitudes needed to address climate change comprehensively and contextually, aligning with global sustainability goals (Acosta Castellanos & Queiruga-Dios, 2021; Aznar et al., 2018; Bergman et al., 2018; Glavič, 2020).

Some Institutional Approaches and Curriculum Models

Institutional approaches and curriculum models across Africa and other continents exhibit diverse educational goals, cultural contexts, and policy frameworks. In Africa, there is a strong focus on integrating indigenous knowledge into curriculum development. This helps connect learners with their lived experiences and encourages societal change through local ideas. It also promotes advancements in science, technology, and business on the continent (Ezeanya-Esiobu, 2019). Efforts to decolonize knowledge and curriculum have gained attention, especially in South African higher education. There are calls to give African perspectives equal weight to Eurocentric views, with the aim of reshaping the knowledge model and making the curriculum more relevant to African contexts (Heleta, 2018). The idea of Ubuntu-Currere has been introduced to rethink teaching and learning. It emphasizes context, democratic differences, and global perspectives, addressing the specific socio-political realities of South African higher education (Hlatshwayo & Shawa, 2020).

Challenges such as limited resources, language barriers, and pressures from neoliberal policies imposed by international bodies shape the development of African higher education policies. This highlights the need for curricula that support local knowledge and tackle the continent's specific developmental issues (Brock-Utne, 2003). Curriculum models that promote Universal Design for Learning (UDL) have been recommended to support inclusive education policy in South Africa. They connect the need for inclusive education to practical classroom support and curriculum differentiation (McKenzie & Dalton, 2020).

In broader international contexts, educational approaches often include international schools. These schools adopt global curricula and assessment models, balancing quality with inclusivity, teacher recruitment, and professional development in various cultural settings (International Schools and International Education, 2013). Nursing education provides examples of curriculum development that spans continents, demonstrating innovation in meeting changing healthcare education needs. This includes African, UK, European, and US contexts (Curriculum Development in Nursing, 2004).

Additionally, the growing use of e-learning platforms reflects differing institutional approaches across continents. In the US, the integration of technology has made e-learning a key component of higher education, with various platforms and MOOCs used to enhance access and interaction. In Africa, while challenges such as economic and infrastructure issues persist, there are also significant innovations aimed at bridging digital divides and promoting inclusivity (Adeniyi et al., 2024).

Highlighting Global South scholarship strengthens decolonial climate education by focusing on knowledge from Latin America and Africa. Ecopedagogy comes from critical traditions in Latin America, while African scholarship emphasizes relational worldviews such as Ubuntu and Indigenous ecological knowledge systems. These views question Eurocentric beliefs and support a variety of approaches to climate education that are important for global relevance and justice-oriented learning.

Overall, African curriculum models often prioritize indigenous knowledge, decolonization, and inclusive education frameworks that align with the continent's socio-political needs. In contrast, models in other continents may focus more on internationalized curricula, technology use, and teaching innovations that try to balance global standards with local requirements (Adeniyi et al., 2024; Ezeanya-Esiobu, 2019; Heleta, 2018; Hlatshwayo & Shawa, 2020; International Schools and International Education, 2013).

Challenges and Barriers to Effective Climate Pedagogy

Effective climate education faces several challenges, many of which come from cognitive, cultural, institutional, and socio-economic factors. One key challenge is the gap between climate risk and how vulnerable communities perceive it. For instance, in Fiji's Rewa River Delta, many residents acknowledge environmental issues related to climate change but have a limited understanding of its causes. This gap arises from a lack of knowledge and from influences such as short-term planning, spiritual beliefs, and traditional governance. These cultural factors attenuate the impact of climate messaging and education, underscoring the need to empower local leaders and tailor education to community contexts (Lata & Nunn, 2011).

Psychological and social factors also impede effective climate education. People may feel ineffective, hopeless, or constrained by limited time and resources. They also face dilemmas, such as the tragedy of the commons and free riding, which discourage pro-environmental actions. Overcoming these challenges requires changing social norms through education and institutional support. This shows how community psychology can boost engagement and behavioral change toward sustainability (Quimby & Angelique, 2011).

Educational systems, particularly in developing areas, present their own barriers. In Sub-Saharan Africa, quality education is often hindered by insufficient funding, lack of inclusive access, inadequate teacher training, poor curricula, and the physical or social exclusion of marginalized groups. These systemic issues limit the reach and quality of climate education necessary to build climate literacy and action (Zickafoose et al., 2024).

At a broader epistemological level, the emphasis on scientific approaches in climate discussions can marginalize indigenous knowledge systems, particularly in Africa. Ignoring Indigenous knowledge restricts inclusive and effective responses to climate issues. Combining indigenous perspectives through decolonizing education can enhance resilience and sustainability by addressing ethical and cultural dimensions often overlooked in traditional climate education (David, 2024).

Additionally, educational institutions may struggle with teachers who are not prepared to address complex climate issues. There is a need for specialized programs, like blue curricula that focus on water literacy and the hydro-social cycle. Teacher sensitivity and confidence are vital for effective climate education, yet these skills are often not well-developed in teacher training programs (Boon, 2024).

Lastly, political and institutional deadlocks at national and global levels hinder the effectiveness of climate education by creating uncertainty around feasible policies and diminishing motivation for climate action. Disconnects at both the international and institutional levels can lead to weak educational support, while policies may not align with what countries can realistically implement, thus limiting the effectiveness of climate education (D. Victor, 2011; D. G. Victor, 2011).

Table 1. Aligning Barriers with Pedagogical Strategies in Climate Education

Identified Barrier	Description	Pedagogical Strategy	Implementation Example
Cognitive complexity	Students struggle with interdisciplinary and systems-based knowledge	Systems thinking & interdisciplinary learning	Systems mapping exercises, cross-disciplinary modules
Affective disengagement	Anxiety, apathy, or emotional overwhelm	Reflective and affective pedagogies	Journals, storytelling, dialogue-based learning
Knowledge–action gap	Students lack pathways to apply learning	Experiential and action-oriented learning	Community projects, simulations, problem-based learning
Equity and access disparities	Unequal participation due to resources or background	Inclusive and decolonial pedagogies	Flexible assignments, diverse knowledge integration
Faculty capacity constraints	Limited training and time for climate teaching	Collaborative and scalable models	Co-teaching, shared resources, faculty communities
Institutional constraints	Misalignment between policy and teaching	Campus-as-curriculum, embedded models	Using campus sustainability initiatives in teaching
Assessment limitations	Difficulty evaluating complex learning outcomes	Multi-dimensional assessment frameworks	Rubrics for systems thinking, reflective portfolios
Digital divide	Unequal access to technology	Blended and low-tech adaptable approaches	Hybrid teaching tools, accessible resources

PEDAGOGICAL STRATEGIES AND APPROACHES

Teaching strategies in climate education have evolved to meet the growing need for students to develop sustainability literacy and climate action skills. Today, educators emphasize interdisciplinary, hands-on, and innovative methods that help students become informed and responsible leaders who can address complex environmental challenges. Approaches such as blended learning, outdoor activities, experiential projects, and community-based learning are increasingly used to strengthen critical thinking, collaboration, and a sustainability-focused mindset.

Research shows that participatory, experiential, and critical teaching methods are especially effective. For instance, studies published in *The Journal of Environmental Education* report that ecopedagogical and experiential approaches boost students' environmental awareness, critical thinking, and civic engagement by connecting emotional, cognitive, and action-oriented learning (Lloro-Bidart, 2018). These findings highlight the importance of linking theoretical concepts with practical experiences in higher education.

The growing integration of sustainability across disciplines – including STEM and business – also reflects a shift toward active learning strategies like case studies and simulations. These methods help students apply what they learn to real-world contexts and contribute meaningfully to sustainability goals and climate resilience. Institutions are also acknowledging the value of inclusive and justice-oriented pedagogies that incorporate diverse ways of knowing and support transformative, systemic change toward sustainability (Ajaps, 2023; Elsayed et al., 2025; Gamage et al., 2022; Lozano et al., 2019; Lugg, 2007; Scolobig & Balsiger, 2024; Versteijlen & Wals, 2023).

These teaching strategies provide a conceptual response to identified barriers. Their success depends on how they are applied in real settings. Thus, the next section shifts from "what works" to how to implement these methods in practice.

Experiential and Place-Based Learning

Experiential and place-based learning are two key approaches in climate education that emphasize active engagement and understanding of climate change. Experiential learning involves students gaining knowledge through direct experiences and reflecting on those experiences. It goes beyond traditional classroom instruction by engaging students in hands-on activities that simulate real-world climate scenarios. This fosters a deeper understanding and a personal connection to climate concepts. For example, students may participate in climate simulations, conduct field experiments to measure local air or water quality, or undertake projects involving renewable energy installations. This method improves scientific inquiry skills and motivation. It allows students to build knowledge actively through doing and reflecting. Research has shown that inquiry-based experiential learning models enhance conceptual understanding and scientific beliefs among students, demonstrating their effectiveness in climate science education (Cheng et al., 2016; Zhao et al., 2021).

Place-based learning, by contrast, anchors climate education in local environmental and community contexts. It uses learners' surroundings as the primary resource and classroom, making education relevant to local climate impacts and cultural perspectives. This approach includes activities such as ecological restoration projects, local climate impact assessments, and collaboration with indigenous knowledge holders on climate adaptation strategies. By linking learning to a specific place, students develop a stronger sense of responsibility and care for their environment. They also gain insight into the socio-ecological systems involved. This localized focus helps make climate challenges and solutions more personal, increasing student engagement and empowerment.

In practice, experiential and place-based learning often overlap. A climate education program may involve outdoor fieldwork (experiential) within a local ecosystem or community (place-based), promoting active participation along with contextual understanding. For example, a study with 5th- and 6th-graders used structured inquiry activities, supported by video examples, to guide scientific processes in a rural setting. This improved students' questioning skills and ability to analyze evidence relevant to local environmental studies (Solé-Llussà et al., 2020).

Both approaches complement each other in climate education. They encourage critical thinking, scientific inquiry, and personal connections to climate issues. They provide frameworks that equip students with the skills and motivation necessary to understand and tackle complex climate challenges effectively. Experiential learning engages students through active, inquiry-based experiences that enhance understanding and beliefs, while place-based learning ties education to local environmental and cultural contexts, fostering care and relevance. Together, they are vital components of effective climate education supported by current research (Cheng et al., 2016; Solé-Llussà et al., 2020; Zhao et al., 2021).

Collaborative and Problem-Based Learning

Collaborative learning and Problem-Based Learning (PBL) both prioritize student engagement, teamwork, and practical problem-solving. In collaborative learning, students work in groups toward shared objectives, enhancing interaction, communication, and peer support. This method strengthens teamwork, leadership, and communication skills while deepening understanding through dialogue. In disciplines such as economics and the social sciences, collaborative projects further improve academic performance and interpersonal competencies. (Martinez & Gomez, 2025).

PBL requires students to address open-ended, real-world problems in small groups, promoting self-directed learning, critical thinking, and the application of knowledge across disciplines. It is widely used in fields including medicine, engineering, and health education, where it effectively prepares learners for complex problem-solving and lifelong learning. (Fukuzawa et al., 2017; Martinez & Gomez, 2025).

In health education, PBL often appears within blended learning environments that integrate traditional instruction with online problem-solving activities. Students may collaborate in virtual classrooms or participate in discussions and wikis to analyze clinical cases. This approach increases flexibility, sustains interaction across distances, and supports positive learning outcomes by combining active, collaborative learning with digital tools. (De Jong et al., 2014).

Studies also show that integrating PBL with collaborative strategies—such as in flipped classrooms—enhances student motivation and performance. Learners placed in structured, diverse groups demonstrate stronger gains in understanding and engagement than those working in unstructured or randomly assigned groups, underscoring the value of combining PBL with well-designed collaborative methods. (Chang et al., 2022).

Technology-enhanced PBL environments also offer significant benefits. For instance, web-based PBL combined with collaborative learning in an acid-based physiology course resulted in higher test scores and greater student satisfaction than traditional PBL alone. Using web tools enabled flexible, student-centered small-group problem solving that leverages both PBL and collaborative learning in a blended context (Taradi et al., 2005).

Additionally, virtual reality (VR) has been studied as a new tool for PBL. It simulates real-world problems and encourages group work skills and self-directed learning by presenting students with realistic, open-ended challenges. This allows them to collaborate and actively engage in the learning process (Abdullah et al., 2019).

The learning process in PBL is cyclical, typically involving phases of collaborative problem analysis, followed by self-directed learning and group reporting. These phases build on each other, highlighting how collaboration and individual learning contribute to deep understanding and skill development (Yew et al., 2010).

Despite its many benefits, PBL can face challenges. Ineffective tutor guidance or poorly structured problems may limit its success. Careful design that adheres to the principles of constructive, self-directed, collaborative, and contextual learning is essential. Research emphasizes the need for theory-based practice and continuous redesign to optimize PBL implementation in various educational environments (Dolmans, 2019; Dolmans et al., 2005).

Collaborative Learning and Problem-Based Learning are connected methods that encourage student engagement, critical thinking, and teamwork through real-world problems and group interactions. They are enhanced by blended teaching models and new technologies, showing positive effects across various subjects and learning environments

Incorporating Indigenous and Local Knowledge

Incorporating Indigenous and Local Knowledge (ILK) into climate education enhances learning by connecting traditional wisdom with scientific inquiry, developing broader understanding of climate change and sustainable development. ILK comprises culturally rooted knowledge from Indigenous peoples and local communities, acquired through generations of environmental observation.

Teaching practices that include ILK involve Indigenous instructional materials, knowledge keepers' engagement, hands-on learning, and discussions connecting scientific and traditional viewpoints (Ogegbo & Ramnarain, 2024). In Morocco, science curricula incorporating Indigenous Knowledge improved students' understanding of ecosystems while promoting cultural diversity and environmental stewardship (Yazidi & Rijal, 2024).

Research demonstrates ILK's value in understanding climate impacts. The Tsimane' people of the Bolivian Amazon possess rich *ethnoclimatological* knowledge that sometimes challenges scientific frameworks (Fernández-Llamazares et al., 2015). Inuit communities in Nunavik provide detailed observations of climate effects, complementing scientific data for adaptive decision-making (Cuerrier et al., 2015).

Southeast Asian coastal communities use traditional ecological knowledge to track environmental changes, employing local materials and customs to address climate impacts (Hiwasaki et al., 2014). The 2014 US National Climate Assessment incorporated Indigenous knowledge alongside scientific data, highlighting the need for Indigenous perspectives in climate education (Maldonado et al., 2015).

The Indigenous Knowledge Systems-Based Ethnoscientific Model promotes eco-conscious education drawing from Indigenous worldviews, emphasizing sustainable resource use through multicultural environmental education (Slikkerveer & Gellaerts, 2024).

Despite broad agreement on ILK's importance for identifying climate change effects, its integration into policies and research remains limited. Indigenous and local communities' observations extend beyond scientific metrics, providing perspectives on environmental changes and their socio-political context (García-Del-Amo et al., 2020; Reyes-García et al., 2024). This necessitates including diverse knowledge forms in climate education.

The Bunun Indigenous group in Taiwan implements programs to connect learners with natural environments through Indigenous ecological knowledge, enhancing environmental learning while supporting Indigenous identities and decolonization efforts (Nesterova, 2020).

Integrating Indigenous and Local Knowledge into climate education enriches learning with localized understandings of climate change, promoting sustainability skills and inclusive climate literacy. Successful implementation requires collaboration with Indigenous knowledge holders, context-specific materials, and dialogue between knowledge systems to prepare students for environmental challenges (Maldonado et al., 2015; Ogegbo & Ramnarain, 2024; Slikkerveer & Gellaerts, 2024; Yazidi & Rijal, 2024).

DEVELOPING CLIMATE LITERACY AND COMPETENCIES

Climate pedagogy in higher education goes beyond just sharing knowledge. It intentionally builds climate literacy and skills that help learners understand, evaluate, and respond to climate change in informed, ethical, and proactive ways. Climate literacy is increasingly understood as a concept that includes cognitive, emotional, and behavioral aspects, supported by critical and systems thinking, and based on principles of climate justice and ethical responsibility (Brundiers et al., 2022; UNESCO, 2020).

Implementing climate pedagogy also raises key questions about evaluating learning. Traditional assessment methods often fall short in capturing complex outcomes like systems thinking and agency. The next section examines assessment frameworks and methods aligned with climate pedagogy.

Cognitive, Affective, and Behavioral Dimensions

The cognitive aspect of climate literacy includes developing students' understanding of climate science. This covers climate systems, human causes of climate change, mitigation and adaptation strategies, and the social and economic effects of climate impacts. Research shows that knowledge alone does not lead to meaningful engagement or behavior change (Kollmuss & Agyeman, 2002). Therefore, effective climate pedagogy includes emotional and behavioral aspects.

The emotional aspect focuses on students' values, attitudes, feelings, and sense of responsibility toward climate change. Higher education institutions recognize the need to address climate anxiety, eco-grief, and feelings of helplessness, as these can harm students' motivation and sense of agency (Ojala, 2016). Teaching approaches like reflective journaling, dialogue-based learning, and narrative inquiry help students process their emotions while building hope, confidence, and commitment.

The behavioral aspect stresses the need to turn knowledge and concern into action. This includes encouraging environmentally friendly behaviors, civic engagement, and professional practices aligned with sustainability. Service-learning, community projects, and campus sustainability initiatives provide students with real-world contexts in which to apply their knowledge, bridging the gap between classroom learning and climate action (Sterling et al., 2013). Together, these three aspects define climate literacy as a holistic educational outcome rather than just a cognitive achievement.

Recent research shows that climate education should focus on emotional aspects like eco-anxiety, grief, and moral distress. New studies emphasize the importance of teaching climate emotions, which impact learners' agency, engagement, and resilience (Pihkala, 2020; Talks, 2024). Emotion-informed teaching methods highlight collective processing, storytelling, and hope-based action to fight disengagement and paralysis (Ojala, 2023; Verlie et al., 2021). These approaches see emotions as drivers of climate action and moral responsibility.

Fostering Critical Thinking and Systems Thinking

Climate change is a complex problem marked by uncertainty, nonlinearity, and connections across ecological, social, political, and economic systems. As a result, climate pedagogy must focus on developing critical thinking and systems thinking skills (Meadows, 2008; Wiek et al., 2011).

Critical thinking allows students to question climate data, assess different perspectives, review policy responses, and recognise power dynamics in climate discussions. This skill is vital in a time of climate misinformation, where students must navigate conflicting claims. Teaching strategies such as case studies, policy debates, and critical media literacy can help develop analytical skills and humility.

Systems thinking enhances critical thinking by helping learners grasp feedback loops, trade-offs, and unintended consequences in climate systems. Utilizing tools like systems mapping, scenario analysis, and modelling exercises, students learn to move beyond linear thinking and understand the interconnected nature of climate issues. These skills are increasingly recognized as essential for preparing graduates to work across fields and sectors to address climate change (Brundiers et al., 2022).

Although systems thinking is seen as an important skill for sustainability, recent research in teaching suggests that clear instructional design is necessary to build this skill (Peretz, 2025; Vuorio et al., 2023). Techniques such as systems mapping, scenario modeling, and feedback loop analysis have helped learners better understand complex interactions in socio-ecological systems and the unintended outcomes that can arise. Studies in environmental education indicate that organized approaches to systems thinking greatly improve problem-solving and sustainability skills (Brundiers & Wiek, 2017).

Promoting Climate Justice and Ethical Responsibility

A key feature of modern climate pedagogy is its focus on climate justice and ethics. Climate change disproportionately impacts marginalized communities, especially in the Global South, where those who are least responsible for emissions often face the greatest challenges (IPCC, 2022). Higher education has an ethical duty to prioritize issues of equity, responsibility, and power in climate education.

Climate justice-oriented teaching encourages students to examine the historical and structural factors that contribute to vulnerability, such as colonial histories, exploitative economic systems, and unequal access to resources. Including insights from environmental justice research, Indigenous knowledge, and perspectives from the Global South disrupts technical and depoliticized approaches to climate teaching (Whyte, 2017). These methods help promote ethical reasoning, solidarity, and commitment to inclusive climate solutions.

When higher education institutions incorporate justice and ethics into climate education, they help shape graduates who are not only technically skilled but also socially responsible and morally engaged. This aligns climate education with broader goals of democratic citizenship, human rights, and sustainable development.

CASE STUDIES AND BEST PRACTICES

Examining real-world examples of climate pedagogy in practice provides valuable insights into how theoretical ideas are implemented across different higher education settings. Case studies and best practices showcase effective models, institutional strategies, and collaborative efforts that expand climate education.

Building on these insights, we can identify broader directions for the future of climate pedagogy. The final section outlines new trends and upcoming priorities, especially regarding equity, scalability, and institutional change.

Examples of Successful Climate Pedagogy Programs

Many universities have developed comprehensive climate pedagogy programs that combine teaching, research, and community engagement. For instance, interdisciplinary climate majors and minors have been introduced at institutions throughout Europe, North America, and parts of the Global South. These programs blend natural sciences, social sciences, policy studies, and ethics. They emphasize hands-on learning, problem-solving projects, and stakeholder engagement, enabling students to address real-world climate challenges (Leichenko & O'Brien, 2020).

Another effective approach is integrating climate learning goals into general education curricula. This ensures that all students, regardless of their discipline, achieve a baseline level of climate literacy. Such institution-wide strategies prevent climate education from becoming a niche topic and instead establish it as a vital part of higher education's mission (Tilbury & Tilbury, 2011).

Cross-Institutional and International Initiatives

Collaborations across institutions and countries play a crucial role in expanding climate pedagogy and sharing knowledge. Networks such as UNESCO's Education for Sustainable Development (ESD) initiatives and regional university partnerships encourage curriculum innovation, faculty training, and the sharing of resources (UNESCO, 2020).

International efforts are especially valuable for incorporating perspectives from the Global South and promoting diversity in climate education. Joint projects among universities across regions facilitate comparative learning, collaborative research, and student exchanges focused on climate adaptation, resilience, and sustainable development. These collaborations help challenge Eurocentric biases and support more inclusive climate pedagogy.

Lessons Learned and Transferable Insights

Across various settings, several key lessons emerge. First, institutional leadership and policy backing are crucial for sustaining climate education programs. Initiatives that depend on individual leaders' alone risk fragmentation and burnout. Second, building faculty capabilities is vital; educators need training in teaching methods, interdisciplinary support, and institutional incentives to effectively deliver climate-related content. Third, meaningful student engagement happens when teaching connects climate learning to personal experiences, career paths, and community needs. Finally, effective climate pedagogy is flexible and responsive, continually adapting to new scientific findings, societal needs, and student input. These lessons highlight the need to see climate pedagogy as an ongoing transformation within institutions, aligned with the broader sustainability goals of higher education.

IMPLEMENTING CLIMATE PEDAGOGY IN PRACTICE: TOOLS, ACTIVITIES, AND INSTITUTIONAL MODELS

To support higher education faculty and institutional leaders in implementing climate pedagogy, this section translates the chapter's ideas into practical, flexible teaching strategies and implementation models. Drawing from the interdisciplinary literature outlined earlier, the section provides clear examples that can work across various subjects and educational settings.

Sample Teaching Activities for Climate Education

Systems Mapping Exercise (Cognitive + Behavioral)

The systems mapping exercise is a cognitive-behavioral activity in which students work in small groups to map the relationships between climate drivers (e.g., energy systems), impacts (e.g., food security), and socio-political factors (e.g., inequality). This activity promotes systems thinking and interdisciplinary analysis of climate issues. It can be assessed using a rubric that evaluates complexity, connections, and suggested interventions. The activity is suitable for sciences, social sciences, or business courses.

Climate Reflection Journals (Affective + Reflective)

The climate reflection journals aim to develop students' affective and reflective abilities. In this activity, students are required to keep weekly journals that reflect on course content, their feelings, and the personal importance of what they learn. It aims to promote critical reflection and emotional engagement with climate change and its attendant negative effects. Assessment of this activity can use a rubric that focuses on the depth of reflection and the integration of concepts. This activity's usefulness lies in its ability to offer low-barrier participation for diverse learners.

Community-Based Project (Behavioral + Action-Oriented)

In the community-based project approach, students team up with local organizations to tackle real climate challenges such as urban sustainability and waste reduction. The learning outcome in this activity is to promote applied problem-solving and civic engagement among students. The activity can be assessed using project deliverables and feedback from stakeholders. A potential constraint to the successful implementation of this activity is its requirement for institutional support, which, if not provided, may hamper its success.

Climate Justice Debate or Role-Play

In this activity, students take on roles (e.g., policymakers, Indigenous leaders, industry representatives) to debate climate policy decisions. The aim is to develop critical thinking and perspective-taking among students. The activity can be assessed by determining the quality of arguments, use of evidence, and ethical reasoning.

Ready-to-Use Teaching Resources

To ease faculty workload and help with scalability, educators can use a variety of teaching resources to deliver climate pedagogy. Some of the resources are:

- Open Educational Resources (OER) for climate case studies and datasets
- Digital simulations and interactive tools for modeling climate scenarios
- Interdisciplinary teaching modules that combine climate content with existing courses
- Assessment rubrics for systems thinking, reflection, and action competence

These resources enable the gradual integration of climate pedagogy without requiring the redesign of entire courses.

Institutional Models and Case-Based Approaches

Embedded Curriculum Model

In this model, climate content is woven into existing courses instead of being offered as separate programs. This approach is effective because it expands reach across disciplines, reduces curriculum overload, and encourages faculty ownership

Pilot-to-Scale Model

Here, institutions start with small pilot courses or modules that are later expanded based on evaluation and faculty input. This approach has the advantage of lowering resistance and risk. It also builds institutional momentum towards large-scale rollout.

Campus-as-Curriculum Model

In the campus-as-curriculum model, university operations such as energy systems and waste management serve as learning labs for students. This model promotes experiential and applied learning. It can be applied to carbon audits and sustainability reporting projects.

Partnership Model

In this model, universities work with external organizations such as NGOs, government, and industry to tackle climate challenges. The model is useful because it offers internship opportunities to students, gives them real-world project experience, and provides academic institutions and the partner organizations with the opportunity to share resources.

Faculty Implementation Pathways

Considering time and resource limitations, faculty can choose tiered implementation strategies such as:

a) **Low-Intensity (Immediate Integration):** This may involve the addition of climate case studies to lectures, or the introduction of one reflective assignment, or the use of existing datasets or examples.
b) **Medium-Intensity (Course Redesign):** In this strategy, academic institutions may redesign their existing courses to incorporate project-based learning or introduce interdisciplinary modules or use structured assessment rubrics.
c) **High-Intensity (Program-Level Change):** This strategy calls for the development of new courses or minors, the formation of community partnerships, or the integration of experiential learning across curricula.

This tiered approach helps educators match their goals with their capacity, making the implementation of climate pedagogy more practical across different contexts.

Linking Practice to Institutional Constraints

While these strategies offer practical starting points, their success depends on institutional conditions. Faculty need to navigate potential barriers like limited time and resources, disciplinary expectations, and assessment and accreditation requirements.

Therefore, effective implementation involves strategic adjustments, such as:

a) Matching activities with existing learning outcomes
b) Using scalable and reusable resources
c) Collaborating with colleagues to share the workload

Integrating Student Voice into Practice

Student feedback is crucial for improving climate pedagogy. Practical ways to gather student feedback include mid-course feedback surveys, co-designed assignments, and student-led project components. These methods help ensure that teaching strategies respond to diverse learning needs and experiences.

ASSESSMENT AND EVALUATION IN CLIMATE PEDAGOGY

Assessment and evaluation in climate pedagogy are crucial to the uptake of climate education within the higher education landscape. Therefore, assessment frameworks for climate pedagogy must go beyond the traditional measurement of cognitive abilities (DeCamp, 2024). It is important that all climate education-related assessments within the higher education landscape forge a close alignment with the multifaceted aims of climate pedagogy in developing well informed critical and action-oriented graduates who can address climate crisis and issues of social justice. To this end, assessment frameworks governing climate pedagogy must be authentic, multidisciplinary and participatory (Karami, Shobeiri & Jafari, 2017). Assessment must also reflect core sustainability competencies (anticipatory thinking, strategic and normative competence, collaboration and problem solving). In response to the increased need for transformative responses to climate change (UNESCO) institutions of higher learning must be strategic in evaluating not only students' knowledge about climate change but also their attitude and behavior in readiness towards a positive change.

Despite the increasing emphasis on sustainability skills, climate education assessment still lacks validity and standardization. New research highlights the need to use indicators, competency frameworks, and mixed-method approaches to measure learning outcomes effectively. Using indicators aligned with global frameworks, such as the SDGs and UNESCO ESD goals, can help systematically monitor and evaluate sustainability in curricula (Madden et al., 2020). This approach improves the reliability and comparability of assessment practices. It also promotes institutional accountability and continuous improvement.

Methods of Measuring Learning Outcomes

A pluralistic and competency-based approach is required in measuring learning outcomes in climate pedagogy. These involve learning outcomes representing interconnected domains such as cognitive, affective and action-oriented behavior (DeCamp, 2024). In the cognitive domain, assessment must target learning outcomes that measure understanding of climate issues, policy frameworks and strategies of mitigation and adaptation. This will generally ensure students' awareness of climate science. In addition, measuring affective outcomes will ensure civic and ethical commitment, stakeholders' attitudes to and perceived importance of the subject. Studies have identified varied levels of awareness of climate issues among participants across different countries and regions. What is alarming, however, is that even places where awareness levels seem to be higher, there is less agency towards a positive change (Karami, Shobeiri & Jafari, 2017). This underscores the need for action backed behavioral outcomes. These outcomes may include embarking on sustainability initiatives, and the ability to communicate and champion advocacy for climate policies.

Reflective and Participatory Assessment Techniques

To keep abreast of the transformative goals of climate pedagogy, a focus on reflective and participatory assessment techniques is imperative. These innovative techniques address the affective and ethical dimensions of assessment that traditional modes of assessment fail to recognize (Gellermann, Michel & Harms, 2025). Participatory assessment techniques foster collaborative climate actions that co-create knowledge with students and ensure accountability and lifelong learning (DeCamp, 2024). These techniques may include reflective journals which reveal students' evidence of learning, lived experiences and transformation towards climate oriented best practices. Also, assessment can involve peer engagement and collaborative efforts to foster collective solutions through a demonstration of shared and accountable leadership. Portfolios, especially e-portfolios, can serve as demonstrable techniques to consider in assessing students. Portfolios provide concrete evidence of the students' reflection, multimedia advocacy outputs and project participation. They also showcase students' growth throughout the course and emphasize learner agency.

Continuous Improvement and Feedback Mechanisms

Considering the importance of safe climate to the sustainability of humans, assessment in climate pedagogy must be a continuous future driven process instead of a one-time summative attempt. This formative approach deepens learning, agency and responsiveness to climate related issues. In addition, there ought to be an effective feedback system where students continuously revise their work and refine their thinking.

Continuous improvement and feedback mechanisms do not only take place at the classroom level but also at the curriculum and institutional levels. At the curriculum level, there must be continuous review of the curriculum to accommodate current research and emerging trends to tackle the climate crisis (Gellermann, Michel & Harms, 2025). It is equally imperative for higher education quality assurance units to integrate evaluation of climate pedagogy into their annual institutional audits. Universities can also receive feedback from their alumni through tracer studies involving their graduates to obtain feedback on sustainability of their climate pedagogy. These institutional level initiatives will ensure a holistic approach to aligning institutional mandates to sustainability commitments (Karami, Shobeiri & Jafari, 2017).

FUTURE DIRECTIONS AND INNOVATIONS

Future directions and innovations in climate education are pivotal for addressing the evolving challenges of climate change and preparing learners to engage effectively in sustainable solutions. This section examines emerging trends, novel pedagogical approaches, and technological advancements that are shaping the future landscape of climate education. It emphasizes the integration of interdisciplinary knowledge, experiential learning, and digital tools to enhance accessibility, engagement, and impact. By exploring innovative frameworks and forward-looking strategies, this section aims to inspire transformative practices that not only deepen understanding but also empower learners to become proactive agents of change in the face of a global climate crisis.

Emerging Trends in Climate Education

Emerging trends in climate education reflect a multidisciplinary and transformative approach aimed at equipping learners with the knowledge, skills, and agency to effectively address climate change challenges. One significant trend is the emphasis on transformative educational practices that go beyond knowledge transmission to foster active community engagement and policy influence, highlighting the role of youth as change agents and integrating energy literacy as a core component (Muccione et al., 2025). This shift prioritizes climate education not merely as awareness raising but as a driver of behavioral and societal change.

Another emerging focus is the integration of climate sustainability education within community settings, promoting grassroots participation to empower individuals and communities to adopt sustainable practices and advocate for climate action. This includes overcoming educational and promotional challenges through unified, cooperative strategies that blend theoretical knowledge with practical, locally relevant approaches (Ramesh Kumar et al., 2024).

Professional development for educators, particularly science teachers, is also gaining prominence as a trend. With frameworks like the Next Generation Science

Standards in the U.S. incorporating climate topics explicitly for the first time, considerable attention is being given to preparing educators through research-based professional development. This includes integrating technology, providing standards-aligned curriculum resources, and fostering regionally relevant, observation-based learning progressions to make climate change personally meaningful for students (Hestness et al., 2014).

In formal education, particularly higher education institutions, there is a growing recognition of their role in creating ecosystems of climate education. Universities are called to partner with schools and non-formal education providers to support context-specific climate strategies, develop institutional capacity, and use multidisciplinary approaches that address cultural, psychological, and political factors. Such engagement aims to facilitate deeper learning and behavioral shifts distinct from didactic knowledge transmission (Reimers, 2020).

There is also a critical trend toward shifting the outcome of climate education from mere climate literacy — understanding climate change — to climate agency, which empowers learners to take meaningful civic and climate action. This approach emphasizes overcoming barriers related to emotional responses and disciplinary siloes by promoting holistic, action-oriented pedagogies within existing curricular frameworks (Janney et al., 2024).

Furthermore, challenges and opportunities are recognized in specific regions such as East Africa, where climate change education is embedded in curricula but lacks coherent strategic approaches to leverage education effectively for adaptation and mitigation. Addressing these gaps involves capacity building and aligning education to local climate vulnerabilities and sustainable development goals (Apollo & Mbah, 2021).

Lastly, climate change education increasingly highlights the importance of addressing affective dimensions, such as educator experiences and emotional challenges when teaching climate topics, as well as developing pedagogies that support resilience and action without necessitating radical systemic changes (Beasy et al., 2023).

Recommendations for Policy and Institutional Support

Implementable recommendations for climate pedagogy in the context of policy and institutional support should address both practical educational methods and systemic frameworks fostering sustainable learning environments. Firstly, embedding holistic, interdisciplinary, and transformative learning approaches that integrate climate change and sustainability content into curricula is vital, though institutional constraints such as class size and managerial support can hinder progress (Cotton et al., 2009). To counteract these constraints, policies should encourage flexible and adaptive pedagogy, recognizing the value of incremental or "second-best" approaches to integrate sustainability education effectively.

Secondly, open educational practices that blend in-person teaching with technological tools promote more accessible, collaborative, and participatory learning, conducive to sustainability goals. Institutional frameworks must support faculty and student scaffolding in open pedagogy while investing in digital infrastructure and training to enhance pedagogical innovation and resilience—especially relevant given recent shifts towards hybrid learning modalities (Halder Adhya et al., 2024).

Effective climate change education should encompass multidisciplinary content including environmental, social, and economic dimensions with a focus on developing local, tangible, and actionable knowledge for mitigation and adaptation. Policy and institutional support should ensure education systems themselves become climate-proofed, resilient, and green by fostering sustainable school environments that reinforce the educational content (Anderson, 2012).

Developing climate pedagogy also requires participatory teaching that motivates and empowers learners to act as future leaders for sustainability. Policies must prioritize pedagogical approaches that encourage critical thinking, problem-solving, and collaborative learning to build social and ecological awareness among students (Reza, 2016).

Philosophical frameworks, such as Ubuntu, underline community-centred education and policymaking, emphasizing collective responsibility and human-ecosystem interconnectedness. Institutions should incorporate such relational models to foster inclusive environmental education and participatory policy formulation that is culturally relevant and community-oriented (Kyei-Nuamah & Peng, 2024).

In addition, the following examples could improve the adoption of climate pedagogy in higher education for those attempting to introduce it:

Reducing barriers to ESD coverage through partnerships for professional development has proven effective in U.S. models. Professional development enhances educators' climate knowledge and pedagogical skills, improving their confidence to teach climate topics. Training models targeting teacher self-efficacy benefit those inexperienced in climate education, demonstrating that partnerships can lower preparedness barriers (Ennes et al., 2021; Li & Monroe, 2019).

Case studies of successful resistance overcoming support adoption through institutional reforms such as New Jersey's reforms starting 2024-2025 show how systemic changes enable climate pedagogy integration. These reveal how distributed leadership facilitates interdisciplinary collaboration needed for climate education (Davison et al., 2013).

Additionally, assessment methods aligned with transformative goals are crucial for monitoring outcomes. Tools like action competence rubrics and community impact metrics evaluate students' engagement with climate issues, ensuring effective understanding (Brumann et al., 2022).

Integration of concepts like "kaitiakitanga" and traditional practices demonstrates practical climate strategies while promoting inclusive content (Eisenack et al., 2014). Pragmatic guidance helps educators address constraints through low-resource activities and cross-disciplinary collaborations, enabling broader implementation of climate education (Davison et al., 2013; Ennes et al., 2021).

These approaches emphasize distributed leadership, capacity building, institutional reform, and culturally relevant pedagogy for effective climate education. Further, governance roles of institutions and local governments need clarity to support innovative education policies that align with climate change mitigation strategies. Institutional support must facilitate strategic action, conflict management, and coalition-building to implement workable governance frameworks that strengthen community-led sustainability initiatives (Hoppe & Van Bueren, 2015).

Finally, continuous professional development for educators, adequate resource allocation, and national frameworks accommodating regional diversity support the effective adoption of innovative climate pedagogy. These institutional supports are critical to overcoming systemic barriers and ensuring educational reform enhances equity, learning quality, and climate literacy (Fitriyah et al., 2024).

CONCLUSION

With climate change being an ongoing phenomenon that is volatile and varies across regions, educators must also be versatile and adaptable to develop and choose relevant pedagogy that can transform learners. The chapter examines theoretical foundations as well as practical methods of teaching climate change in higher educational institutions.

Final Reflections for Educators:

Climate pedagogy plays a pivotal role in shaping sustainable futures by equipping learners with the knowledge, skills, and ethical frameworks necessary to address complex climate challenges. It fosters climate literacy that spans cognitive, affective, and behavioral dimensions, enabling students to critically engage with climate issues through systems thinking and transformative learning approaches. By integrating interdisciplinary content and diverse knowledge systems—including Indigenous and local perspectives—climate pedagogy promotes holistic understanding and inclusivity.

Moreover, experiential, collaborative, and problem-based learning strategies empower students to apply theoretical insights to real-world contexts, nurturing a sense of agency and responsibility. The emphasis on climate justice and ethical responsibility further aligns education with broader sustainability goals, encouraging learners to advocate for equitable climate action.

Institutional adoption of climate pedagogy, supported by reflective assessment and continuous feedback, ensures curricula remain responsive and impactful. As higher education increasingly positions itself as a driver of climate action and advocacy, climate pedagogy becomes essential for cultivating informed, motivated graduates who can contribute meaningfully to sustainable development.

To realize these outcomes, ongoing innovation, policy support, and cross-institutional collaboration are vital, enabling climate pedagogy to evolve and expand its influence in fostering resilient and just futures.

REFERENCES

Abdullah, J., Mohd-Isa, W. N., & Samsudin, M. A. (2019). Virtual reality to improve group work skill and self-directed learning in problem-based learning narratives. *Virtual Reality (Waltham Cross)*, *23*(4), 461–471. DOI: 10.1007/s10055-019-00381-1

Abgaryan, H., Matevosyan, A., & Asatryan, S. (2023). Revolutionary changes in higher education with artificial intelligence. *Main Issues of Pedagogy and Psychology*, *10*(1), 1–9. DOI: 10.24234/miopap.v10i1.454

Acosta Castellanos, P. M., & Queiruga-Dios, A. (2021). From environmental education to education for sustainable development in higher education: A systematic review. *International Journal of Sustainability in Higher Education*, *23*(3), 622–644. DOI: 10.1108/IJSHE-04-2021-0167

Adeniyi, I., Hamad, N., Adewusi, O., Unachukwu, C., Osawaru, B., Chilson, O., Omolawal, S., Aliu, A., & David, I. (2024). E-learning platforms in higher education: A comparative review of the USA and Africa. *International Journal of Science and Research Archive*, *11*(1), 1686–1697. DOI: 10.30574/ijsra.2024.11.1.0283

Ajaps, S. (2023). Deconstructing the constraints of justice-based environmental sustainability in higher education. *Teaching in Higher Education*, *28*(5), 1024–1038. DOI: 10.1080/13562517.2023.2198639

Anastasopoulou, E., Tsagri, A., Avramidi, E., Lourida, K., Mitroyanni, E., Tsogka, D., & Katsikis, I. (2024). The Impact of ICT on Education. *Technium Social Sciences Journal*, *58*, 48–55. DOI: 10.47577/tssj.v58i1.11144

Anderson, A. (2012). Climate Change Education for Mitigation and Adaptation. *Journal of Education for Sustainable Development*, *6*(2), 191–206. DOI: 10.1177/0973408212475199

Ayanwale, M. A., Molefi, R. R., & Oyeniran, S. (2024). Analyzing the evolution of machine learning integration in educational research: A bibliometric perspective. *Discover Education*, *3*(1), 47. DOI: 10.1007/s44217-024-00119-5

Aznar, P., Calero, M., Martínez-Agut, M. P., Mayoral, O., Ull, À., Vázquez-Verdera, V., & Vilches, A. (2018). Training Secondary Education Teachers through the Prism of Sustainability: The Case of the Universitat de València. *Sustainability (Basel)*, *10*(11), 4170. DOI: 10.3390/su10114170

Barth, M., Michelsen, G., Rieckmann, M., & Thomas, I. (2023). Digital transformation and sustainability in higher education: Opportunities and challenges for climate pedagogy. *Journal of Cleaner Production*, *384*, 135493.

Bellino, M. E., & Adams, J. D. (2017). A critical urban environmental pedagogy: Relevant urban environmental education for and by youth. *The Journal of Environmental Education, 48*(4), 270–284. DOI: 10.1080/00958964.2017.1336976

Bergman, Z., Bergman, M. M., Fernandes, K., Grossrieder, D., & Schneider, L. (2018). The Contribution of UNESCO Chairs toward Achieving the UN Sustainable Development Goals. *Sustainability (Basel), 10*(12), 4471. DOI: 10.3390/su10124471

Bharti, R., Pomal, K., Ahmed, M., & Singh, C. (2024). Transformative Impact of ICT on Education: Leveraging Technology and Communication to Enhance Teaching and Learning. *Feedback International Journal of Communication, 1*(3), 131–141. DOI: 10.62569/fijc.v1i3.39

Bleazby, J., Thornton, S., Burgh, G., & Graham, M. (2022). Responding to climate change 'controversy' in schools: Philosophy for Children, place-responsive pedagogies & Critical Indigenous Pedagogy. *Educational Philosophy and Theory, 55*(10), 1096–1108. DOI: 10.1080/00131857.2022.2132933

Boon, H. J. (2024). A Comprehensive Approach to Water Literacy in the Context of Climate Change. *Education Sciences, 14*(6), 564. DOI: 10.3390/educsci14060564

Breunig, M. (2005). Turning Experiential Education and Critical Pedagogy Theory into Praxis. *Journal of Experiential Education, 28*(2), 106–122. DOI: 10.1177/105382590502800205

Brock-Utne, B. (2003). 2 - Formulating Higher Education Policies in Africa: The Pressure from External Forces and the Neoliberal Agenda. *Journal of Higher Education in Africa, 1*(1), 24–56. DOI: 10.57054/jhea.v1i1.1689

Brumann, S., Ohl, U., & Schulz, J. (2022). Inquiry-based learning on climate change in upper secondary education: A design-based approach. *Sustainability (Basel), 14*(6), 3544. DOI: 10.3390/su14063544

Brundiers, K., Barth, M., Cebrián, G., Cohen, M., Diaz, L., Doucette-Remington, S., Dripps, W., Chang, Y.-H., Yan, Y.-C., & Lu, Y.-T. (2022). Effects of Combining Different Collaborative Learning Strategies with Problem-Based Learning in a Flipped Classroom on Program Language Learning. *Sustainability (Basel), 14*(9), 5282. DOI: 10.3390/su14095282

Brundiers, K., & Wiek, A. (2017). Beyond interpersonal competence: Teaching and learning professional skills in sustainability. *Education Sciences, 7*(1), 39. DOI: 10.3390/educsci7010039

Chen, S.-Y., Lin, P.-H., Lai, Y.-H., & Liu, C.-J. (2024). Enhancing Education on Aurora Astronomy and Climate Science Awareness through Augmented Reality Technology and Mobile Learning. *Sustainability (Basel), 16*(13), 5465. DOI: 10.3390/su16135465

Chen, S.-Y., & Liu, S.-Y. (2020). Developing Students' Action Competence for a Sustainable Future: A Review of Educational Research. *Sustainability (Basel), 12*(4), 1374. DOI: 10.3390/su12041374

Cheng, P.-H., Yang, Y.-T. C., Chang, S.-H. G., & Kuo, F.-R. R. (2016). 5E Mobile Inquiry Learning Approach for Enhancing Learning Motivation and Scientific Inquiry Ability of University Students. *IEEE Transactions on Education, 59*(2), 147–153. DOI: 10.1109/TE.2015.2467352

Cheong, S. M., Sankaran, K., & Bastani, H. (2022). Artificial intelligence for climate change adaptation. *Wiley Interdisciplinary Reviews. Data Mining and Knowledge Discovery, 12*(5), e1459. DOI: 10.1002/widm.1459

Cuerrier, A., Brunet, N. D., Gérin-Lajoie, J., Downing, A., & Lévesque, E. (2015). The Study of Inuit Knowledge of Climate Change in Nunavik, Quebec: A Mixed Methods Approach. *Human Ecology: an Interdisciplinary Journal, 43*(3), 379–394. DOI: 10.1007/s10745-015-9750-4

Curriculum Development in Nursing. (2004). *Routledge*. DOI: 10.4324/9780203313343

David, J. O. (2024). Decolonizing climate change response: African indigenous knowledge and sustainable development. *Frontiers in Sociology, 9*, 1456871. Advance online publication. DOI: 10.3389/fsoc.2024.1456871 PMID: 39640009

Davison, A., Brown, P., Pharo, E., Warr, K., McGregor, H., Terkes, S., Boyd, D., & Abuodha, P. (2013). Distributed leadership: Building capacity for interdisciplinary climate change teaching at four universities. *International Journal of Sustainability in Higher Education, 15*(1), 98–110. DOI: 10.1108/IJSHE-10-2012-0091

De Jong, N., Savin-Baden, M., Cunningham, A. M., & Verstegen, D. M. L. (2014). Blended learning in health education: Three case studies. *Perspectives on Medical Education, 3*(4), 278–288. DOI: 10.1007/S40037-014-0108-1 PMID: 24458338

DeCamp, E. (2024). Integrating climate change across the disciplines: Review of a faculty learning community and student climate literacy assessment model. *Environmental Education Research, 30*(12), 1–19. DOI: 10.1080/13504622.2024.2309588

Diepolder, C. S., Weitzel, H., & Huwer, J. (2021). Competence Frameworks of Sustainable Entrepreneurship: A Systematic Review. *Sustainability (Basel), 13*(24), 13734. DOI: 10.3390/su132413734

Dolmans, D. H. J. M. (2019). How theory and design-based research can mature PBL practice and research. *Advances in Health Sciences Education : Theory and Practice*, *24*(5), 879–891. DOI: 10.1007/s10459-019-09940-2 PMID: 31720879

Dolmans, D. H. J. M., De Grave, W., Wolfhagen, I. H. A. P., & Van Der Vleuten, C. P. M. (2005). Problem-based learning: Future challenges for educational practice and research. *Medical Education*, *39*(7), 732–741. DOI: 10.1111/j.1365-2929.2005.02205.x PMID: 15960794

Elsayed, A. H., Pajuelo, M. L., Almaghaireh, I., Chaaban, K., Homsi, I., & Elmassri, M. (2025). Fostering sustainability leadership through SDG 13 integration in business curricula. *Sustainability (Basel)*, *17*(18), 8297. DOI: 10.3390/su17188297

Ennes, M., Lawson, D. F., Stevenson, K. T., Peterson, M. N., & Jones, M. G. (2021). It's about time: Perceived barriers to in-service teacher climate change professional development. *Environmental Education Research*, *27*(5), 762–778. DOI: 10.1080/13504622.2021.1909708

Ezeanya-Esiobu, C. (2019). *Indigenous Knowledge and Education in Africa*. Springer Singapore., DOI: 10.1007/978-981-13-6635-2

Fernández-Llamazares, Á., Méndez-López, M. E., Díaz-Reviriego, I., Mcbride, M. F., Pyhälä, A., Rosell-Melé, A., & Reyes-García, V. (2015). Links between media communication and local perceptions of climate change in an indigenous society. *Climatic Change*, *131*(2), 307–320. DOI: 10.1007/s10584-015-1381-7 PMID: 26166919

Filho, W. L., Wall, T., Salvia, A. L., Dinis, M. A. P., & Mifsud, M. (2023). The central role of climate action in achieving the United Nations' Sustainable Development Goals. *Scientific Reports*, *13*(1), 20582. Advance online publication. DOI: 10.1038/s41598-023-47746-w PMID: 37996539

Fix, G. M., Rikkerink, M., Ritzen, H. T. M., Pieters, J. M., & Kuiper, W. A. J. M. (2020). Learning within sustainable educational innovation: An analysis of teachers\u2019 perceptions and leadership practice. *Journal of Educational Change*, *22*(1), 131–145. DOI: 10.1007/s10833-020-09410-2

Fukuzawa, S., Boyd, C., & Cahn, J. (2017). Student Motivation in Response to Problem-based Learning. *Collected Essays on Learning and Teaching*, *10*, 175–188. DOI: 10.22329/celt.v10i0.4748

Gamage, K. A. A., Ekanayake, S. Y., & Dehideniya, S. C. P. (2022). Embedding Sustainability in Learning and Teaching: Lessons Learned and Moving Forward—Approaches in STEM Higher Education Programmes. *Education Sciences*, *12*(3), 225. DOI: 10.3390/educsci12030225

García-Del-Amo, D., Mortyn, P. G., & Reyes-García, V. (2020). Including Indigenous and local knowledge in climate research. An assessment of the opinion of Spanish climate change researchers. *Climatic Change*, *160*(1), 67–88. DOI: 10.1007/s10584-019-02628-x PMID: 32457557

Gellermann, G., Michel, H., & Harms, U. (2025). Between Two Worlds: Locating Climate Literacy between Modern Educational Frameworks and Assessment Needs. *Mind, Brain and Education : The Official Journal of the International Mind, Brain, and Education Society*, *19*(2), 61–72. Advance online publication. DOI: 10.1111/mbe.70003

Gero, A., Méheux, K., & Dominey-Howes, D. (2011). Integrating community based disaster risk reduction and climate change adaptation: Examples from the Pacific. *Natural Hazards and Earth System Sciences*, *11*(1), 101–113. DOI: 10.5194/nhess-11-101-2011

Glavič, P. (2020). Identifying Key Issues of Education for Sustainable Development. *Sustainability (Basel)*, *12*(16), 6500. DOI: 10.3390/su12166500

Heeks, R. (2017). *Information and Communication Technology for Development* (ICT4D). Routledge. DOI: 10.4324/9781315652603

Heleta, S. (2018). Decolonizing Knowledge in South Africa: Dismantling the 'pedagogy of big lies.' Ufahamu. *Journal of African Studies*, *40*(2). Advance online publication. DOI: 10.5070/F7402040942

Henderson, J. A., Bieler, A., & Mckenzie, M. (2017). Climate Change and the Canadian Higher Education System: An Institutional Policy Analysis. *Canadian Journal of Higher Education*, *47*(1), 1–26. DOI: 10.47678/cjhe.v47i1.187451

Hickman, C., Marks, E., Pihkala, P., Clayton, S., Lewandowski, R. E., Mayall, E. E., Wray, B., Mellor, C., & van Susteren, L. (2021). Climate anxiety in children and young people and their beliefs about government responses to climate change: A global survey. *The Lancet. Planetary Health*, *5*(12), e863–e873. DOI: 10.1016/S2542-5196(21)00278-3 PMID: 34895496

Hiwasaki, L., Luna, E., Syamsidik, S., & Marçal, J. A. (2014). Local and indigenous knowledge on climate-related hazards of coastal and small island communities in Southeast Asia. *Climatic Change, 128*(1–2), 35–56. DOI: 10.1007/s10584-014-1288-8

Hlatshwayo, M. N., & Shawa, L. B. (2020). Towards a critical re-conceptualization of the purpose of higher education: The role of Ubuntu-Currere in re-imagining teaching and learning in South African higher education. *Higher Education Research & Development, 39*(1), 26–38. DOI: 10.1080/07294360.2019.1670146

Holdsworth, S., & Thomas, I. (2020). Competencies or capabilities in the Australian higher education landscape and its implications for the development and delivery of sustainability education. *Higher Education Research & Development, 40*(7), 1466–1481. https://doi.org/DOI: 10.1080/07294360.2020.1830038

IPCC. (2022). *Climate change 2022: Impacts, adaptation and vulnerability.* Cambridge University Press., DOI: 10.1017/9781009325844

Kagawa, F., & Selby, D. (2012). Ready for the Storm: Education for Disaster Risk Reduction and Climate Change Adaptation and Mitigation1. *Journal of Education for Sustainable Development, 6*(2), 207–217. DOI: 10.1177/0973408212475200

Kaklauskaite, U., & Streimikiene, D. (2024). Climate justice from the lens of sustainable development. *Sustainable Development (Bradford), 32*(6), 6788–6800. DOI: 10.1002/sd.3062

Kang, M., Cholakis-Kolysko, K., & Dehghan, N. (2024). Sustainability teaching in higher education: Assessing arts and design faculty perceptions and attitudes. *International Journal of Sustainability in Higher Education, 25*(8), 1751–1769. DOI: 10.1108/IJSHE-03-2023-0091

Karami, S., Shobeiri, S. M., Jafari, H., & Jafari, H. (2017). Assessment of knowledge, attitudes, and practices (KAP) towards climate change education (CCE) among lower secondary teachers in Tehran, Iran. *International Journal of Climate Change Strategies and Management, 9*(3), 402–415. DOI: 10.1108/IJCCSM-04-2016-0043

Katiyar, N., Khare, M. D., Kumar, J., Sharma, A., Rawat, S., & Srivastav, J. (2024). Intelligent e-learning platform consolidating Web of Things and ChatGPT. In *Internet of things enabled machine learning for biomedical applications* (pp. 202–221). CRC Press. DOI: 10.1201/9781003487647-12

Kinol, A., Miller, E., Axtell, H., Hirschfeld, I., Leggett, S., Si, Y., & Stephens, J. C. (2023). Climate justice in higher education: A proposed paradigm shift towards a transformative role for colleges and universities. *Climatic Change*, *176*(2), 15. Advance online publication. DOI: 10.1007/s10584-023-03486-4 PMID: 36785806

Kollmuss, A., & Agyeman, J. (2002). Mind the gap: Why do people act environmentally, and what are the barriers to pro-environmental behaviour? *Environmental Education Research*, *8*(3), 239–260. DOI: 10.1080/13504620220145401

Lata, S., & Nunn, P. (2011). Misperceptions of climate-change risk as barriers to climate-change adaptation: A case study from the Rewa Delta, Fiji. *Climatic Change*, *110*(1–2), 169–186. DOI: 10.1007/s10584-011-0062-4

Leichenko, R., & O'Brien, K. (2020). *Climate and society: Transforming the future*. Polity Press.

Li, C. J., & Monroe, M. C. (2019). Exploring the essential psychological factors in fostering hope concerning climate change. *Environmental Education Research*, *25*(6), 936–954. DOI: 10.1080/13504622.2017.1367916

Lozano, R., Barreiro-Gen, M., Lozano, F. J., & Sammalisto, K. (2019). Teaching sustainability in European higher education institutions: Assessing the connections between competences and pedagogical approaches. *Sustainability (Basel)*, *11*(6), 1602. DOI: 10.3390/su11061602

Lozano, R., Barreiro-Gen, M., Pietikäinen, J., Gago-Cortes, C., Favi, C., Jimenez Munguia, M. T., Monus, F., Simão, J., Benayas, J., Desha, C., Bostanci, S., Djekic, I., Moneva, J. M., Sáenz, O., Awuzie, B., & Gladysz, B. (2021). Adopting sustainability competence-based education in academic disciplines: Insights from 13 higher education institutions. *Sustainable Development (Bradford)*, *30*(4), 620–635. DOI: 10.1002/sd.2253

Lugg, A. (2007). Developing sustainability-literate citizens through outdoor learning: Possibilities for outdoor education in Higher Education. *Journal of Adventure Education and Outdoor Learning*, *7*(2), 97–112. DOI: 10.1080/14729670701609456

Mahmoud, C. F., & Sørensen, J. T. (2024). Artificial intelligence in personalized learning with a focus on current developments and future prospects. *Research and Advances in Education*, *3*(8), 25–31. DOI: 10.56397/RAE.2024.08.04

Maldonado, J., Bennett, T. M. B., Chief, K., Cochran, P., Cozzetto, K., Gough, B., Redsteer, M. H., Lynn, K., Maynard, N., & Voggesser, G. (2015). Engagement with indigenous peoples and honoring traditional knowledge systems. *Climatic Change*, *135*(1), 111–126. DOI: 10.1007/s10584-015-1535-7 PMID: 26346998

Martin, G., Nakata, V., Nakata, M., & Day, A. (2015). Promoting the persistence of Indigenous students through teaching at the Cultural Interface. *Studies in Higher Education*, *42*(7), 1158–1173. DOI: 10.1080/03075079.2015.1083001

Martinez, M. E., & Gomez, V. (2025). Active Learning Strategies: A Mini Review of Evidence-Based Approaches. *Acta Pedagogia Asiana*, *4*(1), 43–54. DOI: 10.53623/apga.v4i1.555

Mccarty, T., & Lee, T. (2014). Critical Culturally Sustaining/Revitalizing Pedagogy and Indigenous Education Sovereignty. *Harvard Educational Review*, *84*(1), 101–124. DOI: 10.17763/haer.84.1.q83746nl5pj34216

Mckenzie, J. A., & Dalton, E. M. (2020). Universal design for learning in inclusive education policy in South Africa. *African Journal of Disability*, *9*(3). Advance online publication. DOI: 10.4102/ajod.v9i0.776 PMID: 33392062

Meadows, D. H. (2008). *Thinking in systems: A primer*. Chelsea Green Publishing.

Misiaszek, G. W. (2015). Ecopedagogy and citizenship in the age of globalization: Connections between environmental and global citizenship education to save the planet. *European Journal of Education*, *50*(3), 280–292. DOI: 10.1111/ejed.12138

Montrose, L. (2002). International Study and Experiential Learning: The Academic Context. *Frontiers: The Interdisciplinary Journal of Study Abroad*, *8*(1), 1–15. DOI: 10.36366/frontiers.v8i1.91

Moodley, D. E., & Chetty, R. (2024). conceptualizing critical thinking pedagogy in teacher education. *Childhood & Philosophy, 20*, 01–23. DOI: 10.12957/childphilo.2024.81799

Muccione, V., Ewen, T., & Vaghefi, S. A. (2025). A scoping review on climate change education. *PLOS Climate*, *4*(1), e0000356. DOI: 10.1371/journal.pclm.0000356

Nesterova, Y. (2020). Rethinking Environmental Education with the Help of Indigenous Ways of Knowing and Traditional Ecological Knowledge. *Journal of Philosophy of Education*, *54*(4), 1047–1052. DOI: 10.1111/1467-9752.12471

Nugraha, A. E., Wibowo, D., & Hendrawan, B. (2024). Paulo Freire's Critical Pedagogy Analysis Of Educational Transformation. *MSJ : Majority Science Journal*, *2*(2), 220–228. DOI: 10.61942/msj.v2i2.157

Oberman, R., & Sainz, G. M. (2021). *Critical thinking, critical pedagogy and climate change education*. Routledge., DOI: 10.4324/9781003003021-5

Ogegbo, A. A., & Ramnarain, U. (2024). A Systematic Review of Pedagogical Practices for Integrating Indigenous Knowledge Systems in Science Teaching. *African Journal of Research in Mathematics, Science and Technology Education, ahead-of-print(ahead-of-print),* 343–361. DOI: 10.1080/18117295.2024.2374133

Ojala, M. (2016). Facing anxiety in climate change education: From therapeutic practice to hopeful transgressive learning. *Canadian Journal of Environmental Education, 21,* 41–56.

Ojala, M. (2023). Hope and climate-change engagement from a psychological perspective. *Current Opinion in Psychology, 49,* 101514. DOI: 10.1016/j.copsyc.2022.101514 PMID: 36502586

Peretz, R. (2025). Integrating systems thinking into sustainability education: An overview with educator-focused guidance. *Education Sciences, 15*(12), 1685. DOI: 10.3390/educsci15121685

Pihkala, P. (2020). Eco-anxiety and environmental education. *Sustainability (Basel), 12*(23), 10149. DOI: 10.3390/su122310149

Priatna, D., & Khan, S. M. (2024). The importance of education and role of educational institutions in climate change mitigation and achieving UN SDG 13 Climate Action. *Indonesian Journal of Applied Environmental Studies, 5*(1), 1–5. DOI: 10.33751/injast.v5i1.10559

Quimby, C. C., & Angelique, H. (2011). Identifying Barriers and Catalysts to Fostering Pro-Environmental Behavior: Opportunities and Challenges for Community Psychology. *American Journal of Community Psychology, 47*(3–4), 388–396. DOI: 10.1007/s10464-010-9389-7 PMID: 21203832

Reyes-García, V., García-Del-Amo, D., Porcuna-Ferrer, A., Schlingmann, A., Abazeri, M., Attoh, E. M. N. A. N., Vieira Da Cunha Ávila, J., Ayanlade, A., Babai, D., Benyei, P., Calvet-Mir, L., Carmona, R., Caviedes, J., Chah, J., Chakauya, R., Cuní-Sanchez, A., Fernández-Llamazares, Á., Galappaththi, E. K., Gerkey, D., & Zant, M. (2024). Local studies provide a global perspective of the impacts of climate change on Indigenous Peoples and local communities. *Sustainable Earth, 7*(1), 1. Advance online publication. DOI: 10.1186/s42055-023-00063-6

Rushton, E. A. C., & Walshe, N. (2025). Curriculum making and climate change and sustainability education: A case study of school teachers' practices from England, UK. *Environmental Education Research, 31*(5), 1083–1097. DOI: 10.1080/13504622.2025.2471990

Sajja, R., Sermet, Y., Cwiertny, D., & Demir, I. (2025). Integrating AI and learning analytics for data-driven pedagogical decisions and personalized interventions in education. *Technology, knowledge and learning, 1*-31.

Schindel Dimick, A. (2016). Exploring the Potential and Complexity of a Critical Pedagogy of Place in Urban Science Education. *Science Education, 100*(5), 814–836. DOI: 10.1002/sce.21233

Scolobig, A., & Balsiger, J. (2024). Emerging trends in disaster risk reduction and climate change adaptation higher education. *International Journal of Disaster Risk Reduction, 105*, 104383. DOI: 10.1016/j.ijdrr.2024.104383

Sen, G., Chau, H.-W., Tariq, M. A. U. R., Muttil, N., & Ng, A. W. M. (2021). Achieving Sustainability and Carbon Neutrality in Higher Education Institutions: A Review. *Sustainability (Basel), 14*(1), 222. DOI: 10.3390/su14010222

Shih, Y.-H. (2018). Some Critical Thinking on Paulo Freire's Critical Pedagogy and Its Educational Implications. *International Education Studies, 11*(9), 64. DOI: 10.5539/ies.v11n9p64

Slikkerveer, L. J., & Gellaerts, S. L. (2024). *The Indigenous Knowledge Systems-Based Ethnoscientific Model of Integrated Eco-Education (IEE).* IGI Global., DOI: 10.4018/979-8-3693-2577-3.ch008

Solé-Llussà, A., Aguilar, D., & Ibáñez, M. (2020). Video-worked examples to support the development of elementary students' science process skills: A case study in an inquiry activity on electrical circuits. *Research in Science & Technological Education, 40*(2), 251–271. DOI: 10.1080/02635143.2020.1786361

Sterling, S., Maxey, L., & Luna, H. (Eds.). (2013). *The sustainable university: Progress and prospects.* DOI: 10.4324/9780203101780

Suklun, H., & Bengü, E. (2024). Raising Awareness of Sustainable Development Goals in Higher Education Institutions. *Research in Educational Administration and Leadership, 9*(1), 39–72. DOI: 10.30828/real.1357661

Svarstad, H. (2021). Critical climate education: Studying climate justice in time and space. *International Studies in Sociology of Education, 30*(1–2), 214–232. DOI: 10.1080/09620214.2020.1855463

Talks, I. (2024, November 6). *Hope, affect, and emotion in environmental education.* North American Association for Environmental Education. https://eepro.naaee.org/research/collections/hope-affect-and-emotion-ee

Taradi, S. K., Taradi, M., Radić, K., & Pokrajac, N. (2005). Blending problem-based learning with Web technology positively impacts student learning outcomes in acid-base physiology. *Advances in Physiology Education*, *29*(1), 35–39. DOI: 10.1152/advan.00026.2004 PMID: 15718381

Tarlau, R. (2014). From a Language to a Theory of Resistance: Critical Pedagogy, the Limits of "Framing," and Social Change. *Educational Theory*, *64*(4), 369–392. DOI: 10.1111/edth.12067

Tilbury, D., & Tilbury, D. (2011). '*Education for Sustainable Development: An Expert Review of Processes and Learning*' Paris: UNESCO. Available in Spanish, French and English. ED-2010/WS/46.

Trott, C. D. (2024). Activism as education in and through the youth climate justice movement. *British Educational Research Journal*, berj.4082. Advance online publication. DOI: 10.1002/berj.4082

UNESCO. (2020). *Education for sustainable development: A roadmap*. UNESCO Publishing.

Vashishth, T. K., Sharma, V., Sharma, K. K., Kumar, B., Chaudhary, S., & Panwar, R. (2024). Transforming classroom dynamics: The social impact of AI in teaching and learning. In *AI-enhanced teaching methods* (pp. 322–346). IGI Global Scientific Publishing. DOI: 10.4018/979-8-3693-2728-9.ch015

Verlie, B., Clark, E., Jarrett, T., & Supriyono, E. (2021). Educators' experiences and strategies for responding to ecological distress. *Australian Journal of Environmental Education*, *37*(2), 132–146. DOI: 10.1017/aee.2020.34

Versteijlen, M., & Wals, A. E. J. (2023). Developing design principles for sustainability-oriented blended learning in higher education. *Sustainability (Basel)*, *15*(10), 8150. DOI: 10.3390/su15108150

Vesterinen, M., & Ratinen, I. (2023). Sustainability competences in primary school education – a systematic literature review. *Environmental Education Research*, *30*(1), 56–67. DOI: 10.1080/13504622.2023.2170984

Victor, D. (2011). Global warming gridlock: Creating more effective strategies for protecting the planet. *Choice (Chicago, Ill.)*, *49*(03), 49–1484. DOI: 10.5860/CHOICE.49-1484

Victor, D. G. (2011). *Global Warming Gridlock*. Cambridge University., DOI: 10.1017/CBO9780511975714

Vuorio, J., Ristimäki, E., & Aksela, M. (2023). Supporting systems thinking in chemistry education through instructional design and modeling approaches. *Journal of Chemical Education, 100*(3), 1234–1243. DOI: 10.1021/acs.jchemed.5c00196

Whyte, K. (2017, March 1). Anthropocene. *English Language Notes, 55*(1–2), 153–162. DOI: 10.1215/00138282-55.1-2.153

Whyte, K. (2017). Indigenous climate change studies: Indigenizing futures, decolonizing the Anthropocene. *English Language Notes, 55*(1), 153–162. DOI: 10.1215/00138282-55.1-2.153

Wiek, A., Withycombe, L., & Redman, C. L. (2011). Key competencies in sustainability: A reference framework for academic program development. *Sustainability Science, 6*(2), 203–218. DOI: 10.1007/s11625-011-0132-6

Yazidi, R. E., & Rijal, K. (2024). Science Learning in the Context of "Indigenous Knowledge" for Sustainable Development. *International Journal of Ethnoscience and Technology in Education, 1*(1), 28. DOI: 10.33394/ijete.v1i1.10880

Yew, E. H. J., Chng, E., & Schmidt, H. G. (2010). Is learning in problem-based learning cumulative? *Advances in Health Sciences Education : Theory and Practice, 16*(4), 449–464. DOI: 10.1007/s10459-010-9267-y PMID: 21161374

Yusuf, M. O., & Balogun, M. R. (2011). Student-Teachers' Competence and Attitude towards Information and Communication Technology: A Case Study in a Nigerian University. *Contemporary Educational Technology, 2*(1). Advance online publication. DOI: 10.30935/cedtech/6041

Zhai, X., & Krajcik, J. (2024). Artificial Intelligence-Based STEM. *Uses of artificial intelligence in STEM education, 1*(10.1093).

Zhao, L., He, W., Liu, X., Tai, K.-H., & Hong, J.-C. (2021). Exploring the effects on fifth graders' concept achievement and scientific epistemological beliefs: Applying the prediction-observation-explanation inquiry-based learning model in science education. *Journal of Baltic Science Education, 20*(4), 664–676. DOI: 10.33225/jbse/21.20.664

Zickafoose, A., Ilesanmi, O., Diaz-Manrique, M., Adeyemi, A. E., Walumbe, B., Strong, R., Wingenbach, G., Rodriguez, M. T., & Dooley, K. (2024). Barriers and Challenges Affecting Quality Education (Sustainable Development Goal #4) in Sub-Saharan Africa by 2030. *Sustainability (Basel), 16*(7), 2657. DOI: 10.3390/su16072657

Chapter 2

Climate Change Education Policy and Pedagogy for Training Teachers' Practices in England

Jo-Anna Reed-Johnson
http://orcid.org/0000-0002-0247-4555
University of Reading, UK

Kirsty Shakespeare
University of Reading, UK

Hong Yang
University of Reading, UK

ABSTRACT

This chapter focuses on climate education policy, pedagogy and practice and its implementation in one Higher Education Institution in England. By exploring the Climate Change Education Framework developed for pre-service teaching in England, United Kingdom in 2022 this chapter will draw on insights from teacher educators and pre-service teachers during the initial implementation phase, 2022-23, and highlighting some of the challenges and benefits of the framework. In doing this it emphasises the potential for the theoretical framework underpinning the initial teacher education framework as a theory of change model for embedding climate and sustainability education throughout teacher education. This chapter also offers insight into further work the university are engaged in in relation to nature education

DOI: 10.4018/979-8-3373-6975-4.ch002

and carbon dioxide monitoring in schools. Further scrutiny of these approaches in relation the framework alignment with UK Government 'Curriculum and Assessment Review' (DfE, 2025) for England highlights the importance of aligning policy, pedagogy and practice.

INTRODUCTION

This chapter outlines the development and implementation of a Climate Change Education Framework across Initial Teacher Education (ITE) programmes at a Higher Education Institution in England. It situates this work within the wider policy context, examining how national strategies, curriculum expectations, and assessment structures shape opportunities for climate and sustainability education.

The authors introduce the theoretical foundations underpinning the Framework; drawing on the principles of teacher learning, interdisciplinary pedagogy, and place-based, project-based approaches. These ideas are illustrated through practice examples across the Early Years, Primary, and Secondary phases, as well as a cross-phase project designed to embed climate education meaningfully within ITE. Student and teacher educator reflections provide insight into how the Framework is experienced in practice.

By exploring the implementation challenges and barriers encountered; including curriculum constraints, time pressures, and varying levels of teacher readiness and expertise; the authors can identify the strategies that supported progress whilst highlighting lessons learnt through the process. The synthesising of insights across policy, theory, and practice provide a conclusion. This conclusion outlines implications for the future of teacher education and identifies areas for further research that might strengthen climate change education across the system.

Context

The Intergovernmental Panel on Climate Change (IPCC, 2022) has made clear that high-quality climate education is essential for addressing the accelerating challenges posed by climate change. As the impacts of climate disruption intensify, education systems must equip young people with the knowledge, skills, and dispositions needed to navigate an uncertain future. However, research consistently shows that many teachers feel under-prepared for this task, lacking both the confidence and competence to teach climate and sustainability topics effectively (Majid et al., 2023; SOS-UK, 2021; Teach the Future, 2022). This gap in teacher preparedness places significant responsibility on initial teacher education (ITE) providers to ensure that new teachers enter the profession ready to engage learners with climate-related issues in meaningful, empowering ways.

Days prior to COP 26, the UK Secretary of State for Education announced the publication of a draft 'Sustainability and climate change strategy for the education and children's services systems' (Department for Education, 2021), which put 'climate change at the heart of education' as part of the UK's pledge, The Glasgow Climate Pact. This landmark strategy marked the first time England's education system would formally address climate change and sustainability as core priorities. Prior to this strategy, climate and sustainability had been addressed in some areas of the national curriculum, namely science and geography (Dawson et al., 2022) and climate change education sat amongst policy texts originating from multiple ministries whose responsibilities spanned education, environment and climate change, but no unified education strategy or policy existed (Greer et al., 2022).

In response to this need, one university in England undertook the development of an initial teacher education framework; that drew on these core components of ESD, competencies and pedagogies that foster agency and critical thinking; to embed climate and sustainability education across all pre-service teacher education programmes. This work formed part of the National Climate Education Action Plan (NCEAP, 2021), a sector-wide initiative launched in the wake of COP-26 and aligned with the Department for Education's (DfE) Sustainability and Climate Change Strategy (DfE, 2022). The pilot phase ran from September 2022 to June 2023, followed by national roll-out from October 2023.

UNESCO positions climate education as a core component of Education for Sustainable Development (ESD) (UNESCO, 2017). Emphasising that it must be interdisciplinary, action-oriented, and values-driven. Rather than focusing solely on the transmission of scientific facts, ESD encourages pedagogies that foster agency, critical thinking, and ethical engagement. Climate-change education, in this context, needs to be transformative with an emphasis on social and political dimensions as well as the ability to questions values and belief systems (Lotz-Sisitka et al., 2015; Walshe et al., 2025). The aim being to develop learners' competences to respond to the challenges of a rapidly changing world. These competences include problem-solving, the ability to "dilemma-flip" when faced with complex or competing priorities, collaborative working, and the capacity to generate innovative, contextually relevant solutions leading to advocacy and agency. Preparing teachers to nurture these capabilities requires a systemic and embedded approach within teacher-education programmes, which this framework aimed to do.

The framework developed provides a structured, research-informed approach to embedding climate and sustainability education across subject specialisms and phases. It directly supports the UK Government's policy paper Sustainability and climate change: a strategy for the education and children's services systems (DfE,

2023), offering a practical mechanism to address the widespread gaps in climate-education provision across the sector. Collaboration was central to the project's success, involving partnerships with schools, NGOs, national bodies, and student-teachers themselves. The significance of this work was recognised nationally when the project team received the Advance Higher Education Collaborative Award for Teaching Excellence (CATE) in 2024, highlighting its contribution to sector-wide innovation and capacity building.

Policy: Climate Change Education in England

This section turns to England's climate change education policy that is built around the Department for Education's Sustainability and Climate Change Strategy (DfE, 2022),which aims to make the education sector in England a world leader in climate literacy and green skills by 2030. It focuses on curriculum development, teacher training, sustainable school operations, and preparing young people for a net-zero future. The core national strategy (2022-2030) applies to early years, schools, Further Education, Higher Education, and children's social care services. The key foci include:

- Reducing the environmental footprint of education settings
- Embedding climate literacy across subjects
- Developing green skills pathways
- Improving estates and digital infrastructure
- Supporting teachers and leaders with training and resources

From a policy perspective the DfE is currently updating its strategy to ensure teachers understand available support and can confidently teach sustainability; as well as ensuring that professional development is being expanded to help educators integrate climate themes across subjects. For the educating of teachers through pre-service teaching it is possible to draw on some of the examples shown in the 'practice' section below. The authors highlight how climate literacy across subjects; developing understanding of green careers; working with estates and providing outdoor experiential learning opportunities; developing teaching resources might be approached. What this policy does not emphasis is the need for interdisciplinary, project and place-based experiential learning.

Figure 1 outlines the key policy drivers that support these changes.

Figure 1. Key Policy Drivers for delivering embedded Climate and Sustainability Education in England

Global Drivers

- IPCC (2022): Climate education as essential for climate resilience
- UNESCO (ESD): Interdisciplinary, action-oriented, values-based learning

National Policy Frameworks

- National Climate Education Action Plan (NCEAP, 2021)
  - 9-point action plan
  - Mandate to design an ITE framework
- DfE Sustainability & Climate Change Strategy (2022; 2023 update)
  - Climate literacy, green skills, whole-system change

University and HEI Framework Development

- Research-informed ITE framework (2022)
- Embedding climate & sustainability across all pre-service programmes
- Pilot phase (Sept 2022–June 2023)
- National roll-out (Oct 2023)
- Collaboration with schools, NGOs, national bodies, student teachers
- Recognised with Advance HE CATE Award (2024)

Transformation of Initial Teacher Education

- Increased teacher confidence & competence
- Interdisciplinary, action-oriented pedagogies
- Values-based and agency-building approaches
- Development of key competences:
  - Problem-solving
  - Dilemma-flipping
  - Collaboration
  - Innovation & contextual relevance

Learner (Pupil) Outcomes

- Climate literacy
- Empowerment and agency
- Ability to navigate climate-related challenges
- Skills for a climate-changing world
- Contribution to a sustainable and just future

This chapter showcases some of the work taking place to embed climate and sustainability education across initial teacher education in one higher education institute in England, in schools and through a nature project, place-based and interdisciplinary approaches.

DfE Strategies (DfE, 2022 and 2023)

As part of this policy approach, whilst this strategy was welcomed by many as a step in the right direction, research by Dunlop and Rushton (2022) highlighted clear differences between the priorities of the government compared to students and educators. They identified several areas of additional action needed beyond those identified by the DfE strategy, including teacher training and an action oriented, rather than a knowledge based focus to learning.

The Department for Education (2022) strategy states 'by 2025, all education settings will have nominated a sustainability lead and put in place a climate action plan'. It is now 2026 and this is not yet the case, but the Climate Ambassador Programme (DfE, 2023) are making steps to address this. Non-statutory guidance has been published to support settings and the DfE have funded three key initiatives to support education settings in taking a whole school and evidence based approach to respond to climate change. Sustainability Support for Education (SSFE) is an online digital resource and knowledge hub, providing advice to help schools progress with tailored sustainability actions. The Climate Ambassadors Scheme connects trained voluntary climate and sustainability experts with schools to support them in developing their climate action plans and delivering climate education. Whilst the National Education Nature Park, provides resources and activities for schools to encourage children to monitor and plan how to increase biodiversity in their grounds and to map it on a digital version of their school site, which connects to a broader national nature park map. A fourth Department for Education supported project called Let's Go Zero (funded by Green Future Investments Limited), works in partnership with the three DfE funded initiatives to provide support through their Climate Action Advisors, who can connect schools with funding opportunities and trusted suppliers and contractors to complete decarbonisation site works. To date over 12,000 schools have signed up to receive support from the various Department for Education-funded and supported sustainability programmes according to data from the Climate Ambassadors scheme.

CURRICULUM AND ASSESSMENT REVIEW, ENGLAND (DFE, 2025)

England's 2025 Curriculum and Assessment Review (DfE, 2025) represents the most significant re-orientation of curriculum policy since the 2014 National Curriculum reforms. Teachers have said that the lack of visibility and emphasis on climate education within the national curriculum is a barrier to creating an effective curriculum on climate education. 16-19 providers have also shared evidence of the need for stronger provision (DfE, 2025, p.40). this has resulted in the publishing (November 2025) of a Review that sets out a comprehensive vision for a "world-class curriculum for all," emphasising updated subject content, clearer progression pathways, and modernised approaches to assessment. Its proposals extend across all key stages, in England, and are intended to guide curriculum development throughout 2026 and beyond. During 2026, curriculum drafters will translate the Review's high-level principles into detailed programmes of study, with a full public consultation scheduled for Summer 2026. This timeline positions the Review as a foundational driver of system-wide reform, shaping not only what pupils learn but also the broader policy architecture that supports teaching and learning in England and future review of the ITTECF (Initial Teacher Training and Early Careers Framework) that determines what training teachers 'learn' in terms of 'learn to' and 'learn that' around 8 teacher professional standards (DfE, 2024).

Theory: ITE Climate Change Education Framework

Climate change education sits at the intersection of several influential educational theories, each offering insights into how teachers can be prepared to engage learners with complex socio-environmental challenges. Social constructivism, teacher-agency theory, and the Education for Sustainable Development (ESD) competences framework all provide conceptual grounding for the development of climate-responsive initial teacher education (ITE) and underpins this work.

Social constructivism positions learning as a socially mediated process in which knowledge is co-constructed through dialogue, collaboration, and interaction with cultural tools. Social constructivism argues that learning happens through social interaction, dialogue, and shared meaning-making. The framework draws on this by posing questions with the pre-service teachers that they may have to model in their own classrooms. It's rooted in the work of Lev Vygotsky, who believed that thinking develops first between people, and only later within the individual. The thinking between people helps others to understand and consider different perspectives, whilst then reflecting to re-consider their own understandings. The key thinking is that learning is social, involving talking, collaborating, questioning, and negotiating meaning. Vygotsky's emphasis on the Zone of Proximal Development (ZPD) highlights the importance of guided participation and scaffolding in enabling learners to engage with challenging concepts. Climate change—characterised by uncertainty, complexity, and contested knowledge—requires precisely these dialogic and collaborative approaches. Climate-change education therefore aligns closely with social constructivist pedagogy, which encourages learners to negotiate meaning, explore multiple perspectives, and develop shared understandings of global issues. Embedding climate and sustainability education within ITE programmes supports future teachers to adopt these pedagogical orientations, enabling them to facilitate inquiry-based, participatory learning that empowers pupils to make sense of climate challenges collectively.

Teacher-agency theory further strengthens the rationale for embedding climate education across ITE. This draws on the work of Priestley, Biesta and Robinson (2015) who wrote about the importance of engaging teachers in curriculum review and advocate and ecological understanding of agency. Agency is understood not as an individual trait but as a capacity that emerges through the interplay of personal, structural, and relational factors. Teachers require both the confidence and the professional autonomy to adapt curricula, respond to local contexts, and engage learners in transformative sustainability practices. However, as national surveys indicate, many teachers currently lack the competence or confidence to teach climate change effectively (Teach the Future, 2021; SOS-UK, 2021). ITE programmes therefore play a crucial role in cultivating teacher agency by providing the knowledge base, pedagogical tools, and reflective spaces needed to engage critically with climate issues. When teachers develop a sense of professional purpose and efficacy in this domain, they are more likely to enact climate-responsive pedagogies and advocate for sustainability within their schools.

The ESD competences framework (UNESCO, 2017; Reichmann, 2018) provides a third theoretical lens, emphasising the development of values, dispositions, and action-oriented capabilities. UNESCO's framing of ESD highlights competences such as critical thinking, collaboration, problem-solving, and the ability to navigate ethical dilemmas—skills essential for responding to climate change. These competences align with the aims of the ITE framework developed by the English HEI, which emphasises problem-solving, "dilemma-flipping," innovation, and contextual relevance. By embedding ESD competences across pre-service programmes, the framework supports teachers in developing pedagogies that move beyond knowledge transmission toward transformative learning experiences that build learner agency and resilience.

For climate education to be effective in enabling action it needs to be focused on competencies that enables agency (Reed Johnson et al., 2025). The framework (see Figure 2) was developed around three interconnecting pillars to empower and develop young people's self-efficacy to enable them to understand the global picture whilst being able to take action at the local level. It is known that current teachers wanting to address concerns related to children's eco-anxiety around understanding climate change there still remains a teacher lack of knowledge, tools and time to deliver high-quality climate-education. The framework for embedding climate and sustainability education across pre-service teacher education programmes was designed to address this. It provides training teachers the opportunity to understand how to cultivate climate literacy across subjects in their learners. It gets the training teachers to reflect on developing their own understanding of green careers; how to engage in outdoor learning opportunities; providing opportunities for them to develop teaching resources and artifacts or protypes for their own place based teaching. Whilst at the same time moving towards actions through developed agency and consideration of the global and local perspectives through their reflections on climate justice.

Taken together, these theoretical perspectives reinforce the importance of embedding climate and sustainability education within ITE. Social constructivism provides the pedagogical foundation, teacher-agency theory highlights the professional conditions required for enactment, and ESD competences articulate the capabilities needed for both teachers and learners to respond effectively to a climate-changing world. The HEI's framework, developed in alignment with national policy initiatives such as the NCEAP (2021) and the DfE Sustainability and Climate Change Strategy (2022; 2023), operationalises these theoretical commitments, offering a coherent model for preparing the next generation of teachers to lead meaningful climate education across the sector.

Figure 2. ITE CASE Framework Summary

University of Reading

INSTITUTE OF EDUCATION

CLIMATE EDUCATION AND SUSTAINABILITY INITIAL TEACHER EDUCATION (ITE) FRAMEWORK

Introduction

Climate change is one of the biggest issues facing the planet. To tackle this threat, we must work collaboratively in helping everyone to understand our joint responsibility in making change happen. Together, we can be empowered to find solutions.

The University of Reading's Climate Education work is leading in driving this change. The National Climate Education Action Plan has provided a clear vision to consider how climate education can be delivered at scale. This framework delivers on action point 2 of that plan, with developments made during the 2022–23 pilot phase.

Aims and learning outcomes

Adapted from Thew et al. (2022)

Knowledge

The causes and consequences of climate change (IPCC, 2021a, 2021b, 2021c).

Climate change's consequences on a local and global scale, with an emphasis on different aspects of the problem having different levels of confidence and certainty e.g. it is much easier to say things at global level rather than local.

Attitudes, values and behaviours

To develop affective and behavioural skills that enable engagement with climate change and its impacts, i.e. 'approaches that cultivate integrated knowledge and global citizenship, while preparing students for curious, well-informed big-hearted lives' (UNESCO, 2017).

Competences and capabilities

Enabling the learner to act and work with others to solve problems.

Developing systems thinking, strategic thinking, collaborative competency, critical thinking, self-awareness and integrated problem-solving (Ojala, 2016).

Session overview/aims

Session A

What is climate education & why is it important?

Considering ones' own view of climate education (positionality).

What do you know about climate change? How does it make you feel? Have you seen it being taught and how?

Session B

What is climate justice & how can it be explored?

Exploring what is climate justice? The intersectionality of climate justice to be explored and the impact of this on individuals (UN, 2015).

Session C

Action & developing self-efficacy

Encouraging personal action from students and pupils. How personal action can ripple out into collective local, national, and international action (DFE, 2021).

Expression of Interest

If you are an ITE provider you can apply to use this ITE Framework under licence (free of charge). Please complete the Expression of Interest form here:
reading.ac.uk/planet/climate-education/initial-teacher-education

Programme directors drew on a range of teaching resources focused around the three pillars of climate education, climate justice and climate actions (Figure 3)

Figure 3. Three Pillars Teaching Resources

Good examples emerged from the practices developed through the early years, primary and secondary programmes and are shared below.

Practice: Climate Change Education in Pre-Service Teaching

'All teacher trainers and initial teacher trainees should be able to access training that empowers them to effectively incorporate climate education within their teaching across all levels and all subjects. This is to enable teachers to understand where and how to access verified information, and to consider the potential impact of a changing climate within the context of their level/subject of expertise such as climate education can be appropriately and meaningfully be incorporated in their teaching without exacerbating eco-anxiety.'

(NCEAP, 2022 - Lead: University. Contributions: Royal Meteorological Society, Met office, SoS-UK, Green Schools Project, Ashden, Education and Training Foundation, ClimaTalk, STEM learning, Outwood Grange Academies Trust)

Through examples of teaching strategies from different age phases the authors saw emerging a range of strategies being drawn on from the ITE Framework that were presented in the previous 'theory' section. These practices focused on climate education, climate justice and climate action to varying degrees as well as adopting social constructivist, teacher-agency theory, and the Education for Sustainable Development (ESD) competences framework led approaches.

These examples of practice included:

- Early Years – project, place and play based approaches;
- Primary – group work and game design with teacher resources/artifacts produced;
- Secondary – group work, experimental and experiential learning, outdoors learning, project based, interdisciplinary approaches;
- Project work demonstrated through the nature park and carbon dioxide monitoring in schools.

Early Years practitioners took a project, place, play based approach to learning about understanding the world. They set a task to establish gardening and sustainability initiatives in early years settings. The purpose being to set up a nature based learning environment that supported children's physical, social and emotional cognitive development whilst promoting awareness of sustainability practices and engaging the nursery's wider community. The projected linked in with the professional standard of promoting equality and opportunity (8.1), leading practice (8.4) and contributing to the wider life and ethos of the setting (8.2) (National College Teaching and Learnership [NCTL], 2013). It linked to the aspects of physical development, and understanding of the world. The trainee, through observation, interviews and anecdotal records (Figure 3 below) showed that the parents and staff demonstrate being overwhelmed by the emphasis on environmental education and the practical tips provided to parents through the 'parent corner' developed through this project helped reduce this. The participants said they could see now how the learning of the children had been extended at home. Staff members also reported that the project had enhance children's learning whilst fostering good relationships between staff, children and parents.

Figure 4. Early Years Gardening and Recycling Project

(Source: Early Years Trainee Teacher)

Primary Education focus came through group work on a first-year BA Primary Education (3 year programme) module. The module 'Global Inclusion and Sustainability'. The students remit was to design a classroom board game to support climate education that linked to the National Curriculum and Sustainable Development Goals. The game board and playing cards, rules and objectives also needed to consider the needs of pupils with Special Educational Needs. The target audience are children aged 8-9 years old in Year 4 of primary school and would be able to meet some of the National Curriculum requirements for science and geography (DfE, 2013).

Figure 4 above shows an overview of one group of students' game they designed and went on to use school teaching whilst on school placement in primary school settings.

Figure 5. Example of Primary Project Climate Education Games – Climate Tipping Points (image courtesy of BA Primary Education Trainee Teacher)

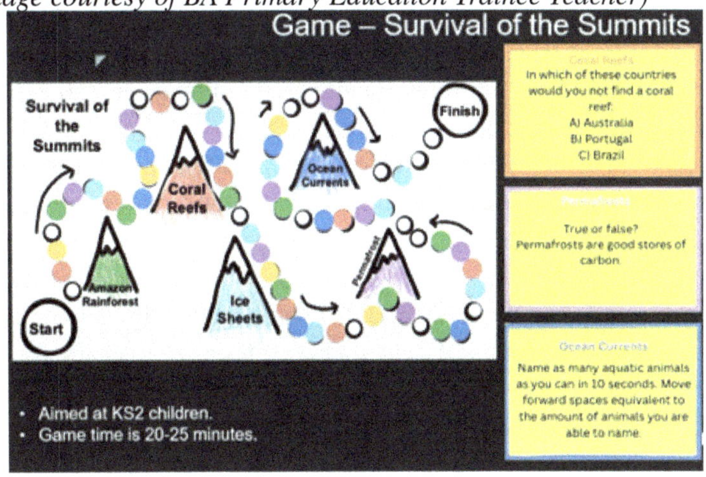

The focus of this game was climate tipping points; the causes and effects, and how they can be avoided. Through climate literacy the learners develop an understanding of climate tipping points, referring to the point at which a critical limit is reached, which if passed can cause dramatic and potentially irreversible damage to Earth's system (Met Office, 2023). Tipping points that are currently causing the most concern include the melting of ice sheets, thawing of the permafrost, shifting ocean currents, collapsing coral reef ecosystems and the decline of the Amazon Rainforest (World Wildlife Fund, 2025). Therefore it is about helping young children to understand that it is never too late to take action against tipping points and there are things individuals can do to combat the climate crisis (World Wildlife Fund, 2025). The game is focused on helping children to maintain a positive outlook and reduce eco-anxiety. The Sustainable Development Goals (SDGs) goals focused on were 12 responsible consumption, 13 climate action, 14 life below water and 15 life on land (UN, 2025).

Secondary Education practitioners took on a variety of approaches from fieldwork, experimental and experiential learning, outdoor learning, project based and interdisciplinary. A few examples of these are shared below from a range of subjects: Geography, Art, Design & Technology, Science, History.

Geography postgraduate certificate in education (PGCE) Secondary Geography student teachers took on a project and experimental based approach. They were asked to design and deliver fieldwork to a group of 13-year old pupils visiting the University campus. The PGCE students guided the pupils to visit a number of possible field locations around the University campus to develop relevant geographical fieldwork skills. They collected data on key variables such as soil p.H., air temperature and soil infiltration capacity. These trainee teachers experience was valuable as it enabled them to design and lead their own fieldwork - something required when they qualify as teachers and work as professionals in the future. It also satisfied a number of competencies such as collaboration, strategic group work, critical thinking, self-awareness and integrated problem solving. The work aligned with the SDGs 3 good health and well-being, 4 quality education, 11 sustainable cities and communities, 15 life on land and 13 climate action.

The pupils were taught by the pre-service teachers how to present their data using Geographic Information System (GIS) software to help them to inform their decision making. The experiential fieldwork was designed to challenge the school pupils to investigate where a new flower bed should be situated, using an enquiry approach (Roberts, 2003). The fieldwork reinforced the learning that the pupils had already done in school on the concepts of sustainability and climate change.

Art, Design Technology & Science (Physics, Chemistry and Biology) trainee teachers worked in interdisciplinary ways by sharing different expertise through a project based approach. They were asked to consider how to tackle eco-anxiety through hope, empathy, action and resilience. In designing an artifact through this collaborative problem solving the different subject specialists provided their own perspectives and thus produced a model to use in their own teaching.

The pre-service teachers needed to learn how to communicate, collaborate and co-operate across subject specialisms. They needed to be creative and innovative in their thinking whilst learning about possible ways to promote sustainable futures for their pupils in schools. The trainees were asked to draw on the SDGs in particular life below water (SDG 13). They were presented with a series of images (examples shown in Figure 5) from various artists or designers, as well as to consider how it might fit in with the school national curriculum and specific awareness days such as ‘ world water day’, ‘world bee day’, ‘national marine week’ etc…

The task set was to draw on the expertise and requirements of the national curriculum for art, design and technology and the sciences. Each subject specialist had to have an input in terms of how the product or exhibit was produced, the science behind it and the artistic value in relation to draw on thinking from other artists.

Figure 6. Example slides for trainee teacher presentation

Trainee teachers presented their work, as an artifact/exhibit to showcase the end outcome (Figure 6 and 7), and sequence of lessons that supported this for a specific age range in schools.

Figure 7. Example artifact produced

Secondary Interdisciplinary Project 'The Bee Meadow Project' drew on a range of subject specialisms, to work on an interdisciplinary project. Working with the Museum of English Rural Life artifacts to relate their subject teaching to the history of agriculture. With the Bee Meadow Project being an outdoor stimulus to enable the pre-service teachers to develop a range of teaching and learning resources and activities to use on school teaching placements. The following year another group of Science pre-service teachers engaged in the 'Pollinator Project'. This involved a collaboration with the School of Agriculture and Pollinating London Together, bringing in a range of expertise. Students used a research tool 'Fit Count' app; developed a range of teaching and learning activities and resources for future initial teacher education educators and teachers in schools to use. Poster Presentations were produced (Figure 8).

Figure 8. Poster Presentations

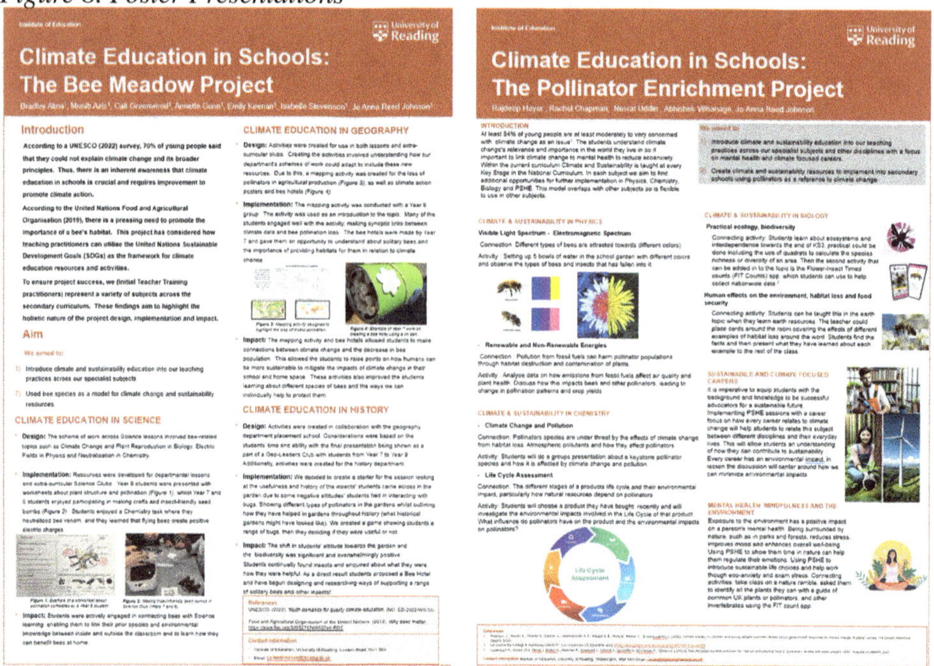

The University Nature Park Project

A discovery learning and green skills University wide project. A multidisciplinary project designed with input from key academic schools across the university including our School of Health and Life Science, School of Agriculture Policy and Development, School of Geographical and Environmental Sciences, Institute of Education,

School of Meteorology, School of Built Environment, School of Humanities many of whom will be embedding the space into their own student learning opportunities.

The aim of the University of Reading Nature Education Park Project is to help pupils from the local community build a connection to nature through discovery and learning, whilst developing their green skills such as:

· Species identification and ecological knowledge
· Carrying out practical fieldwork and research
· Recording and interpreting data
· Creative thinking and decision making
· Communication
· Environmental stewardship

According to research by WWF-UK in their 2024 'Schools For Nature' report, which surveyed teachers from 1885 primary and secondary schools across the UK, only 27% of schools have outdoor learning integrated into their curriculum for all pupils. 35% of schools surveyed excluded outdoor education entirely and for secondary schools, 56% offer no outdoor learning opportunities for their pupils. Children from low socioeconomic backgrounds are even less likely to have access to nature, with only 18% pupils attending schools with a high percentage of children in receipt of free school meals (a long standing measure of family circumstances in England) given daily opportunities to experience nature.

In Reading this equated to 21.1% of pupils in 2024. Barriers for pupils accessing nature in local schools include lack of teacher confidence in delivering climate, sustainability and nature education, particularly in outdoor settings, a lack of funding to support school trips and a lack of access to nature and green space in their own school grounds due to the urban setting.

By enabling school groups to engage in hands on experiences to explore and investigate biodiversity, sustainability and climate on the estate, the University encourages school groups to apply the skills and knowledge gained through this experience to their own settings; to help increase biodiversity and improve climate resilience, both in targets outlined in the DfE strategy (2023). Schools are encouraged to engage with the DfE funded National Education Nature Park scheme, to help the schools keep track of the improvements they make and see how these contribute to biodiversity net gain at a national scale. The University campus also offers access to specialist education facilities such as the Cole Museum of Zoology and the Atmospheric Observatory which have been used to further enhance climate and environmental learning opportunities for local schools.

This project is still in its' infancy stage, and currently activities being offered are linked to the national curriculum and span a range of subjects including science, geography, English, music, art, design and technology and religious education. This subject range models to participating schools and teachers that climate education and fostering a connection with nature are both holistic and interdisciplinary and do not need to be confined to science and geography where it is most explicitly included in the national curriculum. The duration of sessions being run has ranged from one hour for schools who are able to walk to campus, to a whole day for schools travelling from further afield.

Examples of activities being delivered include: estimating the amount of carbon stored by trees, habitat mapping and biodiversity sampling, exploring plant and animal adaptations to different climate zones and habitats, exploring the role of pollinators in an ecosystem, creating stories and soundscapes inspired by nature with natural instruments (rocks, sticks, leaves, grasses etc) and learning what insects need to survive to help design insect hotels. Through the sessions, pupils discuss how our environments are being impacted by human activity, as well as considering how more sustainable alternatives and actions can help to increase biodiversity and reduce our contributions to climate change.

Figure 9 shows primary school children carrying out practical fieldwork using sweep nets to investigate how invertebrate biodiversity varies in our Bee Meadow vs a patch of mown grass and considering how different grounds management practices influence biodiversity.

Figure 9. Children Sweep Netting on IoE Campus Sustainability Garden

Figure 10 shows children learning about the role of pollinators in our ecosystems. By creating a physical web of connections between different living and non- living parts of a local ecosystem (e.g. sun, soil, rain, pond, plants, trees and animals) they can explore the relationships (e.g. predator, prey, producer, consumer, pollinator, habitat) and consider what happens when human driven activity, such as using pesticides or extreme drought due to climate change, impacts the ecosystem and connections.

Figure 10. Children learning about the role of pollinators in our ecosystems

The response from local schools during the pilot has been very positive, with over six hundred school children taking part in activities to date (Feb 2026) and by the end of the pilot period (July 2026) we expect to have had close to one thousand five hundred children taking part. Further research is needed to demonstrate the impact of this work.

The Learning Garden consists of five curriculum-linked zones designed to educate and empower learners of all ages to understand environmental challenges and take meaningful action.

Figure 11. The University of Reading Learning Garden (opening in summer 2026)

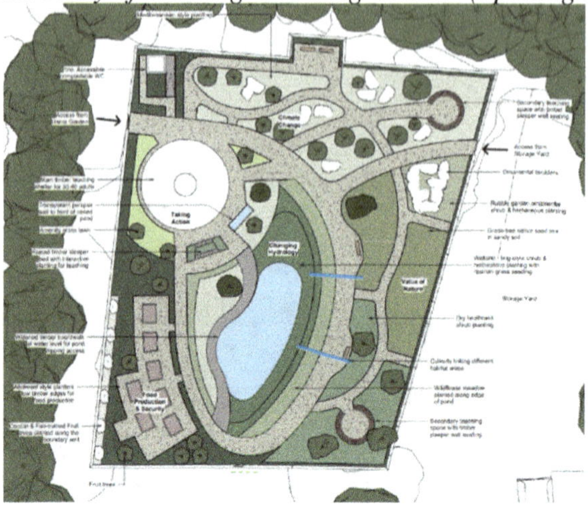

- Climate Change Zone: Explores causes of climate change using UoR research and historic data, Mediterranean-style planting, weather-monitoring instruments, and examples of renewable energy technologies.
- Food Production & Security Zone: Focuses on plant life cycles, pollinators, soil health, sustainable food systems, and the impacts of climate and biodiversity loss on food security, supported by simple comparative growing experiments.
- Changing Hydrology Zone: Examines climate-driven changes in water cycles, flooding, drought, plant adaptations, and the role of natural solutions in managing water, runoff, and soil erosion.
- Value of Biodiversity Zone: Showcases local habitats and species, highlighting biodiversity importance, vulnerability, and the effects of land-use change, pollution, and unsustainable development.
- Taking Action Zone: Demonstrates practical steps schools and communities can take to build climate-resilient, biodiverse spaces, including green roofs, vertical gardens, ponds, wildlife habitats, and a community tree nursery.

The carbon dioxide monitoring project is a schools based, experiential learning project taking experts in carbon dioxide monitoring into schools. Within the broader context of climate change education, carbon dioxide monitoring in schools classrooms offers a tangible, data-driven mechanism through which abstract global processes can be made locally visible. Carbon dioxide is central to discussions of anthropogenic climate change, as atmospheric accumulation from fossil fuel combustion and land-use change is the primary driver of global warming (IPCC, 2021). Yet for many pupils, atmospheric carbon cycles remain conceptually distant. By measuring carbon dioxide concentrations within their own classroom environment, learners can observe real-time fluctuations linked to occupancy, ventilation practices, and building management decisions. This creates opportunities to connect personal behaviour, built environment systems, and global carbon dynamics. Such experiential learning aligns closely with social constructivist pedagogies and Education for Sustainable Development (ESD) competences, which emphasise inquiry, systems thinking, and action-oriented learning (UNESCO, 2017). In the English policy context, the Department for Education's Sustainability and Climate Change Strategy (DfE, 2022) emphasises improving school estates and environmental performance alongside embedding climate literacy across subjects. Classroom carbon dioxide monitoring sits at the intersection of these priorities: it supports evidence-based estate management while simultaneously providing authentic curriculum material for science, geography, and citizenship. Moreover, it offers teachers a concrete example of climate-related data.

In this project, we monitored carbon dioxide in a school in South England in July 2025 (Figure 12). In the classroom, the mean CO_2 concentration was 455 part per million (ppm), with a minimum recorded value of 410 ppm and a peak value of 885 ppm. These figures are close to typical outdoor background concentrations, which currently average approximately 420 ppm (IPCC, 2021). CO_2 concentrations exceeded 800 ppm for only four recorded hours, representing just 0.14% of the total monitoring period, and did not exceed 1000 ppm at any point. The modest peak of 885 ppm corresponds to periods of higher occupancy (i.e., during concentrated classroom activity), reflecting patterns of classroom use.

Figure 12. Hourly carbon dioxide (CO_2) concentrations recorded in a school classroom in South England.

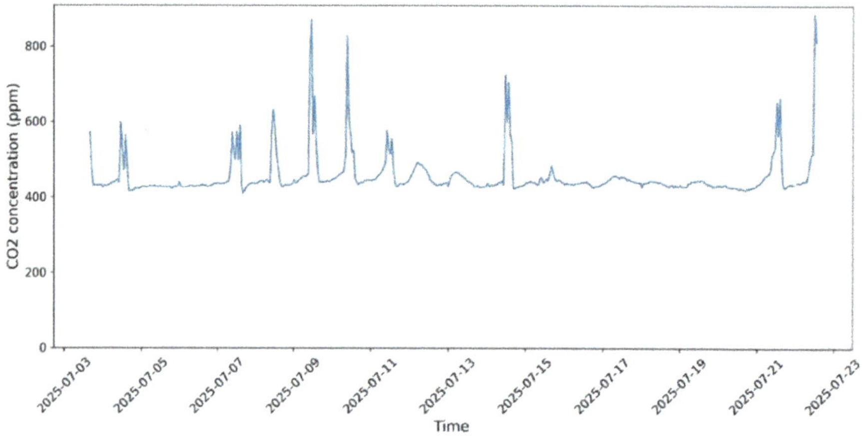

Embedding classroom carbon dioxide measurement within climate change education aligns with contemporary pedagogical frameworks and national policy drivers. First, it provides a concrete entry point for interdisciplinary curriculum integration. In science, pupils can investigate the carbon cycle, respiration, and combustion, using classroom data to explore how human activity alters atmospheric composition. In geography, discussions can extend from local indoor air quality to global greenhouse gas concentrations and their role in radiative forcing. In citizenship education, pupils can debate policy responses to climate change and consider how building design and community action contribute to sustainability goals. Such integration reflects UNESCO's (2017) emphasis on systems thinking and critical reflection as core ESD competences. Measuring carbon dioxide locally allows pupils to understand that climate change is not solely a distant phenomenon but a process connected to everyday activities, energy use, and ventilation practices. The act of monitoring transforms learners from passive recipients of information into active investigators. This aligns with inquiry-based and dialogic pedagogies rooted in social constructivism, where knowledge is co-constructed through collaborative interpretation of evidence. For teachers, incorporating carbon dioxide monitoring into school placements provides a powerful model of teacher agency in action. Rather than relying exclusively on prescribed curriculum materials, teachers can design data-informed learning experiences that respond to their immediate environment.

Priestley et al. (2015) suggest that teacher agency emerges through the interplay of professional knowledge, contextual understanding, and practical action. Implementing carbon dioxide monitoring exemplifies this process: teachers interpret environmental data, make pedagogical decisions, and potentially influence school operational practices. In doing so, they move beyond compliance with policy to enact informed, context-sensitive leadership in sustainability education. Furthermore, integrating carbon dioxide monitoring supports whole-school approaches to sustainability. The DfE strategy (2022) encourages schools to reduce carbon footprints and improve estate management. By involving pupils in monitoring and interpreting indoor carbon dioxide, schools foster a culture of shared responsibility. Students may propose evaluate energy trade-offs, ventilation strategies, or design awareness campaigns linking carbon emissions and air quality . Such activities develop problem-solving and "dilemma-flipping" competences, as learners navigate tensions between energy conservation, thermal comfort, and ventilation.

- Student and Teacher reflections

Early Years practitioners shared that their parents became more engaged and could see how their children had change the activities they engaged with at home due to this work.

Primary trainees said that they had enjoyed the challenge set and could see the benefits of developing a climate change education game that was also inclusive for special educational needs students.

The impact was noted by the president to the Geographical Association, for the Geography site experimental work, who said: '

'There was emphasis on enabling students to take part in outdoor learning as well as exploring the data collected through GIS technologies. This collaboration was very successful... '

'This project also helped support our whole-school focus on skill development including leadership and presentation skills, as well as providing an aspirational experience in a higher education setting.'

A teacher educator observed taught sessions (Spring, 2022) said:

'... the students seemed to engage in some really deep conversation about education on a global scale. The teacher educators modelled excellence in teaching (questioning, passion, group work, reflection) through the lens of climate and sustainability. Resources were great too - all very thought-provoking.'

Trainee Teacher (Steering Committee Member) commented:

'Working with the university (as a student)... gave me vital insight into the most engaging, comprehensive, and advantageous applications for D&T and Art. It is vital that trainee teachers experience the CASE project provided by the UoR, IoE, and I am excited to be contributing back to HE.' (2023)

Teacher Educators have said that:

'using the framework and teaching resources has provided a great starting point'
'...it has allowed me to further research areas of knowledge that were limited'
'..enabling a range of teacher educators to tackle this in a variety of ways to ensure that the work undertaken with out students becomes embedded rather than tokenistic'.

The Nature Park pilot project has demonstrated anecdotal opportunities to enhance university student learning where several students interested in careers in environmental education and science communication, or are in ITE have volunteered to support sessions. This has enhanced their practical experience and knowledge of working with schools, allowing them to observe how more experienced staff deliver outdoor education activities as well as being mentored to be able to develop and lead sessions themselves. Expanding their understanding that climate education and inspiring a connection to nature can be delivered in variety of ways across a range of subjects. But research needs to be conducted to understand any significant benefits.

- Implementation Challenges and Barriers

Challenges and barriers for implementation include time, funding, the academic contracts of educators, lack of knowledge and understanding, and lack of engagement in practitioner research. As well as national curriculum constraints. The next section explores some of these challenges and barriers in more detail.

- Curriculum constraints and time limitations

National Curriculum can create a barrier as teacher educators and absent themselves from engagement as the teachers in schools may say that it is not their responsibility. Often teacher educators, not in the traditional subjects associated with climate and sustainability education (geography, sciences, citizenship) would say that it does not relate to the content of their curriculum.

Due to time constraints it may also be that those engaged in delivering their subjects would argue that if this is not being examined on then schools will not support this work.

- Teacher readiness and expertise

This work highlighted the initial engagement with teacher educators who need to build their confidence and competence to firstly design and deliver this work as well as to empower their pre-service teachers.

- Strategies to overcome challenges & lessons learnt

The Curriculum and Assessment Review report acknowledges the knock on impact of this Strategy for exam boards and publishers who are beginning to development schemes of work. The challenges of time, finance and expertise can be supported through the other DfE strategies through the national education nature park projects and climate ambassador schemes. For pre-service teaching teacher education programmes can take steps to make changes to their delivery without extensive spending and time. It is about finding ways for programme directors and their teams to imagine what is possible with their existing provision and then reflecting the ITE framework into that. It is about developing flexible approaches that the ITE framework afforded opportunities for. Thenga et al. (2020) found that rigid approaches to ESD through the Fundisa for Change project did not have the impact that is expected. This requires scheduling time to discuss gaps in expertise; and building understanding through CPD. It was also about scaffolding that understanding and providing some starting point teach resources and exemplars. It is about connecting up with other expertise across your own university and wider community. In the future, when the DfE complete their review of the ITTECF framework there will be further opportunities for supporting this work and driving further change through ITE.

Policies need to be interlinked with practices, as found by Meekaew and Saenhum (2025) in their bridging policy and practice in climate change education for teacher training in Northeast Thailand. Both Meekaew and Saenhum (2025) and Assaraf et al. (2023) noted the importance of policy makers directly working with practitioners with governments needing to take leadership on developing evidence base climate change curriculum and providing appropriate professional development (Assaraf et al., 2023 p.209). Climate change education needs to be holistic and a social learning process to avoid ‘death by formal curriculum’ (Kagawa & Selby, 2009).

There are clear implications for research in initial teacher education. These include enabling practitioner researchers to demonstrate the impact of their implementation work. Teacher Educators needs to be supported to engage in research, or work in collaboration with research communities. The research demonstrating impact is essential to drive policy change and wider national and local support. Further research to explore the impact of this work through a longitudinal study, in different context, settings and phases would be beneficial for driving policy.

CONCLUSION

The chapter has showcases different ways for implementation and highlights some of the challenges and barriers encountered. These include curriculum constraints, time pressures, and varying levels of teacher readiness and expertise that connect to both the policy and practice boundaries. The chapter identifies strategies that supported progress for embedding approaches to climate and sustainability education across initial teacher education and highlights lessons learnt through that process.

Policy

Alongside these policy drivers, outlined in Figure 1, the practices outlined demonstrating how institutional resources play a critical role through leadership commitment within higher education institutions, staff expertise, continuous professional development, partnerships with schools and NGOs, and the allocation of funding, time, and curriculum space. This all requires engagement through policy linked with practice.

The national roll-out of the ITE framework is still on-going, with informal dissemination through organisations such as Department for Education, Universities College for Educating Teachers, Conferences, NCEAP group, and Royal Meteorological Society. Sixteen higher education teacher training institutions and five school centred teacher training institutions have joined the community of practice from across the four nations of the United Kingdom. By 2025 the framework gained international recognition with a growing number of higher education institutions signing up for the international community of practice.

The embedding of climate and sustainability education within initial teacher education is a new and emerging field strengthened by the various Department for Education initiatives in England. The theoretical framework and pedagogical approach that underpin this ITE Framework is social constructivism, teacher agency theory and ESD competencies. These pedagogical approaches are aligned to the training of teachers as well as pedagogies around ESD.

The limiting aspects for engagement of teachers, training teachers and teacher educators are the content of the ITTECF curriculum and national curriculum. The re-drafting of the National Curriculum in England offers opportunities across for further engagement in this important area of learning. Secondary education, in particular will have explicit climate and sustainability education explicitly mentioned within the subjects: Geography, Physics, Chemistry, Biology and Design Technology. Further work needs to be done to demonstrate the benefits and opportunities for interdisciplinary work across all phases of learning. The Nature Park project also offers opportunity for further connections with our local schools and outdoor education opportunities.

The current landscape of the policy in England is guidance; with the pending Sustainability and Climate Education Policy refresh; the National Curriculum review; Climate Ambassadors Scheme; National Education Nature Parks all providing openings for support and funding at the schools level. The next steps review of the

ITTECF framework might open up opportunities for a policy level drive to support this at the initial teacher education level. The authors remain optimistic.

Theory

This approach could be viewed as a theory of change model for embedding climate and sustainability education in initial teacher education (ITE). As well as an evidence-informed pathway for transforming teacher preparation in response to escalating global environmental challenges. The evidence base, including surveys highlighting teacher confidence gaps and research on ESD, social constructivism, and teacher agency, further legitimises the need for systemic change (Reed Johnson et al., 2025).

The theory of change model anticipates short-term outcomes for pre-service teachers, including increased climate knowledge, greater confidence to teach climate change, which necessitates adoption of social constructivist, dialogic, and action-oriented pedagogies. Pre-service teachers develop ESD-aligned competences such as critical thinking, ethical reasoning, collaboration, and context-sensitive innovation. In the medium term, these pre-service teacher-level changes translate into school-level impacts: more frequent and higher-quality climate education, cross-curricular integration, school-based sustainability projects, and strengthened teacher agency to lead climate initiatives. Ultimately, the Theory of Change aims toward long-term system and societal impacts. These include a teaching workforce equipped for a climate-changing world, learners with climate literacy and resilience, contributions to national sustainability goals, and a broader cultural shift toward just and sustainable futures within the education system.

Others who have written about the possibilities of theory of change to tackle the climate crisis in a UK university are Bookbinder, Mdee and Roelich (2024). They argue that a flexible and democratic approach to change is essential to mitigate tensions that arise that might block engagement in climate change education. These tensions include time, access to resources, professional development and requirements stated in the national curriculum and ITEECF (initial teacher education).

Practice

These case stories offer opportunities for hope and possibilities of what can be achieved in embedding our climate change education throughout our teacher education programmes, schools programmes and through more innovative ways of learning, such as outdoors. Connecting us to nature is a powerful mechanism for

REFERENCES

Assaraf, O., Dawson, V., Eilam, E., Gokpinar, T., Goldman, D., Naugauker, N., Agung, G., Eka Putri, P., Wijaya Subiantoro, A., Tolppanen, S., White, P., Widdop Quinton, H., & Dillon, J. (2023). Climate change education implementation: The voices of policymakers, professional development providers, and teachers in five countries. *International Journal of Science Education*, *47*(2), 191–213. DOI: 10.1080/09500693.2024.2314572

Bookbinder, R., Mdee, A., & Roelich, K. (2024). The possibility of a theory of change to tackle the climate crisis in a UK university. *International Journal of Sustainability in Higher Education*, *25*(8), 1929–1944. DOI: 10.1108/IJSHE-05-2023-0185

Dawson, V., Eilam, E., Tolppanen, S., Assaraf, O. B. Z., Gokpinar, T., Goldman, D., Putri, G. A. P. E., Subiantoro, A. W., White, P., & Widdop Quinton, H. (2022). A cross-country comparison of climate change in middle school science and geography curricula. *International Journal of Science Education*, *44*(9), 1379–1398. DOI: 10.1080/09500693.2022.2078011

Department for Education. (2013). The national curriculum in England key stages 1 and 2 framework document. In Gov.UK (pp. 1–201). Gov.UK. https://assets.publishing.service.gov.uk/media/5a81a9abe5274a2e8ab55319/PRIMARY_national_curriculum.pdf

Department for Education. (2021). Sustainability & climate change: A draft strategy for the education & children's services systems. https://assets.publishing.service.gov.uk/government

Department for Education. (2022). Sustainability and climate change strategy. London: Crown Publication Retrieved from https://www.gov.uk/government/publications/sustainability-and-climate-change-strategy/sustainability-andclimate-change-a-strategy-for-the-education-and-childrens-services-systems

Department for Education. (2023). *Sustainability and climate change: A strategy for the education and children's services systems (Updated policy paper)*. Crown Publishing.

Department for Education (DfE). (2024). *Initial Teacher Training and Early Careers Framework (ITTECF)*. Crown Publishing.

Department for Education (DfE). (2025). *Curriculum and Assessment Review: Building World Class Curriculum for All Final Report*. Crown Publishing.

Dunlop, L., & Rushton, E. A. C. (2022). Putting climate change at the heart of education: Is England's strategy a placebo for policy? *British Educational Research Journal, 48*(6), 1083–1101. DOI: 10.1002/berj.3816

Greer, K., Kitson, A., Rushton, E. A. C., Walshe, N., & Dillon, J. (2022). Teaching climate change and sustainability in England: Committed individuals and the prevalence of 'self-taught' professional learning. *Professional Development in Education*, 1–23.

Intergovernmental Panel on Climate Change (IPCC). (2021). *Climate change 2021: The physical science basis.* Cambridge University Press., DOI: 10.1017/9781009157896

IPCC. (2022a). Mitigation of Climate Change. UNEP. WMO. 2022. Available online: https://report.ipcc.ch/ar6wg3/pdf/IPCC_AR6_WGIII_FinalDraft_FullReport.pdf (accessed on 15 January 2023).

IPCC. (2022b). [Impacts, Adaptation and Vulnerability. Intergovernmental Panel on Climate Change.]. *Climatic Change*, 2022.

Johansen, B. (2007). *Getting There Early*. Institute of the Future.

Kagawa, F., & Selby, D. (Eds.). (2009). *Education and Climate Change: Living and Learning in interesting times*. Taylor & Francis Group.

Lotz-Sisitka, H., Wals, A., Kronlid, D., & McGarry, D. (2015). Transformative, transgressive social learning: rethinking higher education pedagogy in times of systemic global dysfunction, Current Opinion in Environmental Sustainability, Vol. 16, pp.73-80, ISSN 1877-3435, DOI: 10.1016/j.cosust.2015.07.018

Majid, N., Marston, S., Reed Johnson, J. A., & Happle, A. (2023). Reconceptualising Preservice Teachers' Subject Knowledge in Climate Change and Sustainability Education: A Framework for Initial Teacher Education from England, UK. *Sustainability (Basel), 2023*(15), 12237. https://www.mdpi.com/2429150. DOI: 10.3390/su151612237

Meekaew, N., & Saenkum, D. (2025). Bridging policy and practice in climate change education: Evidence from Northeast Thailand. *Frontiers in Education, 10*, 1561574. DOI: 10.3389/feduc.2025.1561574

Met Office. (2023). What do we mean by a climate tipping point? Met Office. https://www.metoffice.gov.uk/blog/2023/what-do-we-mean-by-a-climate-tipping-point

National Climate Education Action Plan. (2021). *National Climate Education Action Plan*. NCEAP.

National College Teaching and Leadership (NCTL) (2013). Teachers' Standards (Early Years). Reference: NCTL-00108-2013.

Priestley, M., Biesta, G., & Robinson, S. (2015). *Teacher Agency: An Ecological Approach*. Bloomsbury., DOI: 10.4324/9781315678573-15

Priestley, M., Biesta, G., & Robinson, S. (2015). *Teacher agency: An ecological approach* (1st ed.). Bloomsbury Academic., https://www.bloomsbury.com/uk/teacher-agency-9781472534668/ DOI: 10.4324/9781315678573-15

Reed Johnson, J. A., Marston, S., Happle, A., & Majid, N. (2025). *Implementation of a Climate Change and Sustainability Education Framework into pre-service teaching: opportunities for initial teacher education. Cogent Education Journal 2023*

Reichmann, M. (2018). Learing to Transform the World: Key competencies in education for sustainable development Chp.2 In Issues and trends in education for sustainable development, Leicht, A., Heiss, J. and Byun, W.J. (eds). UNESCO DOE .org/DOI: 10.54675/YELO2332

Roberts, M. (2003). *Learning through Geographical Enquiry: Making Sense of the Key Stage 3 Classroom*. The Geographical Association.

SOS-UK. (2021). *Climate education survey report*. Students Organising for Sustainability UK.

Teach the Future. (2022). Curriculum for a Changing Climate: Subjects. 2021. Available online: https://www.teachthefuture.uk/blog/curriculum-for-a-changing-climate-science (accessed on 12 February 2026).

Teacher the Future. (2022). Our Research. https://www.teachthefuture.uk/research

Thenga, M., Goldschagg,P., Ferguson,R., Mandikonza,C. (2020). Teacher Professional Development and Geography Teachers' Pedagogical Practices for Climate Change Education. Southern African Journal of Environmental Education Vol 36.

UNESCO. (2017). Education for sustainable development goals: Learning objectives. UNESCO Publishing. https://doi.org/ (2018). Issues and trends in education for sustainable development Leicht, A., Heiss, J. and Byun, W.J. (eds). UNESCO DOE.org/DOI: 10.54675/CGBA9153UNESCO

United Nations. (2021). COP26 The Glasgow Climate Pact. United Nations Climate Change Conference UK 2021, Glasgow.

United Nations. (2023). Goal 14 | department of economic and social affairs. Sdgs.un.org; United Nations. https://sdgs.un.org/goals/goal14#targets_and_indicators

United Nations. (2024b). Goal 15 | department of economic and social affairs. Sdgs. un.org. https://sdgs.un.org/goals/goal15#targets_and_indicators

United Nations. (2025a). Goal 12 | ensure sustainable consumption and production patterns. Sdgs.un.org; United Nations. https://sdgs.un.org/goals/goal12#targets_and_indicators

United Nations. (2025b). Goal 13 | department of economic and social affairs. United Nations. https://sdgs.un.org/goals/goal13#targets_and_indicators

United Nations Educational Scientific and Cultural Organisation. (2019). Framework for the implementation of education for sustainable development (ESD) beyond 2019. General Conference 40th Session, Paris.

Walshe, N., Perry, J., & Healy, G. (2025). Student perspectives on climate change and sustainability education in England: Experiences and expectations. *UCL Open Environment*, *7*(1). Advance online publication. DOI: 10.14324/111.444/ucloe.3341 PMID: 40661851

World Wildlife Fund. (2025). What are climate tipping points? World Wildlife Fund. https://www.worldwildlife.org/resources/explainers/what-you-need-to-know-about-tipping-points/

Chapter 3
Empowering South African Teachers for Interdisciplinary Climate Change Education

Ntandokamenzi Penelope Dlamini
http://orcid.org/0000-0002-6152-5607
University of South Africa, South Africa

Dumsani Wilfred Mncube
http://orcid.org/0000-0001-5566-2288
University of South Africa, South Africa

ABSTRACT

Climate change education presents unique challenges that demand interdisciplinary approaches and innovative teaching strategies. This chapter explores how South African teachers can be supported and empowered to integrate climate change into their teaching across multiple subjects, including languages, science, geography, and mathematics. Drawing on national curriculum policies and teacher development programmes, it highlights the importance of connecting classroom learning to real-world environmental issues to foster critical thinking, collaboration, and problem-solving among learners. The chapter argues that empowering teachers requires more than enhancing subject knowledge; it involves building confidence, encouraging teamwork, and equipping educators with practical tools to design meaningful lessons.

DOI: 10.4018/979-8-3373-6975-4.ch003

INTRODUCTION

Revisiting the continuous professional development of teachers to address climate change, one of the most urgent humanitarian and environmental crises of our time, continues to gain traction. Studies place greater emphasis on the need for transformation in the education systems. These unexpected changes have become increasingly concerning since 2020, culminating in the identification of climate change education (CCE) as a critical solution that goes beyond technical knowledge to embrace interdisciplinary, action-oriented, and socially responsive pedagogies (Morales-Aguilar et al., 2025). Mounting scientific evidence from the Keep it Cool (KIC) investigation points to a need for Higher Education and Basic education to take on the CCE impact of this shift. In the main, this shift forced the Global South to strengthen resilience strategies through informed and empowered citizenship. Education is a powerful vehicle positioned at the nexus of knowledge production, critical thinking, and community engagement. Therefore, the development of strong capacity-building mechanisms for teachers can become a central pillar of national climate frameworks to support the UN Sustainable Development Goals (SDGs 4 and 13). This goal advocates integrating sustainability and climate action at all levels of schooling (Mthethwa et al., 2025).

Climate change is a complex global phenomenon driven primarily by greenhouse gas emissions from human activities such as fossil fuel combustion, deforestation, and industrial processes (IPCC, 2023; NASA, 2025). These emissions trap heat in the atmosphere, causing global temperatures to rise and producing cascading effects such as sea-level rise, extreme weather events, and biodiversity loss (IPCC, 2023; NOAA, 2024). Understanding these scientific mechanisms is critical for educators, as it provides the foundation for teaching climate change in an integrated and evidence-based way (UNESCO, 2021; Lehtonen et al., 2018). Without a strong grasp of climate science, teachers may struggle to connect ecological, social, and economic dimensions in their classrooms, limiting the effectiveness of interdisciplinary and transformative pedagogies (Darling-Hammond et al., 2017; Naidoo & Heath, 2024).

Grounding teacher preparation in core climate science concepts enables both pre-service and in-service professional development to more effectively support teachers' ability to deliver accurate and integrated climate change education. In response to these challenges, countries across the Global South have strengthened resilience strategies through informed and empowered citizenship. Likewise, international approaches vary: in Europe and North America, teacher education programs often include mandatory climate-focused modules, online professional development platforms, and interdisciplinary school projects that promote experiential learning. In Southeast Asia, governments integrate climate literacy frameworks into national curricula and support teachers through targeted training and regional knowledge-sharing networks

(UNESCO, 2021). Meanwhile, in Latin America, programs emphasize community engagement and the integration of indigenous knowledge to enhance teachers' ability to deliver contextually relevant climate education (Morales-Aguilar et al., 2025). These diverse strategies highlight how teacher preparation, curriculum integration, and systemic support differ internationally, providing a useful comparative lens for understanding the South African context.

The growing trend points to the government's appetite to address the possible threat climate change poses to our environment and human existence. The government solution lies in working with local and international agencies to develop a curriculum policy that implements a multidisciplinary approach integrating climate change content to address impacts such as droughts, flooding, water scarcity, and heatwaves. Studies show that teachers can play a decisive role in developing knowledge, skills, values, and attitudes that shape learners' climate literacy and adaptive competencies. High-stakes assessments and standardized testing regimes further constrain teachers' flexibility to implement integrated climate change education, reinforcing content coverage over critical thinking, systems thinking, and sustainability competencies (Thenga, 2025; Govender et al., 2025). Though empirical studies praise government intervention in addressing policy imperatives, educators and lecturers feel underprepared to implement curriculum reforms. The evidence from the KiC project shows that most teachers fail to deliver climate-responsive lessons that draw on natural, social, and human sciences, geography, economics, and indigenous knowledge systems. In one study, it was clear that teachers struggle to develop content knowledge of climate change and to infuse pedagogical strategies with confidence to integrate it meaningfully across subject areas (Mzilikazi, Khalo & Caga, 2025). Despite the praise KIC has received for its efforts to introduce curriculum-aligned climate change, the lack of confidence, insufficient training, poor policy coordination, and the absence of LTSM in schools all derail effective classroom and school-wide implementation of CCE.

Specific subjects identified in the Curriculum and Assessment Policy Statement (CAPS) can implement CCE within the curriculum. Furthermore, research indicates that while climate change topics appear in South Africa's Geography and Natural Sciences curricula at various levels, subject-specific teachers face challenges in implementing these curricula. These include prescriptive lesson plans, assessment pressures, and a lack of professional autonomy, which have consistently been proven to restrict their agency and creativity in CCE (Thenga, 2025). Such constraints mirror global trends and confirm that education systems must first strengthen teacher development, curriculum flexibility, and cross-disciplinary integration before they are tested on scale. Well-conceived initiatives meant to prepare learners to pass the means test, such as complex socio-ecological challenges (Morales-Aguilar et al., 2025). The South African context also highlights the importance of aligning school-based climate education with community realities, as many learners live in environments directly affected by climate-induced food insecurity, infrastructure damage, and water shortages.

First, this chapter primarily uses a systematic review to assess the feasibility of implementing an interdisciplinary approach to CCE teaching. In essence, practical pedagogical teaching and learning strategies must be investigated and proposed for such an ambitious, noble course in the curriculum system. Most South African scholars continue to formulate hypotheses for integrating CCE. However, some argue that this call is ambitious and that transformative approaches to teacher empowerment emphasise collaborative professional development, community partnerships, and interdisciplinary training models, such as the Keep it Cool – Climate Change Education (KIC-CCE) initiative. This initiative demonstrates how targeted support can strengthen teachers' reflective practice, environmental literacy, and pedagogical adaptability (Govender et al., 2025). Similarly, research on curriculum reform argues that integrating climate education into the Curriculum and Assessment Policy Statement (CAPS) is essential to promote resilience and foster deeper learning across subjects (Matsepe & Maluleke, 2024). Collectively, literature from 2020–2025 underscores a growing consensus: empowering teachers through interdisciplinary climate change education is not merely desirable, it is essential for building a climate-resilient South African society.

LITERATURE REVIEW

Climate change education (CCE) has been widely discussed in international scholarship, particularly within the broader field of education for sustainable development (ESD). Global research highlights that CCE is not only about transferring scientific knowledge but also about developing critical thinking, ethical awareness, and the ability to respond to complex, real-world challenges (UNESCO, 2017; UNESCO, 2020). For the purposes of this chapter, climate literacy is defined as the knowledge, skills, and dispositions necessary to understand climate systems and respond effectively to climate change (UNESCO, 2021). Interdisciplinary learning integrates multiple disciplinary perspectives to address complex socio-ecological problems (Beane, 1997), while transformative pedagogy seeks to shift learners' values and practices toward sustainable outcomes (Sterling, 2010). Systems thinking is another key concept, enabling learners and teachers to explore the relationships between environmental, social, and economic systems rather than viewing knowledge in isolated subject areas (Lehtonen et al., 2018; Tilbury, 2011).

Within this global context, the South African literature reflects similar concerns, particularly regarding the need to support teachers in translating these ideas into classroom practice. Recent studies indicate gaps in teacher knowledge and pedagogical readiness (Mzilikazi, Khalo & Caga, 2025). The Department of Basic Education has developed national policy frameworks to guide integrated mitigation and adaptation strategies, including the National Climate Change Response White Paper (2017), which documents South Africa's vulnerability to climate change and provides crucial implementation guidelines. These policies underscore the importance of linking teacher professional development, curriculum reform, and classroom practice to support meaningful climate education. International theoretical debates further emphasise transformative learning in climate education. Transformative learning theory suggests that meaningful learning occurs when individuals critically reflect on their assumptions and experiences, leading to changes in perspectives and actions (Mezirow, 1997; Taylor, 2007). In the South African context, this highlights the importance of moving beyond subject boundaries to support more connected and meaningful learning experiences that prepare learners not only to understand climate change but also to act on it (UNESCO, 2020; UNESCO, 2021).

This approach led to the establishment of imperatives, led by Fundisa for Change and the KIC team, to develop an integrated climate change response, in which climate change education and new ideas were to be addressed through formal education. Motsepe and Maluleke (2024) argue that the CAPS curriculum is fragmented, resulting in difficulties in adequately embedding climate change concepts and basic literacy necessary to address an interdisciplinary approach.

Teachers' Content Knowledge and Pedagogical Readiness

From an international perspective, teachers play a central role in enabling transformative and interdisciplinary climate change education. Research shows that teachers require not only content knowledge, but also the pedagogical skills and confidence to facilitate critical, participatory, and action-oriented learning (UNESCO, 2021; UNESCO, 2024). This includes the ability to guide discussions, support inquiry-based learning, and connect classroom content to learners' lived experiences (Darling-Hammond et al., 2017). These global insights reinforce the need for sustained professional development and support structures for teachers, particularly in contexts where resources and training opportunities may be limited (Avalos, 2011). These insights resonate strongly with the challenges identified in South African schools, where teachers often require additional support to apply such approaches in practice. Most studies agree that South African teachers are struggling to implement CCE due to curriculum demand and classroom realities (Thenga et al, 2020). In essence, the CAPS curriculum includes a few selected subjects (Natural Science, Life Sciences, Geography FET, and Social Science) that address CC issues directly. However, teachers are uncomfortable infusing climate change during the teaching of these subjects because they were not adequately trained due to a lack of professional development programs. It was clear that in-service teacher development programs attended by teachers significantly improved their practice and lesson delivery (Olawumi et al., 2023). According to Mzilikazi, Khalo and Caga (2025), there exists a potential gap in both teachers' content knowledge and PCK that stifles the delivery of innovative and effective classroom integration of CCE. Linked to this finding is the realisation by teachers that implementing an interdisciplinary curriculum proved more complicated due to limited training. Olawumi et al. (2023) highlighted the critical role of sustained teachers' professional development (TPD) in strengthening both content knowledge and PCK for teachers across the spectrum. This provides teachers with opportunities to reflect on their practices and on the purpose of integrating interdisciplinary teaching methods for climate literacy.

Professional Development Interventions and Teacher Empowerment

Professional development (PD) initiatives consistently serve as a strong mechanism for shaping teacher quality, addressing professional development needs, and enhancing teacher empowerment in contemporary education systems. This study argues that schools have a serious responsibility to place teachers at the heart of curriculum development and implementation. Emphasising teachers' empowerment is likely to enhance their autonomy, self-efficacy, leadership, decision-making power, and professional identity. For example, in South Africa, the establishment of Fundisa for Change came at a time when curriculum transformation was at the crossroads. This same initiative was triggered by teachers' limited conceptual understanding of climate change, due to insufficient professional development. Huoponen (2023) supported the initiative by Fundisa for Change and all government departments to join hands with NGOs in casting a spotlight towards environmental education, where issues of environmental literacy are addressed. In other words, targeted teacher-training initiatives were needed to enhance teachers' knowledge of climate-related curriculum tasks across almost every subject (Thenga, 2025; Naidoo & Heath, 2024). Hence, the need for ongoing professional development in climate change education cannot be overstated; however, practical programs that support all stakeholders offer an opportunity to implement CCE.

Integrating Climate Change Education into Initial Teacher Education

While this chapter has highlighted the importance of in-service professional development in strengthening teachers' capacity for climate change education, less attention has been given to how teachers are prepared before entering the profession. Initial Teacher Education (ITE) programmes offered by universities play a critical role in shaping teachers' knowledge, beliefs, and pedagogical approaches (Darling-Hammond, 2006; Zeichner, 2010). Both authors emphasise that early professional formation strongly influences classroom practice; however, their work pays limited attention to emerging areas such as climate change education. In contrast, more recent policy-oriented studies (UNESCO, 2017; Naidoo & Heath, 2024) explicitly highlight climate change as a gap in teacher preparation, arguing that many teachers enter the profession underprepared to teach climate-related content in integrated and meaningful ways. This suggests a disconnect between general teacher education theory and the evolving demands of sustainability-focused curricula.

One important area for development is the integration of climate change education within Bachelor of Education (B.Ed.) programmes through curriculum mapping. International frameworks (Tilbury, 2011; UNESCO, 2020) strongly advocate for embedding sustainability across the curriculum rather than treating it as an isolated topic. These perspectives align with Lehtonen et al. (2018), who argue that complex issues such as climate change require interconnected and systems-based approaches to teaching. However, while these authors agree on the need for integration, they differ in emphasis: UNESCO (2020) focuses more on policy-level curriculum alignment, whereas Lehtonen et al. (2018) place greater emphasis on pedagogical transformation in classroom practice. Together, they suggest that effective integration requires both structural curriculum reform and shifts in teaching practice. Universities can strengthen teacher preparation by introducing interdisciplinary modules focused specifically on climate change education and sustainability. Sterling (2010) argues that sustainability education requires a fundamental shift in how knowledge is structured and taught, moving toward transformative and holistic learning. Similarly, UNESCO (2021) promotes action-oriented and competency-based approaches. While both perspectives support transformative learning, Evans et al. (2017) provide a more practice-oriented view, showing that without structured opportunities to apply these approaches, student teachers may struggle to translate theory into practice. This highlights a tension between conceptual ambition and practical implementation within teacher education programmes.

University–school partnerships are widely recognised as a key mechanism for bridging this gap between theory and practice. Zeichner (2010) critiques traditional teacher education models for separating university coursework from classroom experience, while Darling-Hammond et al. (2017) advocate for more integrated models that link theory, practice, and reflection. In the context of climate change education, such partnerships become even more important, as they allow student teachers to experiment with climate-focused lesson planning in real-world settings. However, the effectiveness of these partnerships often depends on contextual factors such as school resources and mentor support, particularly in under-resourced environments. There is a need for stronger alignment between initial teacher education programmes and national policy frameworks, particularly those guided by the Department of Higher Education and Training (DHET). While policy frameworks (DHET, 2015; UNESCO, 2020) call for the integration of sustainability into education systems, implementation at the programme level remains uneven. Avalos (2011) argues that teacher development must be viewed as a continuum, linking pre-service and in-service phases; however, in practice, these phases are often poorly coordinated. This lack of alignment limits the effectiveness of both initial preparation and ongoing professional development. Strengthening coherence across these levels would support a more systemic approach to climate change education and ensure that teachers are better prepared from the outset.

Policy Context and National Imperatives

Studies point to several engagements that culminated in streamlined South African national policy landscapes aimed at addressing climate change education through adaptation, mitigation, and resilience strategies (Education Strategy Plan, 2020). The National Climate Change Response White Paper (NCCRP) gave the green light to commence the public participation initiative for South Africa to consider investment in policy alignment. Based on this consideration by NCCRP, the outcome directed that climate change becomes a cross-cutting developmental issue. The directive supported the need for interdisciplinary teaching approaches, which directed teachers to reconsider their pedagogical strategies. All teachers were expected to move beyond siloed subject mentality by embracing systems thinking (Kurent & Avsec, 2026). Public consultation and investment in building policy alignment attracted interest from interested parties. The report, compiled by Fundisa for Change and UNESCO's Sustainability Starts with Teachers programmes, casts a spotlight on teachers' professional development and on curriculum alignment with sustainability priorities. Furthermore, this global framework significantly influenced education reforms in South Africa, resulting in the coordination of the national-level curriculum masterplan to address climate change. Similarly, the National Development Plan 2030 draws a parallel between the need to streamline curriculum policy guidelines and the practical realities highlighted by SDG 4, which aims to advance inclusive and quality education for all (National Planning Commission (NPC), 2013). In essence, Goal 4 calls for dramatic shifts in curriculum implementation, advocating the integration of education for sustainable development (ESD) into local curriculum frameworks, teacher training, and the learning environment (UNEP/UN SDG Global Database, 2025; UNESCO, 2017; UNESCO, 2021; 2022; 2023a). This occurs after the realisation that education is one of the most powerful and proven vehicles for sustainable development.

Reimagining Interdisciplinary and Transformative Pedagogies for CCE

The Education system in South Africa is outcomes-driven, under the Curriculum, Assessment, and Policy Framework (CAPS), which serves as a vehicle for formulating and achieving learning outcomes (DHE, 2019). Therefore, the complexities surrounding the enactment of curriculum demands require assessing the contexts in which the majority of vulnerable communities reside. Most studies point to the need for the development of advanced teaching materials, but also aligned to digital OER to assist educators in supporting integrative learning in order to foster "deep" and flexible thinking that forces them to move away from a silos mentality (Matsepe & Maluleke, 2024; James, 2022). According to Vogel and Schaibold (2015), South Africa can develop transformative climate change education that involves systems thinking and a creative, critical approach. In the main, systems thinking promotes a safe space within the curriculum for learners and educators to engage in addressing climate change as intended. This culminated in a programme called Teaching and Learning for Change, which championed TPD to promote quality, interdisciplinary education aligned with sustainability goals (James, 2022). These international perspectives complement the South African literature by providing a broader theoretical foundation for understanding how interdisciplinary climate change education can be effectively implemented across diverse, resource-constrained contexts (UNESCO, 2020; Tilbury, 2011). Taken together, both global and local perspectives point to the need for stronger alignment between theory, policy, and classroom practice. Education in South Africa aligns with the international education for sustainable development (ESD) agenda, which supports a cross-curricular approach grounded in competencies for sustainable development.

AIM OF THE STUDY

This study aims to explore practical strategies for empowering South African teachers to implement interdisciplinary Climate Change Education. This study specifically focuses on the approaches teachers use to improve climate change education and proposes alternative practices for South African schools.

The objectives of the study are framed to:

- Determine innovative interdisciplinary teaching strategies aligned with the CAPS curriculum to teach climate change education in South Africa.
- To propose possible alternative teaching strategies that can improve the teaching of climate change education.

THEORETICAL FRAMEWORK

A theoretical framework provides a conceptual lens that guides the organisation, analysis, and interpretation of a body of literature. In a systematic review, theory is used not to explain participant behaviour directly, but to structure the synthesis of published research and to support the interpretation of recurring patterns and conceptual relationships across studies (Grant & Osanloo, 2014; Kivunja, 2018). In education research, theory-informed reviews are particularly valuable because they help illuminate how social, cognitive, motivational, and contextual factors are represented in existing scholarship. This systematic review is guided by an integrated framework combining Sociocultural Theory and Teacher Empowerment Theory to interpret research on teacher professional development and climate change education in resource-constrained school contexts.

From a sociocultural perspective (Vygotsky, 1978), teacher learning is understood as socially mediated and contextually situated, developing through collaboration, dialogue, mentoring, and participation in professional communities rather than through isolated individual activity. This lens helps interpret how the reviewed studies conceptualise professional development models such as peer coaching, collaborative lesson design, and communities of practice. The concept of the Zone of Proximal Development (ZPD) further supports analysis of how scaffolding structures, including guided practice, modelling, and structured support, are described across the literature. Sociocultural Theory also foregrounds the importance of context, enabling the synthesis to account for how studies treat environmental constraints such as limited resources, curriculum pressures, and institutional conditions, particularly in rural and under-resourced settings.

Complementing this perspective is Teacher Empowerment Theory, which focuses on professional agency, instructional decision-making, motivation, and confidence (Short & Rinehart, 1992; Tschannen-Moran & Hoy, 2001). This lens supports the interpretation of how the literature describes teachers' capacity to adopt interdisciplinary approaches, adapt curriculum content, and implement innovative pedagogical practices. In this chapter, interdisciplinary learning is understood as integrating knowledge across subjects such as science, geography, and social studies to address climate challenges in an integrated way (Beane, 1997; Tilbury, 2011). Systems thinking supports climate change education by enabling teachers and learners to explore relationships between environmental, social, and economic systems, fostering holistic understanding and problem-solving (Sterling, 2010; Lehtonen et al., 2018). Transformative pedagogy emphasizes reflective, action-oriented learning that reshapes learners' values, attitudes, and practices toward sustainability (Sterling, 2010; Mezirow, 1997). Climate literacy encompasses the knowledge, skills, and dispositions necessary to understand climate systems and respond effectively to climate change (UNESCO, 2021).

These concepts collectively guide the chapter's exploration of teacher preparation, professional development, curriculum integration, and classroom practice for effective climate change education in South African schools. Within the synthesis, empowerment functions as an interpretive lens for understanding how, in the literature, professional development is linked to sustained teacher engagement and practice change. The two theories, together with these conceptual definitions, provide a complementary analytical framework: Sociocultural Theory supports interpreting how professional learning is socially structured and contextually embedded, while Teacher Empowerment Theory aids understanding of agency, motivation, and professional confidence as enabling factors or outcomes. This integrated framework enables a balanced synthesis that accounts for collaborative learning processes, teacher agency, and the interdisciplinary competencies necessary for climate change education. The framework also informed the development of the thematic coding structure used in the review, particularly in categorising forms of professional development, collaborative learning structures, contextual constraints, and teacher agency outcomes, ensuring the synthesis was conceptually grounded while remaining responsive to patterns emerging from the reviewed studies.

RESEARCH METHODOLOGY

This study employs a systematic review methodology (Zawacki-Richter, Kerres, Bedenlier, Bond, & Buntins, 2020) to synthesise existing research on strategies that empower South African teachers for interdisciplinary climate change education. A systematic review is appropriate for identifying patterns, gaps, and insights from diverse academic sources, guiding policy, curriculum development, and pedagogical innovations (Davies, 2010). The review is framed around the following key research questions:

RQ1: What teacher professional development strategies support interdisciplinary climate change education in South Africa?

RQ2: What are the reported challenges and barriers faced by teachers in implementing interdisciplinary climate change education?

RQ3: What evidence-based strategies have been proposed to enhance teacher empowerment, curriculum integration, and pedagogical effectiveness in resource-constrained environments?

To ensure reliability and minimise bias, the study follows the Preferred Reporting Items for Systematic Reviews and Meta-Analyses (PRISMA) guidelines (Page et al., 2021; Heiser, Palalas, & Gollert, 2025).

ELIGIBILITY CRITERIA

The review focused on peer-reviewed journal articles published between 2010 and 2025. Both public and private schooling contexts were considered, with emphasis on South Africa. Studies were selected based on the following inclusion criteria:

Table 1.

CRITERION	*DETAILS*
Type of study	Original research articles presenting empirical findings
Publication type	Peer-reviewed journal articles
Language	English
Publication date range	From 2010 to 2025
Education level	Basic education (primary and high schools)
Region/context	South Africa
Focus	Teacher empowerment, professional development, or capacity building in climate change or sustainability education. Involvement of interdisciplinary or cross-curricular teaching approaches.

Studies were excluded if they focused solely on learners, without considering teachers or teacher-related processes. Publications that addressed climate change only from a scientific or technical perspective, without an educational or pedagogical component, were also excluded. The review further excluded research conducted outside South Africa to maintain contextual relevance. Moreover, opinion pieces, essays, and commentary articles lacking empirical research evidence were not considered. Only English-language publications were included; therefore, studies published in other languages were excluded.

DATA SOURCES AND SEARCH STRATEGY

A comprehensive literature search was conducted using ERIC, DOAJ, Google Scholar, AOSIS Scholarly Book, and Interdisciplinary Journal of Education Research; Scopus and Web of Science; and Springer Nature, selected for their coverage of education, environmental studies, and teacher development research relevant to South African contexts. The search focused on publications from 2010 to 2025 to capture contemporary strategies and challenges in climate change education. Keywords included:

"climate change education", "interdisciplinary teaching", "multi-disciplinary teaching", "Continuous Professional Development", "teacher empowerment", "professional development", "sustainability education", "South Africa", "teacher development", "environmental education", "environmental literacy", "climate literacy", "transdisciplinary learning".

Boolean operators (AND, OR) were used to combine terms. Reference lists of relevant studies were also examined to identify additional sources.

SCREENING AND SELECTION PROCESS

Following PRISMA guidelines, the selection process involved four stages:

Identification: All studies identified through database searches were recorded, and duplicates were removed.

Screening: Titles and abstracts were reviewed to exclude studies that did not meet the inclusion criteria.

Eligibility: Full-text articles were reviewed against inclusion and exclusion criteria.

Inclusion: Studies meeting all criteria were selected for analysis.

A PRISMA flow diagram was used to document the number of studies at each stage, including reasons for exclusion, ensuring transparency and reproducibility.

The second-level screening involved full-text sources that were identified during the first-level screening; twelve items were retrieved for full-text review. After systematic identification and screening, nine articles were selected for coding and data extraction, as shown in Figure 1 and Table 2.

Table 2.

#	AUTHORS	YEAR	STUDY TITLE
1	Raymond Anyanwu, Lesley Le Grange and Peter Beets	2015	Climate change science: The literacy of Geography teachers in the Western Cape Province, South Africa
2	Yvonne Nsubuga, Ingrid Schudel, Zintle Songqwaru, and Helen Fox	2025	Exploring the Development of Teachers' Personal Pedagogical Content Knowledge about Climate Change Education during a Continuing Teacher Professional Development Course
3	Kayode B. Olawumi and Mzuyanda P. Mavuso	2024	In-Service Teacher Training Programmes that Promote Education for Sustainable Development: A Review of Emerging Literature
4	Coleen Vogel and Shanu Misser	2015	'Teaching and Learning for Climate Change' – the Role of Teacher Materials and Curriculum Design in South Africa
5	Nonkanyiso Pamela Shabalala	2025	Challenges, opportunities and strategies of distributed leadership in managing environmental education curriculum in South African secondary schools
6	Msezane Sikhulile	2023	Content and Teaching Methods Regarding Climate Change: South African Curriculum Perspective
7	Bonilla-Jurado, D., Zumba, E., Lucio-Quintana, A., Yerbabuena-Torres, C., Ramírez-Casco, A., & Guevara, C.	2024	Advancing University Education: Exploring the Benefits of Education for Sustainable Development
8	Awuzie and Emuze	2017	Promoting Sustainable Development Implementation in Higher Education: Universities in South Africa
9	Nakidien, Singh and Sayed	2021	Teachers and Teacher Education: Limitations and Possibilities of Attaining SDG 4 in South Africa

CODING AND DATA EXTRACTION

Data from the studies that met the inclusion criteria were extracted using a structured data extraction table explicitly designed for this review (van Wyk, 2025). The table captured important details from each study, including the context of the study (such as province and school type), the teacher professional development strategies described, the interdisciplinary or cross-curricular approaches used, the reported outcomes (for example, changes in teacher knowledge, skills, and pedagogy), as well as the challenges and limitations noted by the authors. Each selected article was carefully read, and in some cases more than once, to ensure the information recorded was accurate and consistent. After the extraction stage, a thematic analysis following Braun and Clarke's (2006) approach was conducted to bring the findings together across studies. This involved coding and grouping recurring ideas to identify common patterns and strategies related to teacher empowerment in resource-constrained contexts, with particular attention to institutional support, curriculum integration methods, pedagogical approaches, and teachers' views on what worked effectively.

LIMITATIONS

This systematic review has several limitations that should be considered when interpreting the findings. First, only English-language publications were included, mainly to ensure consistency in analysis and because of language accessibility constraints; however, this may have excluded relevant studies published in other languages. Future reviews could reduce this bias by including multilingual research or collaborating with translators. Second, the review drew only on selected academic databases and peer-reviewed sources to maintain quality control, thereby excluding grey literature such as policy reports, dissertations, and NGO publications. Future studies could incorporate structured grey literature searches to capture practice-based and policy-level insights that are often underrepresented in journals. Third, the focus on South African studies was intentional to maintain contextual relevance, but it limits the transferability of findings to other regions. Future research could conduct comparative or multi-country reviews to identify cross-context patterns and differences.

Another limitation concerns the variation in research designs, methods, and reporting standards across the included studies, which can make direct comparisons difficult; future primary research would benefit from more consistent reporting frameworks and greater methodological transparency. In addition, although PRISMA procedures were followed to improve rigour and transparency, the screening, coding, and thematic interpretation processes still involve researcher judgement, which introduces some interpretive subjectivity. This could be addressed in future reviews through multiple independent reviewers and inter-coder reliability checks. Finally, because systematic reviews rely on published studies, they may not fully reflect emerging practices or recent innovations that have not yet been documented in academic literature; ongoing review updates or living systematic reviews could help address this gap over time.

Presentation of findings

Theme 1: Teachers' Climate, Content Knowledge, and Pedagogical Readiness

Findings may highlight gaps in the teaching materials used in schools on climate science and how these shortcomings influence practitioners' confidence in teaching climate concepts. (Msezane, 2023). Climate change practitioners in both the public and private sectors, textbook authors, and educational policymakers face the dilemma of how to coordinate climate change response without the full support of the National Government through policy (Vogel & Schaibold, 2015; Shabalala, 2025).

This was evident during the Keep it Cool Climate Change Project, which showed that Natural Sciences and Geography teachers sampled across different provinces continue to struggle with conceptual clarity when integrating climate topics into instruction, thereby constraining interdisciplinary climate education (Mzilikazi, Khalo & Caga, 2025; Govender, Maphalala & Mncube). These findings suggest that teacher educators and policymakers should improve professional development programmes and support interventions to strengthen teachers' knowledge and understanding of climate change concepts, thereby enhancing climate change education in schools.

Teachers' climate, content knowledge, and pedagogical readiness remain weak, stemming directly from the lack of DBE coordination and response to climate change education in South Africa, as recent literature highlights persistent conceptual gaps and uncertainties among educators. Mzilikazi Khalo and Caga (2025) and Ajani and Govender (2025) demonstrated that teachers often struggle to integrate climate change pedagogies into the content knowledge they deliver, highlighting the underlying, unintended, inconsistent, and superficial teaching of climate change concepts at the Senior Phase level. This challenge is compounded by insufficient training and weak pedagogical content knowledge (PCK). At the same time, some contend that a lack of clear curriculum frameworks for integrating climate topics into disciplinary teaching collectively constrains teachers' ability to deliver coherent, scientifically accurate instruction (Mzilikazi, Khalo & Caga, 2025; Thenga, 2025).

The consensus among experts is that teachers' content knowledge remains uneven across subjects, signalling a systemic gap in pre-service and in-service teacher education that impacts interdisciplinary climate pedagogy (Vogel & Schwaibold, 2015). The FET band teachers reinforce this pattern, revealing that while many educators possess an elementary understanding of environmental issues, their ability to deliver interdisciplinary climate change education remains constrained by limited curriculum support (Nsubuga, Schudel, Songqwaru & Fox, 2025; Olawumi & Mavuso, 2024). Thenga's (2025) study agrees that prescriptive lesson plans and assessment-driven instruction impede teachers' flexibility to adapt or deepen climate teaching, limiting their capacity to integrate climate science with socio-economic, ecological, and local knowledge systems. According to Galdames-Calderón et al. (2024), there exist widespread challenges such as teachers' pedagogical readiness, shaped not only by individual knowledge but also by institutional structures that either enable or restrict interdisciplinary engagement. With strong curriculum frameworks and professional autonomy embedded in the Climate Change Policy, teachers cannot effectively link climate content across disciplinary boundaries (Singh & Sayed, 2024).

Systematic review findings from Morales-Aguilar et al. (2025) further emphasise that climate change education requires teachers to possess multi-layered knowledge that bridges scientific, socio-environmental, economic, and indigenous knowledge dimensions—what the authors describe as a progression from disciplinary to transdisciplinary pedagogical competence. These places heightened teachers' readiness demands, requiring them not only to understand climate science but also to facilitate systems thinking, ethical reasoning, and culturally grounded climate learning. South African evidence from the *Keep It Cool – Climate Change Education* (KIC-CCE) initiative suggests that when teachers engage in structured, collaborative professional development, their content knowledge and pedagogical capacity improve substantially, enabling more confident and integrated climate instruction. These findings collectively underscore that strengthening teachers' climate content knowledge and pedagogical readiness is essential for advancing interdisciplinary climate change education, and requires systemic support through curriculum reform, targeted professional development, and access to high-quality teaching resources.

Theme 2: Curriculum Constraints and Structural Barriers

CAPS provides some climate-related content, but ambiguity and fragmentation hinder teachers' ability to apply interdisciplinary approaches. Evidence shows that prescriptive lesson plans, assessment-driven teaching, and top-down instructional tools reduce teacher agency and limit integration of climate themes across subjects.

Furthermore, teachers report that CAPS lacks explicit, sufficient climate change content, making it difficult to achieve deeper learning or to embed climate competency in routine teaching.

For instance, research published in AOSIS-linked literature notes that prescribed lesson plans, standardised testing regimes, and tightly structured syllabi reduce teacher agency and restrict the meaningful integration of climate change across subjects, thereby reinforcing disciplinary silos rather than encouraging cross-curricular learning (Thenga, 2025). Similarly, AOSIS book scholarship emphasises that traditional, fact-based curricula continue to dominate, making it challenging to implement transformative, interdisciplinary climate education that integrates social, scientific and community knowledge systems (Govender et al., 2025). These curriculum-level limitations undermine the goals of interdisciplinary CCE by prioritising examination performance and content coverage over critical thinking, systems thinking and sustainability competencies.

Beyond curriculum design, structural barriers embedded in the education system further constrain teachers' capacity to deliver interdisciplinary climate education. AOSIS literature highlights systemic challenges such as limited teacher preparation, inadequate institutional support, resource shortages and weak policy coordination, all of which restrict practical curriculum innovation. For example, AOSIS-published research on curriculum integration notes that a lack of trained personnel, insufficient resources, resistance to change, and the absence of coherent guiding frameworks act as institutional barriers to implementing new educational priorities within teacher education programmes. In the context of climate change education, these structural constraints intersect with broader inequalities in school infrastructure, professional development access and support networks, resulting in uneven implementation across regions and schools. Structural inequalities and uneven resource distribution across schools exacerbate challenges in climate change education. Teachers in under-resourced settings often lack access to materials, professional networks, and collaborative opportunities, limiting their ability to deliver interdisciplinary and transformative learning (Maphalala, Kutame & Kok, 2025; Naidoo & Heath, 2024). AOSIS scholarship further stresses that without systemic investment in teacher capacity, interdisciplinary collaboration and curriculum reform, climate change education risks remaining fragmented and marginal rather than becoming an integrated, transformative component of South African schooling (Maphalala, Kutame & Kok, 2025).

Theme 3: Professional Development and Capacity-Building as Catalysts for Empowerment

Professional development (PD) initiatives such as *Fundisa for Change* and *Keep It Cool-Climate Change Education (KIC-CCE)* significantly enhance teachers' climate pedagogical content knowledge (PCK), reflective practice, and interdisciplinary teaching skills (Mzilikazi, Khalo & Caga, 2025; Govender, Maphalala & Mncube). Studies show that teachers exposed to these initiatives demonstrate improved confidence, a richer pedagogical climate, and greater readiness to navigate socioecological and interdisciplinary content (Anyanwu et al, 2015; Thenga, 2025).

Systematic review evidence further shows that PD that integrates scientific, socio-environmental, economic, and indigenous knowledge systems enhances teacher empowerment for interdisciplinary climate education.

Professional development (PD) remains a pivotal catalyst in empowering South African teachers to deliver interdisciplinary climate change education, particularly because most teachers enter the profession with limited climate-specific pedagogical content knowledge (PCK). Recent research shows that teachers frequently cite inadequate training, insufficient exposure to climate science, and limited understanding of interdisciplinary pedagogies as significant barriers to integrating climate change into their teaching (Mzilikazi, Khalo & Caga, 2025). In response to these gaps, structured in-service training initiatives—such as the Fundisa for Change programme and Higher Education Institution (HEI)-led workshops have demonstrated significant potential to build teacher competence and agency. For example, studies on Natural Sciences and Geography teachers reveal that targeted professional development programmes substantially improve teachers' conceptual understanding and their confidence in integrating climate change across disciplinary boundaries (Olawumi et al., 2023; Mavuso et al., 2022). These findings underscore that well-designed PD initiatives serve not only to strengthen content mastery but also to foster the pedagogical adaptability required for interdisciplinary climate change education.

Capacity-building efforts play a similarly transformative role by providing ongoing institutional support, collaborative spaces, and system-level mechanisms that embed climate change education into everyday teaching practice. Emerging evidence shows that sustained capacity-building—through short courses, professional learning communities, and structured mentorship—enhances teachers' personal PCK development, enabling them to internalise and apply interdisciplinary climate-teaching strategies more effectively (Nsubuga, Schudel, Songqwaru & Fox, 2025). Furthermore, capacity-building initiatives that involve HEIs as long-term partners amplify impact by aligning teacher education, curriculum design, and community engagement with national climate literacy goals. Higher education institutions have been shown to play a key role in advancing climate change education by expanding teacher access to research-informed training, digital teaching tools, and interdisciplinary teaching frameworks (Maphalala et al., 2025; Mhlabathi & Achieng, 2024). Collectively, the literature demonstrates that capacity-building—when institutionally supported and continuously offered—equips South African teachers with both professional resilience and the interdisciplinary competence necessary to empower learners meaningfully in a rapidly changing climate.

Theme 4: Collaboration, Communities of Practice, and Cross-Sector Partnerships

Interdisciplinary climate change education is strengthened when teachers work collaboratively with peers, climate scientists, NGOs, and local communities. Findings may reveal that such collaborations enhance access to contextual knowledge, teaching resources, and practical examples of climate impacts. Models such as KIC-CCE demonstrate how community-embedded professional learning fosters teacher agency and a reflective pedagogical climate.

This collaborative dimension supports integrated teaching by linking scientific, cultural, and local environmental knowledge.

Collaboration and professional learning communities are increasingly recognised as critical mechanisms for strengthening interdisciplinary Climate Change Education (CCE) (UNESCO, 2023b). High-impact research shows that Communities of Practice (CoPs) provide structured spaces where teachers co-construct knowledge, share pedagogical strategies, and collectively address complex curriculum challenges associated with interdisciplinary teaching (UNESCO, 2021). Evidence indicates that CoPs enhance teacher agency, support interdisciplinary planning, and foster context-responsive professional learning, which is essential for integrative approaches such as climate and sustainability education (de Souza, Greda & Ferreira, 2025). Similarly, large-scale studies demonstrate that professional learning communities are positively associated with teacher innovation and creative self-efficacy, suggesting that collaborative cultures enable educators to experiment with new pedagogies and interdisciplinary content (Liu et al., 2022). Within the climate change education context, collaborative initiatives linking teachers, curriculum specialists and researchers have been shown to strengthen teachers' competences, motivation and capacity to integrate sustainability into classroom practice, highlighting the transformative potential of sustained professional collaboration (De Rivas, Vilches & Mayoral, 2025; UNESCO, 2023).

Beyond school-based collaboration, cross-sector partnerships play a crucial role in advancing interdisciplinary CCE by connecting education with broader social, environmental and economic systems. Research in sustainability education emphasises that partnerships among universities, schools, governments, civil society, and industry enhance knowledge exchange, resource mobilisation, and the real-world relevance of learning, thereby improving educational outcomes in sustainability (Bowser, Ho, Ziebell, Lazendic-Golloway, *2024*). Cross-sectoral collaboration is also associated with more inclusive and integrated educational responses to complex societal challenges, including climate change, because it aligns educational goals with community development, health, and socio-economic priorities (Zhang et al., 2025). Global policy scholarship further stresses that teacher empowerment depends on networked professional development and multi-stakeholder cooperation, where collaboration across institutions and sectors sustains innovation and supports whole-system transformation toward sustainable development (UNESCO, 2024). Collectively, these perspectives suggest that effective interdisciplinary climate change education in South Africa requires not only teacher collaboration within schools, but also sustained partnerships that bridge education, research, policy and community action.

Theme 5: Towards Transformative, Interdisciplinary, and Contextualised Climate Pedagogies

Recent systematic reviews emphasise that effective climate education requires movement beyond discipline-specific teaching toward integrative, transdisciplinary approaches that connect science, ecology, social justice, indigenous knowledge, and sustainability. Findings may show teachers' desire for pedagogical tools that enable contextualised, action-oriented, and learner-centred climate education approaches that align with South Africa's realities and empower learners as climate-resilient citizens. Teacher initiatives that embed reflective practice and deeper learning (e.g., KIC-CCE portfolios) demonstrate the transformative potential of interdisciplinary climate pedagogy (Govender et al., 2025). Transformative, interdisciplinary and contextualised climate pedagogies are increasingly recognised as essential for equipping teachers and learners to respond to complex socio-ecological challenges (UNSCO, 2014; 2025). High-impact scholarship emphasises that transformative climate education moves beyond transmission of facts toward critical, reflective and action-oriented learning that reshapes values, behaviours and social practices. Global research and policy frameworks highlight that transformative pedagogy integrates cognitive, social, and behavioural learning while encouraging reflexive thinking and real-world problem-solving, enabling learners to develop sustainability competencies and agency for change (UNESCO, 2021; 2024). In addition, interdisciplinary approaches are central because climate change is a "wicked problem" that requires the integration of scientific, social, and ethical knowledge systems; pedagogies of interconnectedness help learners understand the complex relationships between environment, society, and economy, thereby fostering systems thinking and a holistic understanding (Lehtonen et al., 2018). Empirical research further identifies key pedagogical principles for effective climate education, including interdisciplinary learning, context-based inquiry, equity, social justice and community action, all of which contribute to culturally responsive and transformative teaching practices (Strachan & Markwick, 2025). These insights suggest that empowering South African teachers requires pedagogical shifts toward learner-centred, inquiry-driven and justice-oriented approaches that support critical engagement with climate and sustainability challenges.

Contextualisation is equally crucial in making climate change education meaningful and transformative, particularly in African settings shaped by socio-ecological vulnerability and diverse knowledge systems (UNESCO, 2019). Research from African higher education demonstrates that place-based and experiential pedagogies, grounded in local realities and community engagement, foster more profound understanding, environmental responsibility and long-term behavioural change, as learners connect theory with lived socio-ecological challenges (Offei et al., 2025). Similarly, studies on contextualised environmental and sustainability education emphasise the importance of culturally responsive and socially just approaches that link environmental learning with local experiences, community knowledge and everyday realities, thereby enabling meaningful and relevant learning (Reddy, 2021; UNISCO, 2020). Broader climate education research further shows that effective implementation depends on locally adaptive curricula, teacher professional development and equitable resource support, underscoring the need for context-sensitive pedagogical innovation in diverse educational settings (Leite, 2024). Collectively, these perspectives indicate that transformative interdisciplinary climate pedagogies in South Africa must be locally grounded, socially responsive and action-oriented, enabling teachers to bridge global climate knowledge with local realities while fostering critical consciousness, resilience and sustainability competencies.

RECOMMENDATIONS

This study indicates that while South African teachers recognise the importance of climate change education, many do not feel adequately prepared or supported to teach it in an interdisciplinary and meaningful way. Empowerment, therefore, cannot be treated as an individual responsibility; it must be addressed at a systemic level. Strengthening teachers' climate, content knowledge, and pedagogical readiness should be prioritised. The evidence indicates that gaps in conceptual understanding and interdisciplinary teaching approaches continue to limit effective implementation. Professional development initiatives such as Fundisa for Change and Keep It Cool demonstrate that when training is structured, collaborative, and sustained, teachers' confidence and classroom practice improve significantly. These initiatives should be scaled up and embedded into ongoing teacher development frameworks rather than offered as isolated workshops. A clearer curriculum direction is urgently needed. Although CAPS includes climate-related topics, the content remains fragmented and assessment-driven, making interdisciplinary integration difficult. Teachers require practical guidance, flexible frameworks, and space for professional judgement to connect climate concepts across subjects in ways that are relevant to learners' lived realities. Collaboration must be strengthened both within and beyond schools. Professional learning communities provide safe spaces for shared reflection and interdisciplinary planning. Moreover, partnerships between schools, universities, NGOs, and local communities can bridge the gap between theory and practice, ensuring that climate education is grounded in both scientific knowledge and local environmental contexts.

CONCLUSION

This systematic review examined strategies to empower South African teachers to implement interdisciplinary climate change education. The findings reveal an apparent tension: while the need for climate education is widely acknowledged, the structures required to support teachers remain uneven and, in some cases, insufficient. Teachers' challenges are not simply a matter of limited knowledge. They reflect broader systemic issues: fragmented curriculum guidance, constrained professional autonomy, limited institutional coordination, and unequal access to resources. Where structured professional development, collaborative networks, and institutional support are present, teachers demonstrate greater confidence, stronger pedagogical integration, and a more profound commitment to climate education. Notably, the review highlights that climate change education cannot remain confined to isolated content coverage. It requires a shift toward transformative, contextualised, and interdisciplinary pedagogies that connect science with social justice, indigenous knowledge, and community realities. Such approaches position teachers not merely as curriculum implementers, but as active agents of change within their schools and communities. In the South African context, empowering teachers for interdisciplinary climate change education is both an educational and social imperative. Real progress will depend on coherent policy alignment, sustained investment in teacher development, meaningful collaboration, and curriculum reform that moves beyond compliance toward transformation. Only then can climate change education fulfil its potential to cultivate informed, resilient, and critically engaged citizens prepared to navigate an increasingly uncertain environmental future. This chapter provides insights and practical strategies relevant to a wide audience: including teacher educators, pre-service and in-service teachers, policymakers, curriculum developers, researchers, and NGOs. By highlighting interdisciplinary approaches, teacher empowerment, and curriculum integration, it offers guidance for strengthening climate change education in both South African and broader educational contexts.

REFERENCES

Ajani, O. A., & Govender, S. (2025). Evaluating the impact of professional development programmes on curriculum design and implementation for sustainability education in high schools: A scoping review. *Interdisciplinary Journal of Education Research*, *7*(s1), a01. DOI: 10.38140/ijer-2025.vol7.s1.01

Anyanwu, R., Le Grange, L., & Beets, P. (2015, August). Climate change science: The literacy of Geography teachers in the Western Cape Province, South Africa. *South African Journal of Education*, *35*(3), 1–9. DOI: 10.15700/saje.v35n3a1160

Avalos, B. (2011). Teacher professional development in teaching and teacher education over ten years. *Teaching and Teacher Education*, *27*(1), 10–20. DOI: 10.1016/j.tate.2010.08.007

Beane, J. A. (1997). *Curriculum integration: Designing the core of democratic education*. Teachers College Press.

Bonilla-Jurado, D., Zumba, E., Lucio-Quintana, A., Yerbabuena-Torres, C., Ramírez-Casco, A., & Guevara, C. (2024). Advancing University Education: Exploring the Benefits of Education for Sustainable Development. *Sustainability (Basel)*, *16*(17), 7847. DOI: 10.3390/su16177847

Bowser, G., Ho, S. S., Ziebell, A., & Lazendic-Galloway, J. (2024). Networking and collaborating: The role of partnerships across sectors to achieve educational goals in sustainability. *Sustainable Earth*, *7*(1), 17. DOI: 10.1186/s42055-024-00080-z

Braun, V., & Clarke, V. (2006). Using thematic analysis in psychology. *Qualitative Research in Psychology*, *3*(2), 77–101. DOI: 10.1191/1478088706qp063oa

Darling-Hammond, L., Flook, L., Cook-Harvey, C., Barron, B., & Osher, D. (2017). Implications for educational practice of the science of learning and development. *Applied Developmental Science*, *21*(2), 97–140. DOI: 10.1080/10888691.2018.1537791

Darling-Hammond, L., Hyler, M. E., & Gardner, M. (2017). *Effective teacher professional development*. Learning Policy Institute. DOI: 10.54300/122.311

Davies, P. (2010). The relevance of systematic reviews to educational policy and practice. *Oxford Review of Education*, *26*(3–4), 365–378. DOI: 10.1080/713688543

De Rivas, R., Vilches, A., & Mayoral, O. (2025). Bridging the Gap: How Researcher–Teacher Collaboration Is Transforming Climate Change Education in Secondary Schools. *Sustainability (Basel)*, *17*(3), 908. DOI: 10.3390/su17030908

Department of Basic Education (DBE). (2020). *Strategic Plan 2020/21–2024/25.* Department of Basic Education.

Department of Environmental Affairs. (2017). National Climate Change Response – White Paper. URL: https://www.environment. gov.za/sites/default/files/legislations/national_climatechange_response_whitepaper.pdf

Department of Higher Education and Training (DHET). (2015). *Revised policy on the minimum requirements for teacher education qualifications.* DHET.

Evans, N., Stevenson, R. B., Lasen, M., Ferreira, J. A., & Davis, J. (2017). Approaches to embedding sustainability in teacher education: A synthesis of the literature. *Teaching and Teacher Education, 63*, 405–417. DOI: 10.1016/j.tate.2017.01.013

Galdames-Calderón, M., Stavnskær Pedersen, A., & Rodríguez-Gómez, D. (2024). Systematic Review: Revisiting Challenge-Based Learning Teaching Practices in Higher Education. *Education Sciences, 14*(9), 1008. DOI: 10.3390/educsci14091008

Govender, S., Maphalala, M. & Mncube, D.W. (2025). *Education and Climate Change: Discovering Strategies for Individual and Collective Action includes* multiple case studies from the KIC-CCE initiative that demonstrate reflective practice and interdisciplinary climate pedagogy.

Grant, C., & Osanloo, A. (2014). Understanding, selecting, and integrating a theoretical framework in dissertation research: Creating the blueprint for your "house". *Administrative issues journal, 4*(2), 4.

Heiser, R., Palalas, A., & Gollert, A. (2025). Digital well-being begins with inclusion: A systematic review of videoconferencing guidelines for equitable learning. *Australasian Journal of Educational Technology, 41*(1), 18–26. DOI: 10.14742/ajet.9552

Huoponen, A. (2023). Expert Teachers' Perceptions of Their Students' Environmental Literacy and the Practice of Environmental Education. *Discourse and Communication for Sustainable Education, 14*(1), 112–136. DOI: 10.2478/dcse-2023-0010

IPCC. (2023). Climate Change 2023: The Physical Science Basis. Contribution of Working Group I to the Sixth Assessment Report of the Intergovernmental Panel on Climate Change. Cambridge University Press. https://www.ipcc.ch/report/ar6/wg1/

James, A. K. (2022). A review of Teaching and Learning for Change: Education and Sustainability in South Africa, edited by Ingrid Schudel, Zintle Songqwaru, Sirkka Tshiningayamwe and Heila Lotz-Sisitka. *Southern African Journal of Environmental Education, 38.* DOI: 10.4314/sajee.v38i1.06

Kivunja, C. (2018). Distinguishing between theory, theoretical framework, and conceptual framework: A systematic review of lessons from the field. *International Journal of Higher Education*, *7*(6), 44–53. DOI: 10.5430/ijhe.v7n6p44

Kurent, B., & Avsec, S. (2026). Systems Thinking in the Role of Fostering Technological and Engineering Literacy. *Systems*, *14*(1), 5. DOI: 10.3390/systems14010005

Lehtonen, A., Salonen, A., & Cantell, H. & Laura Riuttanen (2018). A pedagogy of interconnectedness for encountering climate change as a wicked sustainability problem, Journal of Cleaner Production, Volume 199, 2018, Pages 860–867, ISSN 0959-6526, https://doi.org/DOI: 10.1016/j.jclepro.2018.07.186

Lehtonen, A., Salonen, A. O., Cantell, H., & Riuttanen, L. (2018). A pedagogy of interconnectedness for encountering climate change as a wicked sustainability problem. *Journal of Cleaner Production*, *199*, 860–867. DOI: 10.1016/j.jclepro.2018.07.186

Leite, S. (2024). Towards a transformative climate change education: Questions and pedagogies. *Environmental Education Research*, *30*(12), 2376–2393. DOI: 10.1080/13504622.2024.2365983

Liu, S., Lu, J., & Yin, H. (2022). Can professional learning communities promote teacher innovation? A multilevel moderated mediation analysis, Teaching and Teacher Education, Volume 109, 2022, 103571, ISSN 0742-051X, https://doi.org/ DOI: 10.1016/j.tate.2021.103571

Liu, Y., Li, Z., & Yin, X. (2022). Environmental regulation, technological innovation, and energy consumption — A cross-regional analysis in China. *Journal of Cleaner Production*, *2018*(203), 885–897.

Maphalala, M. C., Kutame, A. P., & Kok, P. J. (2025). Embedding climate change education into the school curriculum: Enablers and constraints. In Govender, S., Maphalala, M. C., & Mncube, D. W. (Eds.), *Education and climate change: Discovering strategies for individual and collective action* (pp. 135–147). AOSIS Books., DOI: 10.4102/aosis.2025.BK491.08

Matsepe, M., & Maluleke, M. (2024). Infusing Climate Change Education into Curriculum and Assessment Policy Statement (CAPS) as a Resilience Strategy in South Africa: Towards a Theory of Change. *E-Journal of Humanities* [EHASS]. *Arts and Social Sciences*, *5*(9), 73–82.

Mavuso, M.P., Khalo, X., Kafu-Quvane, B.P., & Olawumi, K.B. (2022). Climate Change Education As a Means to Protect the Planet: A Review of the Relevant Literature. eBangi 19 (3), 179–191.https://doi.org/DOI: 10.17576/ebangi.2022.1903.1

Mezirow, J. (1997). Transformative learning: Theory to practice. *New Directions for Adult and Continuing Education, 1997*(74), 5–12. DOI: 10.1002/ace.7401

Mhlabathi, M., & Achieng, M. S. (2024). Digital Technology-Enabled Education for Sustainable Development in South Africa: A Case Study of a University of Technology. In Chigona, W., Kabanda, S., & Seymour, L. F. (Eds.), *Implications of Information and Digital Technologies for Development. ICT4D 2024. IFIP Advances in Information and Communication Technology* (Vol. 708). Springer., DOI: 10.1007/978-3-031-66982-8_14

Morales, R., Aguilar, A., Arenas, D., Cisternas, M., Bascope, D., Salazar, R., Becerra, J., & Solís-Pinilla, C. (2025). Transforming education for a sustainable future: An analysis of teacher education in the context of climate change. *Frontiers in Education, 10*, 1537129. Advance online publication. DOI: 10.3389/feduc.2025.1537129

Msezane, S. B. (2023). Content and Teaching Methods Regarding Climate Change: South African Curricula Perspective. In Leal Filho, W., Sima, M., Lange Salvia, A., Kovaleva, M., & Manolas, E. (Eds.), *University Initiatives on Climate Change Education and Research.* Springer., DOI: 10.1007/978-3-031-25960-9_48-1

Mthethwa, P., Hlatshwayo, S., & Sibandze, N. (2025). A survey on the role of artificial intelligence in transforming English language teaching. *Journal of Research in Education and Pedagogy, 2*(3), 435–446. DOI: 10.70232/jrep.v2i3.78

Mzilikazi, B., Khalo, X., & Caga, N. (2025). Understanding of natural science teachers' approaches to addressing climate change concepts in the curriculum: Evidence from the Fundisa for Change Keep It Cool Climate Change Project. *Interdisciplinary Journal of Education Research, 7*(1), a11. DOI: 10.38140/ijer-2025.vol7.1.11

Naidoo, S., & Heath, G. (2024). *Representation of Climate Change Adaptation in Geography Education Modules at the University of KwaZulu-Natal, The Journal of Geography Education in Africa.* UKZN., DOI: 10.46622/jogea.v7i1.4991

Naidoo, S., & Heath, G. (2024). Representation of Climate Change Adaptation in Geography Education Modules at the University of KwaZulu-Natal. *Journal of Geography Education in Africa, 7*(1). Advance online publication. DOI: 10.46622/jogea.v7i1.4991

Nakidien, T., Singh, M., & Sayed, Y. (2021). Teachers and Teacher Education: Limitations and Possibilities of Attaining SDG 4 in South Africa. *Education Sciences, 2021*(11), 66. DOI: 10.3390/educsci11020066

NASA. (2025). Global Climate Change: Vital Signs of the Planet. National Aeronautics and Space Administration. https://climate.nasa.gov

National Planning Commission (NPC). (2013). *National Development Plan: Vision for 2030*. Government Printers.

NOAA. (2024). Climate Change: Global Temperature. National Oceanic and Atmospheric Administration. https://www.climate.gov

Nsubuga, Y., Schudel, I., Songqwaru, Z., & Fox. H. (2025). Exploring the Development of Teachers' Personal Pedagogical Content Knowledge about Climate Change Education during a Continuing Teacher Professional Development Course, Southern African Journal of Environmental Education; Vol. 40 No. 2 (2025), 2411–5959, 0256-7504

Offei, M. B., Agyeman, B. Y., Henaku, O. E., Enu Kirk, B., & Boasiako, A. A. (2025). Transgressive pedagogies in a climate-constrained Africa: grounding climate change education and sustainability learning in local communities, Frontiers in Education, Volume 10 – 2025, ISSN: 2504-284X, DOI= 2025.1519914DOI: 10.3389/feduc

Olawumi, K., Mavuso, M. P., Khalo, X., Kafu-Quvane, B., & Mzilikazi, B. (2023). Implementation of teacher development programme for integrating climate change education: Natural sciences teachers' views. *International Journal of Environmental, Sustainability, and Social Science*, *4*(3), 788–798. DOI: 10.38142/ijesss.v4i3.534

Olawumi, K. B., & Mavuso, M. P. (2024). In-service teacher training programmes that promote education for sustainable development: A review of emerging literature. *Interdisciplinary Journal of Education Research*, *6*, 1–20. DOI: 10.38140/ijer-2024.vol6.35

Page, M. J., McKenzie, J. E., Bossuyt, P. M., Boutron, I., Hoffmann, T. C., Mulrow, C. D., Shamseer, L., Tetzlaff, J. M., Akl, E. A., Brennan, S. E., Chou, R., Glanville, J., Grimshaw, J. M., Hróbjartsson, A., Lalu, M. M., Li, T., Loder, E. W., Mayo-Wilson, E., McDonald, S., & Moher, D. (2021). The PRISMA 2020 statement: An updated guideline for reporting systematic reviews. *BMJ (Clinical Research Ed.)*, *372*, 71. Advance online publication. DOI: 10.1136/bmj.n71 PMID: 33782057

Reddy, C. (2021). Environmental Education, Social Justice and Teacher Education: Enabling Meaningful Environmental Learning in Local Contexts. *South African Journal of Higher Education*, *35*(1), 161–177. DOI: 10.20853/35-1-4427

Shabalala, N. P. (2025). Challenges, opportunities and strategies of distributed leadership in managing environmental education curriculum in South African secondary schools. *International Journal of Educational Management and Development Studies*, *6*(1), 98–135. DOI: 10.53378/ijemds.353156

Short, P. M., & Rinehart, J. S. (1992). School participant empowerment scale: Assessment of level of empowerment within the school environment. *Educational and Psychological Measurement, 52*(4), 951–960. DOI: 10.1177/0013164492052004018

Souza, D., Greda, I., & Ferreira, H. S. (2025). From Collaboration to Integration: How a Community of Practice Supports Public School Teachers' Understanding of Integrated STEAM Education. *Education Sciences, 15*(11), 1559. DOI: 10.3390/educsci15111559

Sterling, S. (2010). Transformative learning and sustainability: Sketching the conceptual ground. *Learning and Teaching in Higher Education, 5*, 17–33.

Strachan, A., & Markwick, A. (2025). Using a scoping review to inform a planetary-conscious pedagogical approach to primary science education. *Research in Science Education, 55*(4), 817–871. DOI: 10.1007/s11165-025-10280-y

Taylor, E. W. (2007). An update of transformative learning theory: A critical review of the empirical research. *International Journal of Lifelong Education, 26*(2), 173–191. DOI: 10.1080/02601370701219475

Thenga, M. (2025). Barriers and opportunities in implementing climate change education in the FET Phase Geography curriculum in South Africa. *Journal of Geography Education in Africa, 8*(1), 1–15. DOI: 10.46622/jogea.v8i1.5382

Tilbury, D. (2011). *Education for sustainable development: An expert review of processes and learning*. UNESCO.

Tschannen-Moran, M., & Hoy, A. W. (2001). Teacher efficacy: Capturing an elusive construct. *Teaching and Teacher Education, 17*(7), 783–805. DOI: 10.1016/S0742-051X(01)00036-1

UNESCO. (2014). *Roadmap for implementing the Global Action Programme on Education for Sustainable Development.* Paris, UNESCO. https://unesdoc.unesco.org/ark:/48223/pf0000230514

UNESCO. 2017. *Education for Sustainable Development Goals: learning objectives.* https://unesdoc.unesco.org/ark:/48223/pf0000247444

UNESCO. (2017). *Education for sustainable development goals: Learning objectives*. UNESCO.

UNESCO. (2019). *Country Progress on Climate Change Education, Training, and Public Awareness. An analysis of country submissions under the United Nations Framework Convention on Climate Change.* Paris, UNESCO, (2020). *Education for sustainable development: A roadmap*. Paris, UNESCO. https://unesdoc.unesco .org/ark:/48223/pf0000374802

UNESCO. (2020). *Education for sustainable development: A roadmap*. UNESCO.

UNESCO, (2021). *Getting every school climate ready. How countries are integrating climate change issues in education.*

UNESCO. (2021). *Getting every school climate ready: How countries are integrating climate change issues in education*. UNESCO.

UNESCO. (2022). *Youth demands for quality climate change education*. UNESCO.

UNESCO. 2023. *Impact of Climate Change Education on Learning Outcomes.* (unpublished)

UNESCO. (2023a). *Greening the future: Education for sustainable development.* Retrieved April 3, 2024, from https://www.unesco.org/en/sustainable-development/education/greening-future

UNESCO. (2024). *Green School Quality Standard: Greening every learning environment*. UNESCO.

UNESCO. (2024). *Green school quality standard: Greening every learning environment*. UNESCO.

van Wyk, M.M. (2025). AI-Smart Technologies Digitally Transformed Higher Education Institutions: A Systematic Literature Review. *AI Smart-Enabled Architecture and Infrastructure for Higher Education*, 75–100.

Vogel, C., & Schwaibold, U. (2015). Teaching and Learning for Climate Change –the Role of Teacher Materials and Curriculum Design in South Africa, Southern African Journal of Environmental Education, Vol. 31 (2015).

VVOB / Fundisa for Change (2021). *Keep It Cool: Climate Change Education – Teacher Support Set 3.* Provides professional development tools designed to improve interdisciplinary teaching of climate change within curriculum-aligned projects.

Vygotsky, L. S. (1978). *Mind in Society: The Development of Higher Psychological Processes*. Harvard University Press.

Zawacki-Richter, O., Kerres, M., Bedenlier, S., Bond, M., & Buntins, K. (2020). *Systematic reviews in educational research: Methodology, perspectives and application*. Springer Nature., DOI: 10.1007/978-3-658-27602-7

Zeichner, K. (2010). Rethinking the connections between campus courses and field experiences in college- and university-based teacher education. *Journal of Teacher Education, 61*(1–2), 89–99. DOI: 10.1177/0022487109347671

Zhang, J., Capellini, V., Rocha, E., Lamônica, D., & Muñoz-Martínez, Y. (2025). Editorial: Cross-sectoral collaboration in inclusive education, Frontiers in Psychology, Volume 16 – 2025, ISSN=1664-1078, DOI: 2025.1697199DOI: 10.3389/fpsyg

Chapter 4
Climate-Ready Teachers, Resilient Future: Linking Teacher Education, Climate Action, and Societal Resilience

Divyanshu Mishra
http://orcid.org/0009-0005-6382-0596
Central University of Rajasthan, India

Anjali Sharma
http://orcid.org/0000-0001-6297-3113
Central University of Rajasthan, India

ABSTRACT

In 2015, UN member countries also adopted the 2030 Agenda for Sustainable Development a comprehensive global plan of action for 'people, planet and prosperity' comprising 17 SDGs and 169 targets to be achieved by 2030, including SDG 4 Quality Education and SDG13 on climate action (Nerini et al 2019). This chapter is guided by three research questions: First, What theoretical bases explain the role of teacher education in promoting climate action and societal resilience? Second: Identify the gaps related to teacher education in the context of climate change education through previous research studies. And Third: Creating a model for a climate-ready learning environment through teacher preparedness? The paper follows a qualitative approach secondary data sources and references managed with Zotero.To address the first RQ, the theoretical base was discussed and for the second RQ identifies gaps in teacher education related to SDG 13, as described in previous research and for third RQ, the chapter proposed a 6-step model to prepare climate-ready teachers to develop a learning environment.

DOI: 10.4018/979-8-3373-6975-4.ch004

INTRODUCTION

Issues related to climate change have gained more importance in recent years. Global temperatures have risen by 1.1°C relative to preindustrial levels, leading to more frequent, intense, and long-lasting extreme weather and climate events (IPCC, 2023). The increase in temperature, extreme weather conditions, and environmental degradation have impacts on the ecosystems, economies, and lives. Such issues require solutions not only scientific but also social and educational. In this regard, education can play a great part. In this regard, the United Nations Educational, Scientific, and Cultural Organization (UNESCO) has set 17 Sustainable Development Goals, among them SDG 4, which specifically addresses Quality Education, and SDG 13, which specifically addresses Climate Action. To achieve these goals, UNESCO published a 2017 document outlining learning outcomes and suggested topics.

The educational themes suggested by UNESCO include climate change and its impacts on people. It is also training them to react to its effects. Attitudes, values, and actions can be transformed through education that can facilitate sustainability. In the education system, teachers play a key role. They affect how learners think, learn, and behave regarding climate issues. Teacher education is thus a key to successful climate action. Educators need to be ready to respond to climate change in a significant way in the future. This preparation is not to be limited to content knowledge. Educators require competence, self-confidence, and moral insight. They also require pedagogical strategies that stimulate engagement, contemplation, and action. Climate-ready teachers are developed through such preparation. Teachers are not the only climate prepared personnel. They act as agents of change. They enable the learners to relate the knowledge of climate to action in the real world. They promote critical thinking and problem-solving. They encourage environmental and social responsibilities. Teachers can use their practice to impact learners, schools, and communities.

Busch, Henderson, and Stevenson (2018) offer a reflective assessment of the state of climate change education (CCE) research and argue that it is necessary to expand epistemological and methodological strategies. They begin with positionality statements, in which researchers acknowledge the influence of personal experience on their research. Subsequently, the authors develop a typology of CCE research centered on three paradigms: positivism/post-positivism, social constructivism, and critical/transformational approaches. Moreover, modern school research emphasizes the role of theoretical approaches that foster transformative and adaptive learning in the context of complex global issues. Transformative Learning Theory (Mezirow, 2018) focuses on critical reflection and perspective transformation as approaches that allow a learner to challenge current assumptions and formulate new ways of perceiving socio-environmental problems. On the same note, the Resilience Theory emphasizes the ability of individuals and communities to adapt to environmental changes and uncertainties (Folke, 2016). Although these theoretical approaches can be useful for conceptualizing climate change education, how they are incorporated into teacher education curriculum and instruction is under-researched in the existing literature. Hence, a critical discussion of how such pedagogical theoretical underpinnings can be used to prepare teachers and to aid the creation of effective climate change pedagogy is necessary. Whether Education for Sustainable Development (ESD) is an effective framework to deploy Climate Change Education (CCE) is a controversial issue in the literature. Although ESD advances interdisciplinary and sustainability-based learning, scholars believe that in most cases, it pays partial and insular attention to climate change. According Busch et al. (2019), there is a need to expand the methods of climate change education by going beyond generalized sustainability frameworks. Equally, in their systematic review, Monroe et al. (2019) explain that successful climate change education must be action-based, with specific interventions rather than ESD-based. Reid (2019) also criticises the narrowness of existing research on climate change education, suggesting that treating CCE as a subset of ESD could limit its transformative capacities. These viewpoints suggest that, despite the significance of ESD, there is an increased need to frame climate change education as a more defined and targeted field in educational practice.

The concept of societal resilience offers a significant framework for understanding this position. Societal resilience is a concept that implies how people and communities respond to change. It encompasses the ability to manage the risks of climate change and mitigate its effects. Resilience is reinforced through education, awareness, adaptation ability, and social responsibility. These capacities can only be developed by teachers among the learners. The connection between climate resilience and education is also highlighted in global policy strategies—Sustainable Development Goals focus on climate action and quality education. The objectives identify the fact that education facilitates adaptation and sustainability. They also demand increased teacher training in climate-responsive teaching. Teacher education must be aligned with these objectives to have resilient futures. This recognition notwithstanding, there are difficulties. Most teacher education programmes continue to emphasize theoretical knowledge. Minimal focus is made on climate action, green skills, and ethical aspects. Experiential and action-oriented learning opportunities are not usually adequate. Consequently, it may be that teachers are not ready to address climate issues in their professional practice. These are the challenges addressed in this chapter. It dwells on the relationships among teacher education, climate action, and societal resilience. It asserts that teacher development is an eye-opening approach to building climate-resilient societies. The chapter sheds light on how climate-ready teachers can become change agents both in and out of the classroom. Education systems may make a difference in the long run by empowering teacher education on climate action. Teachers who are climate-ready ensure that learners acquire knowledge, skills, and values grounded in sustainability. In the long run, this helps in making people more informed, adaptive, and resilient. By doing so, one of the fundamental ways to build a strong future in education is through teacher education.

RQ-1: What theoretical base explains the role of the teacher in promoting climate action and societal resilience?

It is necessary to construct a conceptual framework to understand the significance of teacher education in promoting climate action and social resilience. Climate change is not a singular phenomenon; it involves environmental, social, and ethical issues, requiring not only factual knowledge but also a transformation of values, perspectives, and professional practices. Teacher education programs are uniquely positioned to address these challenges by cultivating transformative teachers who can foster climate awareness, adaptability, and responsible action among students and communities. Conceptual frameworks such as transformative learning theory, resilience theory, and education for sustainable development help us understand how teacher education can empower teachers to take action, enhance their capacity to cope with climate-related uncertainties, and build resilient and sustainable societies. Therefore, teacher education can be considered a major driving force for social climate action and resilience.

Theoretical Base

A theoretical framework is a collection of theories proposed by scholars in a particular discipline, which you intend to use as the foundation for your research and to provide theoretical support for data analysis and interpretation of results (Kivunja, 2018). This chapter draws on interdisciplinary perspectives on education and social resilience to illustrate the importance of teacher education for climate action and social resilience. Climate change is not only an environmental issue but also a socio-ecological phenomenon that requires action-oriented, value-driven, and informed educational responses. Therefore, teachers, as agents of change, play a strategic role in shaping learners, communities, and future generations. The discussion below demonstrates how the theoretical foundation allows us to understand how teacher education can empower educators to take action in response to the climate crisis.

Table 1. Theoretical base

Education for Sustainable Development	
Transformative Learning Theory (Internal professional change)	Resilience Theory (Sustaining professional capacity)
Critical Reflection Dialogue & Discourse Perspective Transformation Commitment to Climate Action	Adaptability Transformative Capacity Coping with Climate Change
Teacher Development Outcomes (Theory-wise)	
Climate Conscious Climate-Reflective Teacher Perspective Shifter Climate Action Teacher	Climate Adaptive Transformative Sustainability to Climate Action
Integrated Teacher Identity Transformative Teachers (Climate-Ready Teachers)	
Climate Vigilant Eco-Emotionally Resilient Eco-Socially Responsible Adaptive to Changing Climate Action-Oriented Climate Pedagogues Capable of Leading Climate Action in Society	
Societal Resilience	

EDUCATION FOR SUSTAINABLE DEVELOPMENT

One of the key foundations of the dialogue is Education for Sustainable Development (UNESCO). The Johannesburg World Summit on Sustainable Development in 2002 proposed the United Nations Decade of Education for Sustainable Development (ESD) 2005–2014, in which a clear statement highlighted that education is at the heart of sustainable development (Amador et al., 2015). Teacher education is a prerequisite for promoting ESD in schools (Rieckmann, 2019). ESD has traditionally offered a more inclusive and comprehensive approach to a broader range of sustainability problems, including climate change. Hence, climate change education incorporated into ESD guarantees a comprehensive learning experience that can facilitate understanding of climate change as a systemic, intertwined problem of sustainable development (Thanaponrangsuth et al., 2025). ESD focuses on the knowledge, skills, values, and attitudes that must be developed in order to deliver sustainable futures. It not only transmits content but emphasizes critical thinking, systems thinking, and responsible action. In teacher education, ESD offers an approach to the interdisciplinary integration of climate change messages and to the development of teachers' ability to relate global environmental issues to local ones. This approach upholds that climate-prepared teachers are needed to address sustainability and engage in climate solutions.

Let us turn to two basic theories to ground the ESD approach for climate-prepared teachers.

Transformative Learning Theory

Mezirow proposed that profound and lasting learning occurs through reflection and meaning-making processes. This theory primarily targets adult learners. Transformative learning has become a powerful metaphor for understanding how adults learn in the field of adult education (Dirkx, 1998). According to this theory, individuals change their frames of reference through critical reflection and dialogue with others, thereby altering beliefs, attitudes, and behaviours. In the context of climate change, transformative learning focuses on shifting teachers' perspectives from detached, abstract concern to a sense of urgent moral and professional responsibility. Teacher education can promote positive change by engaging teachers in practices such as reflection, experiential learning, and problem-based learning, thereby influencing climate action and enhancing climate resilience behaviours.

Transformative learning theory aims to cultivate the following characteristics in teachers.

Critical Reflection

The process of transformation begins with self-reflection and feeling. Teachers must analyse what they feel and believe, and what they value, assume, and practice regarding climate change and sustainability. They should entail questioning prevailing developmental paradigms, individual consumption habits, and instructional methods to understand how they contribute to or help solve climate crises. For climate-ready teachers, critical reflection can inform ethical and responsible decision-making in both teaching and professional practice.

Dialogue and Discourse

These can foster open, participatory, and evidence-based discussions, enabling teachers to explore different perspectives on climate change. Dialogue, as a tool, helps in sharing scientific knowledge, social experiences, and ethical perspectives, and assists teachers in understanding and clarifying relevant meanings. Interactive discourse promotes collaborative learning, critical thinking, and collective climate action.

Perspective Transformation

The transformation of perspective involves a radical change in teachers' worldviews, values, and professional identities regarding climate change. It represents a shift toward an immediate educational, ethical, and social task rather than treating climate change as an abstract or distant problem. In the case of climate-ready teachers, perspective transformation leads to a sustainability-oriented mindset that shapes curriculum development, pedagogy, and interactions with learners and communities.

Committed to climate action

This commitment means that teachers are unwavering in their willingness and desire to translate climate knowledge and mindset shifts into practical educational and community activities. In the context of climate change education, this includes integrating climate themes into the curriculum, encouraging sustainable development, supporting student-led projects, and participating in local climate change adaptation and mitigation efforts. For climate-aware teachers, practicing transformative learning is a commitment to climate action.

Resilience Theory

Resilience theory provides a foundation for individuals and societies to adapt to and cope with any stress. Originating in ecology, resilience theory was first formally proposed by C. S. Holling (1973). Holling defined resilience as the capacity of a system to absorb shocks and reorganize during change without losing its fundamental structure, function, and characteristics. Unlike traditional approaches that focus primarily on stability and equilibrium, resilience theory emphasizes flexibility, adaptability, and the capacity to transform in response to relevant environmental stressors. Over time, the application of resilience theory has expanded beyond ecological environments to encompass social, economic, and educational domains. Other scholars, such as (Folke, 2006) and (Folke et al., 2010), have focused on social-ecological resilience, arguing that human societies and natural ecosystems are intimately intertwined. Based on this perspective, resilience refers not only to recovery after a shock but also to the capacity for learning, innovation, and proactive responses to future risks. This expanded understanding is particularly relevant in the context of climate change, which is characterized by uncertainty and complexity. In climate change education, resilience is defined as the capacity to prepare for, respond to, and manage climate-related hazards. Resilience theory in teacher education emphasizes cultivating teachers' preparedness, flexibility, and problem-solving skills. By equipping teachers with the capacity to address climate change, they can ensure that learners and communities can cope with its impacts, encourage adaptive and sustainable practices, and provide a climate-resilience-oriented teaching approach.

This theory aims to cultivate the following characteristics among the teachers.

Adaptability

Refers to teachers' knowledge, skills, and attitudes in the context of climate change. Climate change education requires adjusting curricula, teaching methods, and classroom practices based on local climate hazards and current sustainable development needs. Teachers with strong adaptability can more flexibly adjust to and integrate climate solutions appropriate to specific contexts, incorporating them into the teaching process.

Transformative Capacity

Refers to teachers' ability to initiate and facilitate fundamental shifts in teaching and institutional culture, as well as community roles, when existing methods are insufficient to address climate challenges. In the context of climate change, this means reimagining education to make it more sustainable and resilient.

Coping with Climate Change

Refers to teachers' ability to manage the cognitive, emotional, and professional demands arising from climate-related stressors. This helps manage eco-anxiety, uncertainty, and workload, and supports teachers' effectiveness and motivation in various situations. Mechanisms for coping with climate change help teachers maintain long-term engagement in climate change education and avoid burnout.

The theoretical base outlines the characteristics of a teacher ready to address climate change in the classroom.

Climate-Ready Teacher: Based on Transformative Theory

Climate Conscious

Based on transformative learning, climate consciousness develops when teachers critically reflect on their own assumptions and the relationship between humans and the environment, and view climate change as an urgent educational challenge. For climate-ready teachers to address climate change, climate consciousness means understanding the realities of climate change and believing they can address these challenges through education.

Climate-reflective teachers

Teachers who continuously and critically reflect on their beliefs, teaching practices, and professional stances, and connect them to climate change, are called climate-reflective teachers. According to transformative learning theory, these teachers consider the impact of curriculum choices, teaching philosophies, and personal values on learners' climate awareness and behaviour. Reflection enables climate-aware teachers to refine their methods and ensure their teaching aligns with the Sustainable Development Goals.

Perspective Shifters

Perspective shifters are educators who have experienced a shift in perspective regarding climate change. This role is based on transformative learning theory, meaning they have moved from initially viewing climate change as a distant or insignificant issue to seeing it as a core issue for education, morality, and social well-being. Climate-conscious teachers act as perspective shifters because they can reshape their own worldview and guide students to develop a sustainability-oriented perspective.

Climate Action Teachers

Educators who teach about climate change across various subject areas. They are dedicated to fostering students' awareness, values, and sense of responsibility regarding climate change. They practice sustainable development principles, such as reusing, recycling, and reducing waste, and connect climate issues to everyday life. They encourage students to move from understanding climate change to thinking critically about it and then taking practical action. Climate-aware teachers strive to promote climate justice, sustainable development, and responsible citizenship through their teaching practices, curriculum content, and the use of projects, discussions, debates, field trips, and community engagement.

Climate-ready Teacher: Based on Resilience Theory

Climate Adaptive

Teachers who adapt their knowledge, skills, and attitude to address the climate change issues in the classroom with learners. This involves incorporating disaster risk reduction, local adaptation, and sustainability practices into teaching. Climate-ready teachers assist learners and societies in anticipating, responding to, and adapting to climate change.

Transformative

The teacher becomes the transformative agency and takes initiative in the institutional culture and community. The teacher also helps people and learners challenge the existing methods as insufficient and unaddressed. Teachers establish positive relationships with peers, students, communities, and institutions, as collaboration improves collective thinking and problem-solving. Teachers who are climate-ready do so collaboratively to build educational and societal resilience.

Sustainability to Climate Action

Refers to a sustained commitment to climate action, which denotes the commitment of teachers to ensuring that they promote climate awareness, resilience, and sustainability through long-term commitment to education. It entails regular participation in climate-responsive teaching, advocacy, and community involvement. To climate-ready teachers, this promise is resilience, ethical responsibility, and professional purpose.

RQ-2: IDENTIFY THE GAPS RELATED TO TEACHER EDUCATION IN THE CONTEXT OF CLIMATE CHANGE EDUCATION THROUGH PREVIOUS RESEARCH STUDIES

A systematic literature review offers a scientific and transparent method for synthesizing existing knowledge, minimizing bias, and ensuring that all relevant studies are covered (Tranfield et al., 2003). Peer-reviewed research on teacher education, climate action, and societal resilience comprised the systematic literature review used in this chapter. The review was conducted in accordance with established standards for systematic reviews, including a well-defined search strategy, inclusion and exclusion criteria, and the critical assessment of the identified sources (Budgen et al., 2007). The key academic database, Google Scholar, was searched, and reports from international organizations such as UNESCO and the United Nations were used. It focused on literature related to conceptual frameworks, teaching methodologies, and policy considerations in climate change education and teacher preparation. The analysis of secondary sources was systematic, enabling the identification of the predominant themes, theoretical bases, and research gaps. This approach reinforced the chapter's conceptual basis and helped make the discussion grounded in credible, evidence-based scholarship (Snyder, 2019).

Selection Studies

A set of carefully selected keywords and keyword combinations, such as "Action Oriented Pedagogies for Climate Change Education", "Challenges faced by Teachers in classrooms related to climate change content during teaching", "teacher education and climate change education", "Climate change education and societal resilience", "Climate change pedagogy and teacher education", and "prepare teachers for climate change education", were used to refine the search results.

The details of the research studies are given below in Table 2. Showing the Keywords Search of research articles.

Table 2. **Showing the Keywords Search of research articles**

Key words	Initial Results
"Action-Oriented Pedagogies for Climate Change Education"	650
"Challenges faced by Teachers in classrooms related to climate change content during teaching"	460
"Teacher education and climate change education"	380
"Climate change education and societal resilience"	500
"Climate change pedagogy and teacher education"	68
"Prepare teachers for climate change education."	390
	2448

A total of 2,448 studies were initially retrieved across the selected keywords, as shown in Table 2. After applying different inclusion criteria, such as articles in the English language only and articles available in full text. Then, based on the title related to the study and keywords. The systematic literature review was conducted according to PRISMA to ensure transparency and reproducibility. Lastly, 15 studies were included for this analysis. The process of determining the number of articles to start with identified 2,448 articles, which were narrowed down to those using carefully chosen keywords. The retrieved records were all imported into Zotero, and duplicate records were eliminated, resulting in 230 records and 2,218 unique studies. These studies were further filtered by title and abstract, leading to the exclusion of 2,018 studies unrelated to climate change education in the context of teacher education. The remaining 200 studies were then analysed at the full-text level for eligibility. In the full-text screening process, the studies were reviewed according to predetermined inclusion criteria. First, 200 of the studies met the English language criterion. Nevertheless, 165 studies were retained after full-text review, and 35 were excluded due to unavailability. Additional screening for relevance to climate change education, with a specific focus on teacher education, resulted in the inclusion of 95 studies and the elimination of 70. Lastly, articles were evaluated based on whether they focused on pedagogical methods, teaching practices, or models of instruction, and thus 15 articles were selected for final analysis.

Limitation of the Review

Limitation of the review is that studies available in the English language and available in full-text were included. Then, relevant studies were shortlisted based on predetermined keywords and title-abstract screening. The specific perspective on teacher education and the related framework further reduces the limited scope.

Analysis

The 15 research studies reviewed from the perspectives of the gaps in between teacher education's different dimensions and climate change education found the following challenges, which have been discussed as follows-

Professional Development of In-Service Teachers on CCE

As evidence shows, in-service teacher development programs are instrumental in supporting changes in classroom perceptions of climate change. Olawumi et al. (2023) indicate that specialised professional development programmes made a significant difference to the pedagogical behaviour of Natural Sciences teachers. Nevertheless, it was also found that the content on climate change and teachers' pedagogical content knowledge remained lacking. The conclusions make it clear that professional development in climate change education, particularly for in-service teachers, needs to be sustained and ongoing to enhance conceptual knowledge and overall pedagogical skills.

Gaps in Pre-Service Teacher Education Programs

Studies also show that there are structural and pedagogical discontinuities in pre-service teacher education programmes. One of the shortcomings is the lack of a shared vision of the role and purpose of environmental education in pre-service teacher preparation. This lack is also worsened by the lack of agreement on pedagogy and the excessive focus on knowledge transmission rather than active, inquiry-based learning. The design of transdisciplinary curricula must be carefully coordinated in order to realize climate change education in universities. Teacher education programmes need to shift away from disciplinary silos to ensure that pre-service teachers are provided not only with a scientific understanding of climate change, but also with the skills of critical inquiry and the ethical views that may be developed through knowledge from different disciplines (Beach, 2023).

Pedagogical Transformation: From Knowledge to Action

The literature also underlines the need to balance scientific searching and social fantasy, intellectual comprehension and feeling in the future perspectives of climate change education. transformative CCE necessitates pedagogical approaches that incorporate sound scientific investigation and participatory and value-based learning. These methods will ensure that climate action is based on informed knowledge, not on activism lacking scientific grounding. The important mediators in this process are teachers. Nevertheless, a significant number of educators face practical issues, such as inadequate training, time scarcity, resource scarcity, and the inability to address affective issues, such as eco-anxiety, in classrooms (Guimerà-Ballesta et al., 2025).

Transdisciplinary, Interdisciplinary Nature

It is proposed in several studies that climate change education should evolve into interdisciplinary and eventually transdisciplinary. The authors note that it is necessary to incorporate the lens of climate science, ecology, economics, and social sciences, as well as local and indigenous knowledge systems (Morales-Aguilar et al., 2025). This kind of integration enlarges knowledge and adds ethical and moral aspects, without which climate education is meaningless. Feldbacher et al., 2024 makes it clear that, despite general knowledge of climate change, it is important to complement this knowledge with region- and country-specific effects to increase the perceived proximity between causes and effects. The interdisciplinary ecosystem-based approach can strengthen the systems thinking and knowledge of the intricate interactions between climate and society.

Emotional, Social, and Collaborative Dimensions of CCE

In addition to cognitive learning, successful climate change education should deal with emotional and social skills. Teacher education must help teachers manage emotions associated with climate change, such as anxiety and uncertainty, and foster learning communities. It is encouraged that institutions promote interdisciplinary, place-based, and community-oriented strategies that contextualize climate change and increase learners' engagement (Morales-Aguilar et al., 2025).

Institutional Initiatives, Community Engagement, and Policy Alignment

Experience, e.g., the ALEC Project (Mexico) and the ECC Program (Chile), shows that teacher attitudes towards CCE effectiveness improve. Nevertheless, it could be enhanced by lengthening programmes, strengthening practical activities, and strengthening partnerships with local stakeholders. Reyes et al. (2025) emphasize that climate change education is a collective responsibility that requires the cooperation of scientists, educators, institutions, and policymakers. Inquiry-based approaches, interdisciplinary climate action initiatives, and community participation help make school-based CCE consistent with other global agendas, such as the SDGs.

Contextual and Pluralistic Character of Climate Change Education

Favier (2021) points out that climate change education cannot be considered in isolation and assessed based on its quality, because pedagogical efficacy relies on the orientations of educators, institutional, and local conditions. It requires synergy among many stakeholders, including teachers, teacher educators, policymakers, researchers, parents, and local agencies such as community organizations and water boards. Efforts to strengthen CCE can serve as precursors to broader educational changes, including cross-curricular integration, inquiry-based instruction, technological innovation, and greater school-community collaboration. Nevertheless, the change can be either opposed or supported based on the school culture and contextual limitations (Favier et al., 2021).

Curriculum Transformation, Institutional Culture

Simplifying education systems to be climate action agenda consistent, such as the UN Decade of Action, is a challenge in terms of its institutional implementation. While Hindley (2022) proposes that climate change education has not been prioritized in curricula solely due to structural factors, he also suggests that it reflects individual values, organizational culture, and organizational positional identities. Cultural change in educational organizations is often necessary because of limited institutional agency at the individual educator level, which impedes curriculum reform efforts. Some of the recommendations involve institutional performance reviews, and recruitment processes should incorporate climate change education and systems thinking. These steps help ensure that not only disciplinary expertise is used to evaluate educators, but that involvement in climate change education is mainstreamed into the teaching and research processes (Hindley,2024)

Lack of Implication

Almazroa (2024) emphasizes the fact that even though teachers in Riyadh showed rather high scores in climate knowledge, this did not translate into classroom action that strongly, which means that knowledge alone is ineffective. Teaching practices are influenced by factors such as resource availability, opportunities for professional development, institutional support, and personal beliefs. The lack of official requirements for climate change education and the inconsistencies in educational practices also significantly undermine implementation. Such results can lead to the conclusion that specialised professional development, which implies both conceptual clarity and practical implementation, is necessary (Almazroa, 2024).

Lack of Training among Teachers

Integrating climate change and sustainable development education into teaching across disciplines has consistently faced numerous obstacles. Teachers have cited issues such as insufficient time, inadequate training, insufficient incentives, rigid curriculum structures, and weak external support. In some cases, Education for Sustainable Development (ESD) is perceived as a mandatory project that does not align with subject-specific learning objectives (Rivas et al., 2025). Particularly at the secondary education level, teachers frequently report a lack of confidence in integrating sustainable development themes, which is linked to insufficient teacher training. However, research suggests that professional learning and long-term training programs can enhance teachers' motivation, confidence, and competence, leading to more effective implementation of sustainable development education (Rivas et al., 2025). According to Zhang et al. (2025), secondary education settings are predominant, with fewer studies focusing on primary education, and studies are centered in Global North countries. To tackle these holes, more inclusive research agendas and context-sensitive pedagogical models are needed.

Integrative Frameworks of teacher knowledge in climate and environmental education

To encode the knowledge requirements of efficient environmental education, Zhang et al. (2025) suggest the TPAC+E framework - an extended version of the Technological Pedagogical Content Knowledge (TPACK) framework that explicitly incorporates environmental knowledge. This model promotes cross-cutting, integrative methods in climate and environmental education. TPAC+E is an efficient model of teacher education, as it helps make practical choices regarding curriculum, professional development, and research. It adds that proper climate education demands disciplinary knowledge and pedagogy, as well as the ability to integrate technology and environmental awareness across various educational settings.

Gaps

The following gaps related to teacher education in the context of climate change education through reviewed research studies were identified.

- Most studies indicate that teachers lack vigilance regarding climate knowledge.
- Research concludes that teachers lack conceptual and pedagogical content knowledge related to climate change.
- The research reviewed finds a dearth of training programmes on climate change issues.
- The evidence-based research also points out a gap between theory and practice.
- The recent discussion based on the research reviewed indicated that there is less interest in action-oriented pedagogies and ignorance of the affective domain in the climate-related agenda.
- The studies concluded that practice-related barriers exist in climate change education.
- The conclusion of most of the research was that integrative models lacked sufficient empirical support in climate-related pedagogy.

RQ-3 Creating a climate-ready learning environment through teacher preparedness.

Climate-ready learning environment development is now the focus of climate change education (CCE), but achieving this is difficult due to ongoing deficits in teacher preparedness. Even though there is growing demand to teach about climate change as a complex scientific, social, and ethical issue, studies show that most teachers lack sufficient content knowledge and pedagogical content knowledge to incorporate CCE into their instructional activities effectively (Olawumi et al., 2023; Almazroa, 2024). Consequently, climate change is either handled in a shallow, haphazard, or subject-area manner, which has limited its effectiveness in education and weakened efforts to establish climate-responsive learning ecosystems. The teacher education and professional development programs have proven effective in improving pedagogical practices related to CCE. The need is for the current programs to be more inquiry-based, participatory, and action-oriented, rather than fragmented, standalone, and transmissive (Beach, 2023; Rivas et al., 2025). There is also a lack of a common vision of climate and environmental education in the context of teacher preparation courses, which limits the development of consistent teaching methods, particularly those involving inter- and transdisciplinary work on climate change (Favier et al., 2021; Morales-Aguilar et al., 2025). Thus, a widely expressed concern among educators is their inadequate readiness to place climate change in local contexts, to incorporate ethical and moral thinking, and to tap into local and indigenous knowledge that helps him or her in making climate learning interesting and meaningful to students (Feldbacher et al., 2024; Morales-Aguilar et al., 2025). Besides pedagogical and conceptual constraints, there are also affective and institutional constraints to establishing climate-ready learning environments that teachers must overcome. Instructors also complain of a lack of training to support emotional aspects such as eco-anxiety, limited time and resources, inflexible curricular frameworks, and inadequate sustained institutional support, which limit their teaching agency (Guimerà-Ballesta et al., 2025; Hindley, 2022). Even when teachers exhibit relatively high climate knowledge, preparedness does not necessarily translate into classroom performance, as it is influenced by organizational cultures, assessment priorities, and collaboration opportunities (Almazroa, 2024). Even though the necessity to employ interdisciplinary, systems-oriented, and action-based climate pedagogy has been receiving more and more recognition, there is still a lack of empirical and conceptual clarity to understand how teacher preparedness can be cultivated in a multifaceted way, which allows the construction of climate-ready learning settings (Zhang et al., 2025; Reyes et al., 2025). The need to combine scientific rigor with pedagogical innovation, emotional competence, and civic engagement has been observed in existing research, but the ways in which these aspects of preparedness interact to influence the classroom settings are under-investigated. This knowledge gap is the direct antecedent to the research question of the present study, which aims to investigate the potential of teacher preparedness to support the development of climate-ready learning conditions capable of instilling informed knowledge, resilience, and meaningful climate action in learners.

Integrated Teacher Identity: Based on the above-discussed theories and teacher development outcomes, an amalgam of both theories will be reflected in the teachers. Integrated Teacher Identity is a synthesis of all these dimensions (cognitive, emotional, ethical, and action-oriented) needed to provide climate-responsive education. Going beyond discipline-based, piecemeal preparation, the model makes the teacher not only a conveyor of climate but also a reflective practitioner, an emotionally resilient actor, and a socially responsible change agent. It combines transformative learning (critical reflection, dialogue, perspective transformation, and commitment to action) with resilience capacities (adaptive, coping, and transformative capacities), thus forming a climate-vigilant, ethically grounded, contextually responsive, and climate-action-leading teacher identity. This combined identity is the fundamental tool through which teacher education can facilitate the gap between knowledge and long-term engagement with climate.

DESCRIPTION OF THE PROPOSED MODEL

The model directly links teacher preparedness to the establishment of climate-ready learning conditions by integrating knowledge formulation and pedagogical practice, student interaction, and action-based outcomes. The model provides an integrative framework that brings together scientific, social, ethical, and civic aspects of climate change and operationalizes interdisciplinary and transdisciplinary pedagogy in teacher preparedness. Emotional resilience and eco-emotional competence are positioned as fundamental elements of teacher preparedness, since teachers are identified as intermediaries in the learning of cognitive and affective climates. The model introduces institutional support, a professional learning community, and curriculum coherence as facilitating conditions for teacher preparedness and climate-ready learning. The model integrates place-based learning, local and indigenous knowledge, and ethical inquiry to be much more relevant, think critically, and gain agency. The model builds on current frameworks by theorizing teacher preparedness as multidimensional, encompassing knowledge, pedagogy, emotion, ethics, agency, and institutional support, as needed to prepare climate-ready environments. The proposed model focuses on preparing pre-service teachers to create climate-ready learning environments in classrooms through the following six structured steps.

Climate Ready Teachers: A Climate Ready Teacher is no longer a technician responsible for delivering an edifying commodity, nor are they laid out along the foreseeable landscape of a curriculum. Instead, they fulfil the functions of mediator and companion; they are the facilitator of. Self-determination and social emancipation of the students: teacher as change agent. This involves the promotion of emotional intelligence and critical awareness (critical thinking and critical theory) (Hampson & Rich-Tolsma, 2015).

Climate-ready teachers will be equipped with the following.

Step 1-Climate Vigilant

The First step directed towards the development of climate vigilance included an informed critical awareness of climatic scientific changes, policy, and eco-sociological interdependencies. A climate-ready teacher is always vigilant about the changing climate, continuously attending to climate risks, the changing environment, and sustainability issues at local and global scales. Within the transformative learning paradigm, teachers are encouraged to examine assumptions, critically analyze climate discourses, and examine the sequential and structural determinants of environmental degradation. Teacher training programmes can operationalize this through an interdisciplinary approach, integrating modules on climate literacy, inquiry-based learning, and critical analysis of media sources, aligned with globally recommended sustainability priorities, including education policies and international agreements. With the help of all these teachers, they are groomed as critically conscious and climate change reflective practitioners capable of identifying the systematic and justice-oriented dimension of climate change.

Step 2-Eco-Emotionally Resilient

An eco-emotionally resilient teacher is climate-ready and capable of managing emotional responses to climate change, such as feelings of eco-anxiety, uncertainty, and stress.

The field of climate change education is increasingly intertwined with eco-anxiety, grief, and uncertainty. According to the Resilience Theory, the second step determines the development of emotional regulation, adaptation, coping, and psychological preparedness. Mindfulness training programs in teacher education include aspects such as reflection, teacher peer-support circles based on dialogue, and simulation exercises, which help develop adaptive capacity. Place-based and narrative pedagogies allow educators to rebuild environmental identities, grounded in hope and agency rather than despair. By introducing the concept of affective learning into professional development, teachers can become emotionally stable facilitators who can address students' climate-related concerns constructively.

Step 3-Eco-Socially Responsible

The climate-ready teacher is eco-socially accountable and has realized that climate change is a social justice, equity, and collective responsibility. Such teachers are informed by transformative learning and ESD, which fosters ethical reasoning, inclusivity, and community engagement, motivating the learner to think about the social implications of environmental choices.

ESD emphasizes intergenerational justice, sustainability ethics, and democratic participation. Participation in the global sustainability discourse, together with initiatives pursued by UNESCO, strengthens normative commitments. In service training, the involvement of the community and climate justice collaboratively prepares a social responsibility among teachers. This way, teachers emerge as ethically grounded practitioners devoted to equity and environmental justice.

Step 4- Adaptive to Climate Change

A climate-ready educator is responsive to shifting climatic patterns and demonstrates flexibility in pedagogy, curriculum, and instructional approaches. Adaptive capacity can be described as teachers' capacity to respond to disruptions caused by climate change and the restructuring of pedagogical and institutional practices. This dimension is based on Resilience Theory and focuses on anticipatory planning, flexibility, and systems thinking. The adaptability of teacher preparation is encouraged by the curriculum design, the modularity of lesson plans, and sustainability competencies aligned with Education for Sustainable Development. The continuity-of-learning planning, disaster Risk Reduction training, and school risk mapping enhance resiliency to climate emergencies. Problem-based learning and scenario planning boost the ability to think flexibly and think beyond the present. Action research and reflective practice also support iterative professional development. These processes make adaptive capacity a quantifiable professional ability, enabling teachers to maintain the quality of education and spearhead institutional adaptation to changing climatic conditions.

Step-5 Action-Oriented Climate Pedagogues

A climate-responsive educator is a climate pedagogue who acts by transferring climate literacy and changed attitudes into action-based teaching approaches. The concept of action-oriented climate pedagogy is a more effective approach to translate awareness into practice. The Transformative Learning Theory assumes that the ultimate result of real perspective change is action commitment. Based on that, the teachers need to acquire competencies in creating and facilitating action-oriented learning experiences. This phase focuses on climate education through projects, participatory inquiry, students' sustainability endeavours, and problem-solving models based on local environmental issues. The teacher training incorporates experiential, place-based, and solution-focused pedagogies to teach teachers how to incorporate mitigation and adaptation strategies in the classroom. Action competence is the core of ESD, and allows the teachers to transcend their awareness-building efforts to empower students to design, implement, and evaluate climate interventions. In this way, the teachers become enablers of collective efficacy and student empowerment rather than merely transmitters of information.

Step- 6 Capable of Leading Climate Action in Society

Finally, the framework brings the teachers into the limelight as climate leaders who do not reach out to the boundaries of the classroom. At this point, professional identity develops into civic leadership based on ethical responsibility and transformational agency. Leadership preparation also encompasses advocacy training, community partnerships, active participation in governance, and communication with people. Teachers work with stakeholders and assist with institutional sustainability. The last phase makes teachers climate leaders, enabling them to expand their practice beyond the classroom to institutional and societal transformation.

Figure 1. Model for Climate Ready Learning Environment

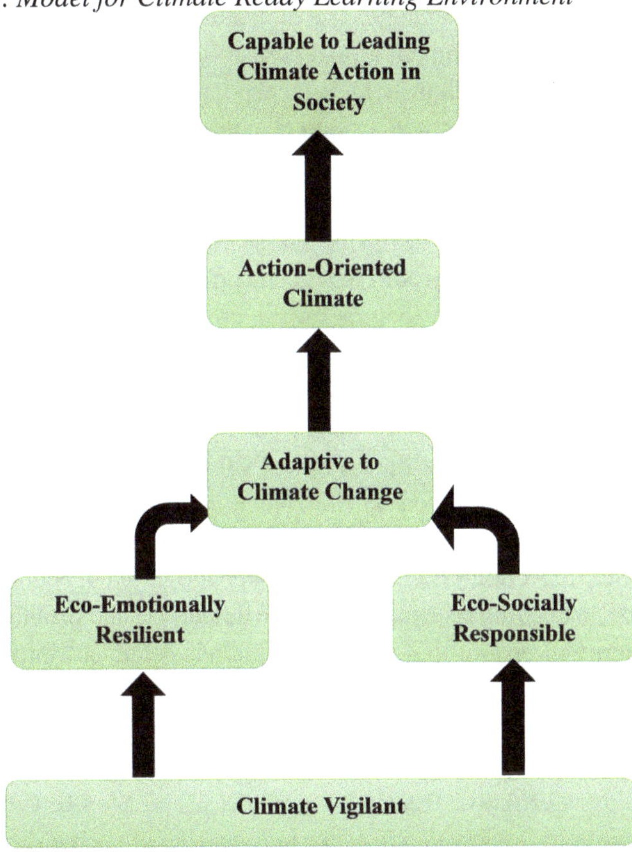

Linkage between Gaps and Proposed Model:

The Climate-Ready Teacher Model (Figure 1) outlined in the suggested 6 steps is a systematic, theoretically grounded response to the major gaps identified in this research. The ongoing lack of conceptual knowledge and pedagogical content knowledge is addressed by Step 1 Climate Vigilant, which builds on interdisciplinary climate literacy and critical engagement with scientific and policy discourses among teachers. It also develops the capacity to contextualize climate knowledge in pedagogical practices. This step is critical because it fosters critical awareness and reflective inquiry among teachers, enabling them to move beyond superficial awareness to pedagogically significant awareness. The given lack of developed training programmes is addressed by the model's very structure, in which Step 1 Climate Vigilant focuses on a changing climate through continuous improvement of professional skills, inquiry-based learning, reflection practice, and adaptive curriculum design. This transforms current teacher preparation models, which are disjointed and overly theoretical, into long-term, competency-centred teacher learning that is attentive to climate change. The disregard for the affective domain, especially regarding eco-anxiety and emotional preparedness, is directly addressed in Step 2: Eco-Emotionally Resilient, which incorporates emotional regulation, resilience-building, and reflective practices into teacher education. It is a step that allows teachers to positively engage with the affective aspects of climate change by integrating mindfulness, peer support, and narrative pedagogies, as well as by promoting personal well-being and nurturing classroom conditions. The lack of connection between theory and classroom practice is effectively addressed by Step 3: Eco-Socially Responsible, which focuses on the ethical and social aspects of climate education typically disregarded and builds values of sustainability, equity, and social justice. It can situate climate change within broader socio-political frames and enhance the teacher's ability to foster ethical thinking, inclusivity, and community involvement, thereby instilling a robust value-based orientation in the pedagogical approach. Practice-related barriers, such as institutional constraints, resource scarcity, and a lack of flexibility, are addressed through Step 4: Adaptive to Climate Change. Step 5: Action-Oriented Climate Pedagogues that operationalize knowledge of climate to encompass experiential, participatory, and problem-based pedagogies. This step will ensure that theoretical knowledge is put into practice through project-based learning, local environmental problem-solving, and student-driven sustainability projects. This way, it also addresses the lack of engagement in action-oriented pedagogies by equipping teachers with the competencies needed to plan and deliver transformative learning experiences. and Step 6: Capable of Leading Climate Action in Society. Such stages help teachers become more adaptive, systems-thinking, and leadership-oriented, enabling them to address structural issues, interact with stakeholders, and participate in institutional and community-wide sustainability initiatives. This would enhance teacher agency and facilitate responsive climate education delivery. Lastly, the empirical weakness of integrative pedagogical models is that they focus on action research, reflective inquiry, and classroom-based implementation in Steps 4 and 5. Finally, the six-step model has incorporated cognitive, affective, ethical, and behavioural aspects of learning, thereby making teachers climate-ready.

CONCLUSION

The three related research questions used to explore the role of teacher education in promoting climate action and resilience in society in this chapter focused on conceptual backgrounds, gaps in the literature, and how climate-ready learning settings can be established with the support of teacher preparedness. Combined, the results highlight the fact that climate change education cannot be the add-on to the existing programs, but it needs to be fundamentally re-oriented to include transformative, interdisciplinary, and action-oriented teacher education. The chapter has responded to the first research question by defining that the conceptual elements of teacher education to address climate action are based on education for sustainable development, transformative learning and resilience theory. All these views point to the role of critical reflection and viewpoint change, adaptability, and ethical involvement in training teachers to act in the face of complexity and uncertainty of climate change. The role of teacher education is therefore critical in cultivating both scientific knowledge as well as agency, civic responsibility, and the ability to assist learners to overcome socio-ecological challenges. These conceptual underpinnings affirm teacher education as the influence of climate action or resilience in society by locating teachers as critical intermediaries between knowledge and action. To answer the second research question, the chapter revealed that there were still gaps in the area of teacher education as witnessed in the literature. These have been thought to be inadequate integration of climate change education into pre-service and in-service programs, disciplinary and divided curricula, excessive dependence on transmissive pedagogies, and insufficient concern on affective, ethical, and contextual aspects of learning climate. Moreover, the literature shows that there is a strong knowledge-action gap, whereby more knowledge about climate change in teachers does not always result in classroom practice, which in many cases is caused by institutional barriers, inadequate coordinated support, and poor professional development. The lack of the following gaps points to the inability of existing teacher education models to treat climate change as a complex, transdisciplinary, and socially embedded issue. Answering the third research question, the chapter claimed that the development of climate-ready learning environments depends on the expanded understanding of teacher preparedness. Teacher preparedness needs to be conceptualized as a multidimensional construct it includes vigilant towards changing climate change, eco- emotionally and eco-socially responsible, content and pedagogical knowledge, emotional and social competence, ethical reasoning, professional agency and institutional support. The climate-ready learning environment occurs when educators are prepared to balance the scientific rigor with participatory and place-based learning, to address challenges behind affective issues like eco-anxiety, and engage learners in informed and meaningful climate action. The chapter revealed that these conditions are not merely a product of isolated teacher competence, but, instead, the consequence of consistent education systems in which teacher education is coordinated with friendly organizational cultures and cooperative professional activities. In the conclusion, this chapter highlights that teacher education is a key area to focus on in promoting climate change education, climate action, and societal resilience. Through these conceptual clarifications, critical gaps, and statements on the program of teacher preparedness in creating climate-ready learning environments. The chapter is adding a more comprehensive picture of how teacher education can be responsive to the climate crisis. Future research, policy, and practice initiatives should therefore focus on integrative and context responsive teacher education models that will enable teachers to no longer focus on knowledge delivery but innovate along transformative climate pedagogy that can equip learners with a rapidly evolving and uncertain world.

REFERENCES

Almazroa, H. (2024). Bridging the Knowledge–Practice Gap: Assessing Climate Change Literacy Among Science Teachers. *Sustainability (Basel)*, *16*(20), 9088. Advance online publication. DOI: 10.3390/su16209088

Beach, R. (2023). Addressing the Challenges of Preparing Teachers to Teach about the Climate Crisis. *Teacher Educator*, *58*(4), 507–522. DOI: 10.1080/08878730.2023.2175401

Budgen, D., Kitchenham, B., Charters, S., Turner, M., Brereton, P., & Linkman, S. (2007, April 1). *Preliminary results of a study of the completeness and clarity of structured abstracts. 11th International Conference on Evaluation and Assessment in Software Engineering (EASE)*. DOI: 10.14236/ewic/EASE2007.7

Busch, K. C., Henderson, J. A., & Stevenson, K. T. (2019). Broadening epistemologies and methodologies in climate change education research. *Environmental Education Research*, *25*(6), 955–971. DOI: 10.1080/13504622.2018.1514588

C. S. Holling (1973). (2022). InFiorella, K. J., Avolio, M. L., Pulver, S., Alexander, S. M., & Burnside, W. R. (Eds.), *Foundations of Socio-Environmental Research: Legacy Readings with Commentaries* (pp. 460–482). Cambridge University Press., DOI: 10.1017/9781009177856.038

Dirkx, J. M. (1998). *Transformative Learning Theory in the Practice of Adult Education: An Overview.*

Favier, T., Van Gorp, B., Cyvin, J. B., & Cyvin, J. (2021). Learning to teach climate change: Students in teacher training and their progression in pedagogical content knowledge. *Journal of Geography in Higher Education*, *45*(4), 594–620. DOI: 10.1080/03098265.2021.1900080

Feldbacher, E., Waberer, M., Campostrini, L., & Weigelhofer, G. (2024). Identifying gaps in climate change education—A case study in Austrian schools. *International Research in Geographical and Environmental Education*, *33*(2), 109–124. DOI: 10.1080/10382046.2023.2214042

Folke, C. (2006). Resilience: The emergence of a perspective for social–ecological systems analyses. *Global Environmental Change, Resilience, Vulnerability, and Adaptation: A Cross-Cutting Theme of the International Human Dimensions Programme on Global Environmental Change, 16*(3), 253–267. DOI: 10.1016/j.gloenvcha.2006.04.002

Folke, C. (2016). Resilience (Republished). *Ecology and Society*, *21*(4), art44. https://www.jstor.org/stable/26269991. DOI: 10.5751/ES-09088-210444

Folke, C., Carpenter, S. R., Walker, B., Scheffer, M., Chapin, T., & Rockström, J. (2010). Resilience Thinking: Integrating Resilience, Adaptability and Transformability. *Ecology and Society*, *15*(4), art20. https://www.jstor.org/stable/26268226. DOI: 10.5751/ES-03610-150420

Fuso Nerini, F., Sovacool, B., Hughes, N., Cozzi, L., Cosgrave, E., Howells, M., Tavoni, M., Tomei, J., Zerriffi, H., & Milligan, B. (2019). Connecting climate action with other Sustainable Development Goals. *Nature Sustainability*, *2*(8), 674–680. DOI: 10.1038/s41893-019-0334-y

Guimerà-Ballesta, G., Calafell-Subirà, G., Jiménez-Valverde, G., & Esparza-Pagès, M. (2025). Climate Change Education in Secondary Schools: Gaps, Challenges and Transformative Pathways. *Encyclopedia*, *6*(1), 8. Advance online publication. DOI: 10.3390/encyclopedia6010008

Hampson, G. P., & Rich-Tolsma, M. (2015). Transformative Learning for Climate Change Engagement: Regenerating Perspectives, Principles, and Practice. *Integral Review*, *11*(3).

Hindley, A. (2022). Understanding the Gap between University Ambitions to Teach and Deliver Climate Change Education. *Sustainability (Basel)*, *14*(21), 13823. Advance online publication. DOI: 10.3390/su142113823

IPCC. (2023). Climate Change 2023: Synthesis Report, Summary for Policymakers. Contribution of Working Groups I, II and III to the Sixth Assessment Report of the Intergovernmental Panel on Climate Change [Core Writing Team, H. Lee and J. Romero (eds.)]. IPCC, Geneva, Switzerland. (pp. 1–34) [Monograph]. Intergovernmental Panel on Climate Change (IPCC). DOI: 10.59327/IPCC/AR6-9789291691647

Kivunja, C. (2018). Distinguishing between Theory, Theoretical Framework, and Conceptual Framework: A Systematic Review of Lessons from the Field. *International Journal of Higher Education*, *7*(6), 44–53. DOI: 10.5430/ijhe.v7n6p44

Mezirow, J. (2018). Transformative learning theory. In *Contemporary Theories of Learning* (2nd ed.). Routledge. DOI: 10.4324/9781315147277-8

Morales-Aguilar, R., Arenas, A., Cisternas, D., Bascopé, M., Salazar, D., Becerra, R., Solís-Pinilla, J., & Merino, C. (2025, May). Transforming education for a sustainable future: An analysis of teacher education in the context of climate change. [). Frontiers Media SA.]. *Frontiers in Education*, *10*, 1537129. DOI: 10.3389/feduc.2025.1537129

Olawumi, K., Mavuso, M. P., Khalo, X., Kafu-Quvane, B., & Mzilikazi, B. (2023). Implementation of Teacher Development Program for Integrating Climate Change Education: Natural Sciences Teachers View. *International Journal of Environmental, Sustainability, and Social Science, 4*(3), 788–798. (none). DOI: 10.38142/ijesss.v4i3.534

Reyes, P. C., Viviani, M., & Robles, C. M. (2025). Strengthening capacities in Latin America by designing teacher professional development programmes for climate change education. *Journal of Professional Capital and Community, 10*(1), 6–19. DOI: 10.1108/JPCC-12-2023-0098

Snyder, H. (2019). Literature review as a research methodology: An overview and guidelines. *Journal of Business Research, 104*, 333–339. DOI: 10.1016/j.jbusres.2019.07.039

Tranfield, D., Denyer, D., & Smart, P. (2003). Towards a Methodology for Developing Evidence-Informed Management Knowledge by Means of Systematic Review. *British Journal of Management, 14*(3), 207–222. DOI: 10.1111/1467-8551.00375

Zhang, X., Jung, W., & Asari, M. (2025). Systematic Review of Environmental Education Teaching Practices in Schools: Trends and Gaps (2015–2024). *Sustainability (Basel), 17*(19), 8561. Advance online publication. DOI: 10.3390/su17198561

Chapter 5
Teacher Training Needs and Capacity Building for Interdisciplinary Climate Change Curriculum Implementation

Kofi Nkonkonya Mpuangnan
http://orcid.org/0000-0001-5566-2288
University of Zululand, South Africa

Samantha Govender
University of Zululand, South Africa

ABSTRACT

The growing urgency of the climate crisis calls for an education system that transcends traditional subject boundaries to promote holistic understanding and sustainable actions. Teachers stand at the centre of this transformation. However, many teachers remain inadequately trained to teach climate change through interdisciplinary frameworks. This raises a critical question: do teachers possess the requisite capacity and confidence to prepare learners for the realities of a climate-impacted world? To address this question, Change Theory in Education theory developed by Fullan in 1982 was used to underpin the study. In this light, conceptualization approach was adopted. This approach involves a systematic process of defining, clarifying, and organizing main concepts to deepen understanding of a research issue.

DOI: 10.4018/979-8-3373-6975-4.ch005

INTRODUCTION

Climate change is no longer just an environmental concern; it has become a complex global challenge that touches nearly every aspect of human life. It affects economies, threatens public health, disrupts food supply, influences education systems, and places growing pressure on social structures (Mpuangnan, 2025). Therefore, education systems have a responsibility to equip learners with the knowledge, skills, and values needed to understand and respond effectively to climate-related challenges. Teachers play a vital role in driving this transformation within education (Assaraf et al., 2025). However, delivering an interdisciplinary climate change curriculum demands skills that extend beyond traditional subject areas. Studies such as (Li & Hu, 2026) and Custodio-Ferrando & Cabero-Fayos (2025), argue that many teachers were originally prepared within single-discipline frameworks. It is imperative to recognize their training needs and support them in building the capacity required to teach climate education in a more integrated and effective way.

Despite the global recognition of climate education as a priority (UNESCO, 2025; Green, 2025), scholarly attention to the specific training needs of secondary school teachers to enhance their instructional capacity remains insufficient. Much of the extant literature has concentrated on curriculum design and initiatives within higher education (Saad et al., 2026; Alimehmeti et al., 2026; Botha, I& Klopper, 2025). This suggests that limited attention is directed to the contextual realities and pedagogical demands encountered by teachers at the secondary level. Meanwhile secondary education represents a pivotal stage in the learning continuum. Secondary education is where learners develop cognitive maturity (Pradhan, 2023) to interrogate complex global challenges and cultivate values that influence long-term environmental responsibility.

Teachers in secondary schools function as essential mediators between foundational knowledge and advanced academic engagement. However, many teachers remain inadequately trained to teach climate change through interdisciplinary frameworks. This raises a critical question: do teachers possess the requisite capacity and confidence to prepare learners for the realities of a climate-impacted world? This study is therefore critical in addressing this question, without adequate professional preparation, the delivery of climate change education risks becoming disjointed and less transformative. Therefore, the inadequate targeted conceptual research on teachers' professional development needs constitutes a significant gap in the literature. Addressing this gap is crucial in preparing teachers with knowledge on interdisciplinary context to successfully implement interdisciplinary climate change curriculum. It can also reinforce interdisciplinary pedagogical approaches and improve the overall quality and effectiveness of climate change instruction within secondary schools.

UNDERSTANDING INTERDISCIPLINARY CLIMATE CHANGE EDUCATION

Interdisciplinary education is an instructional approach that combines knowledge, perspectives, and methods from multiple academic disciplines. Interdisciplinary education helps learners understand and address complex real-world problems (Anand & Singh, 2025). This approach helps learners recognize connections across different fields of study and supports a more holistic understanding of knowledge. It is especially relevant to climate change, which is a complex issue involving scientific, social, economic, and technological dimensions. Engaging with multiple disciplines helps learners develop a thorough understanding of climate-related challenges (Guimerà-Ballesta et al., 2025). It also strengthens their ability to critically evaluate solutions and their real-world impacts on society.

Climate change inherently necessitates an interdisciplinary framework due to its far-reaching impacts across diverse sectors. Within the natural sciences, learners examine climate systems, ecological processes, and biodiversity loss (Heikkilä et al., 2025). Geographic inquiry facilitates analysis of land-use patterns, spatial inequalities, and climate-induced migration. Economic perspectives support the evaluation of resource distribution, development pathways, and sustainability strategies (Wang, et al., 2023), while social studies foreground issues of governance, equity, and public policy (Rodríguez, & Swalwell, 2025). Technological education introduces innovations such as renewable energy and climate mitigation tools. An exclusive focus on any single discipline risks reductionism and constrains students' ability to understand the interconnected drivers and consequences of climate change. An interdisciplinary climate change curriculum integrates these concepts in knowledge production and development of transferable competencies. Such competencies are recognized as essential for preparing young people to respond effectively to contemporary environmental challenges.

Effective implementation of interdisciplinary education depends significantly on teacher preparedness. Teachers must possess both the subject expertise and pedagogical capacity to design integrative learning experiences (Kotsis, 2026). Such competencies help them to facilitate inquiry-driven activities and employ assessment strategies that capture interdisciplinary understanding. Without adequate professional preparation, even well-designed curricula may fail to achieve intended outcomes. Consequently, sustained investment in teacher training and capacity-building is critical to ensuring that climate change education remains rigorous, engaging, and impactful.

THEORETICAL FRAMEWORK

This study is guided by Change Theory in Education. Change Theory in Education by Michael Fullan in 1982. Fullan's Change Theory in Education has influenced educational reform. Change Theory in Education aims to facilitate effective and sustainable educational improvement by guiding successful implementation of meaningful change that enhances teaching and learners' learning (Fullan, 1982). This suggests that Fullan's theory seeks to improve teaching practices and learning outcomes. In this theory, teachers are recognized as central agents within the change process. Fullan (1982) asserts that reforms succeed when teachers receive adequate support. Supports such as professional development, access to resources, collaboration opportunities, help teachers understand new reforms, remained motivated to apply innovation practices.

Furthermore, Change Theory in Education presents interconnected elements necessary for successful educational change. These elements include a clear vision, capacity building, collaboration leadership support and resource availability (Fullan, 1982; Muzakki, et al. 2026). Continuous professional development and ongoing evaluation further reinforce change initiatives. These elements offer a framework for addressing teacher training needs. They support capacity building for interdisciplinary climate change curriculum implementation.

METHODOLOGY

The conceptualization approach was employed to explore interconnected ideas central to this research problem. Conceptualization involves a systematic process of defining, clarifying, and organizing main concepts to deepen understanding of a research issue (Hartmann et al., 2025). This approach integrates diverse perspectives into a coherent analytical framework (Lett et al., 2022). In this study, conceptualization approach helped in reviewing relevant literature, identifying key ideas, and synthesizing scholarly arguments. This process allowed the research to establish relationships among core concepts and build a solid foundation for further analysis. The generated ideas are illustrated in the Figure 1.

Figure 1. Teacher Training Needs

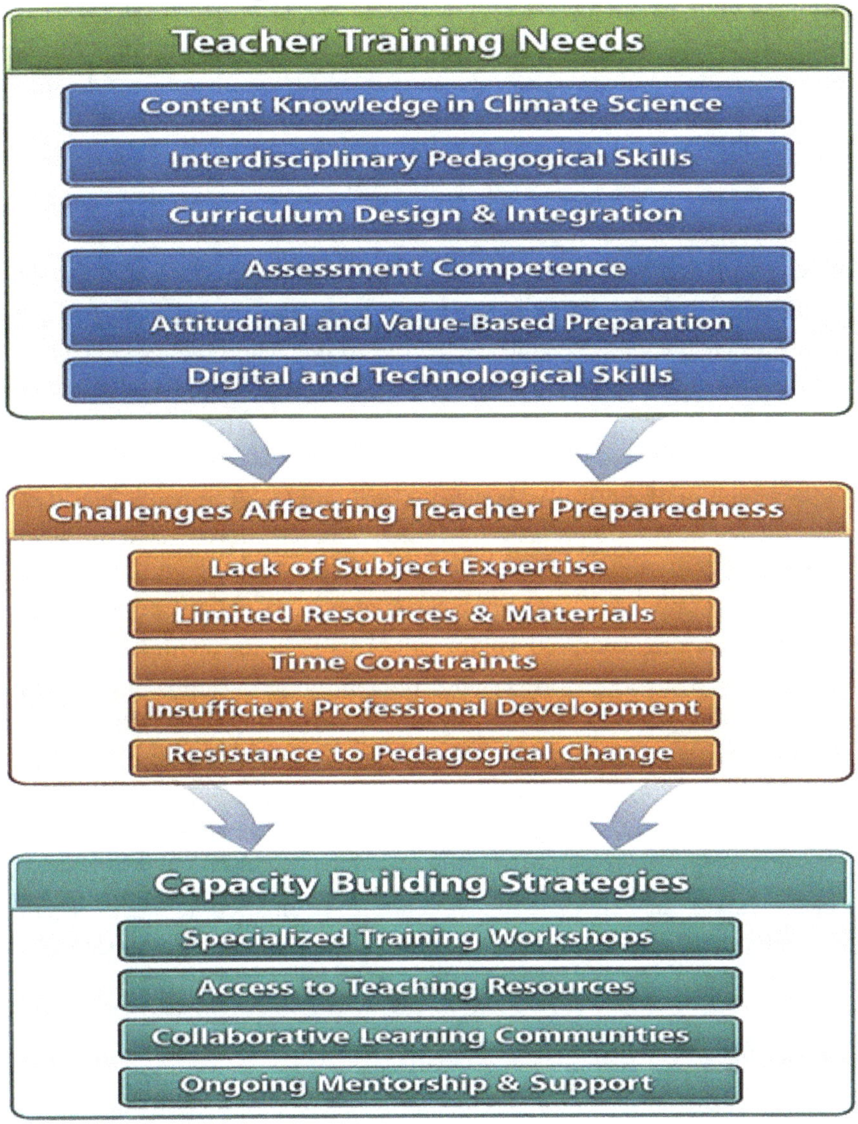

Content Knowledge in Climate Science

Content knowledge in climate science plays a significant role in supporting interdisciplinary climate change curriculum implementation. Romero Ariza et al., (2024) argue that teachers with strong climate literacy demonstrate greater ability to interpret scientific concepts and connect them to sustainability principles across disciplines. Fullan (1982) supports that effective educational reform is contingent upon strengthening teacher capacity, given their role as primary agents of change. According to Newsome, et al., (2023), a strong understanding of climate systems, mitigation strategies, and environmental impacts boosts teachers' confidence in the classroom. It is certain that teacher's knowledge in climate change issues enable them to integrate sustainability concepts effectively into teaching. Such competence contributes to improved learner engagement and fosters the development of environmentally responsible attitudes among students.

On the other hand, inadequate content knowledge can constrain effective curriculum implementation. Additionally, teachers with limited understanding of climate science are often reluctant to employ interdisciplinary approaches (Tang, 2025; Marschall, 2022). This supports Change Theory in Education that reform initiatives often falter when capacity-building mechanisms are insufficient (Fullan, 1982). It can also be said that limited subject expertise can diminish teacher self-efficacy and restrict the adoption of innovative pedagogical strategies. As a result, learners may encounter difficulties in comprehending the complex and interconnected dimensions of sustainability challenges.

Nonetheless, knowledge gaps in climate science education presents opportunities for targeted professional development. This is in line with Change Theory in Education emphasizing the importance of continuous learning and institutional support in sustaining educational improvement. Nassor, et al. (2026) opines that structured training programs can enhance teacher competence and encourage interdisciplinary collaboration. Additionally, training promotes the consistent integration of sustainability principles within instructional processes (Papademetriou et al., 2025). However, failure to address these training needs may impede reform efforts and weaken curriculum effectiveness. Therefore, strengthening teachers' climate science content knowledge remains fundamental to achieving sustainable and effective interdisciplinary curriculum implementation.

Interdisciplinary Pedagogical Skills

Interdisciplinary pedagogical skills are critical for effective implementation of an interdisciplinary climate change curriculum. According to Lam (2025), interdisciplinary pedagogical skills entail the intentional design of instructional practices that enable learners to recognize connections across disciplines, transfer knowledge to diverse contexts, and engage in comprehensive problem-solving. Effectives teachers require these competencies to link climate concepts across disciplines. Such competencies can be achieved through preparing teachers to collaborate, use real-world tasks, integrate perspectives, and adopt learner-centered approaches for deeper understanding (Mpuangnan, 2025). This represents an essential training need that must be prioritized to strengthen teacher capacity in schools.

Moreover, encouraging interdisciplinary teaching has a long-time effect on the students. The students would acquire multiple knowledge and skills as they integrate concepts, skills, and methods from multiple subjects (Melesse et al., 2025). This approach allows students to engage in a collaborative classroom practice which creates an inclusive learning environment. Additionally, Interdisciplinary pedagogical skills further develop critical thinking and analysis of complex environmental issues (Priya, 2025). Such preparedness enhances instructional effectiveness and facilitates students' comprehensive understanding of sustainability.

However, Chamo and Broza (2025) argue that teachers with limited interdisciplinary pedagogy often have trouble synthesizing content across disciplines. This in effect contribute to fragmented instruction and superficial engagement with sustainability themes. Fullan (1982) asserts that reform initiatives are less likely to succeed when teachers perceive themselves as insufficiently prepared for emerging instructional demands. Therefore, addressing these training needs through structured professional development is imperative. this call aligns with Emam and Al-Mahdy (2026) advocating for capacity building and sustained institutional support to enhance teachers' readiness to implement interdisciplinary climate change education effectively.

Curriculum Design and Integration

Teachers' knowledge in curriculum design and integration is critical for the effective implementation of an interdisciplinary climate change curriculum. Curriculum design and integration involve deliberate planning of learning experiences and the purposeful organization of subject areas to ensure meaningful connections among concepts (Vettriselvan & Ramya, 2025). Teachers' knowledge extends to teachers' proficiency in practice teaching. When teachers are proficient in developing integrated lesson plans, they can align climate issues across multiple subjects (Almazroa, 2024). Similarly, when teachers can effectively structure interdisciplinary learning experiences, they are able to deliver coherent lessons (Chen et al., 2025). Their classrooms practices can actively engage students with environmental issues, thereby enhancing learning consistency and promoting meaningful educational outcomes.

On the other hand, inadequate preparation in curriculum design can impede curriculum implementation. Teachers who are not competent in curriculum design often encounter challenges in embedding climate change content within existing curricula (Mavuso et al., 2022). This incompetencies can result in fragmented lessons and limited coverage of sustainability concepts. In this sense, it for this reason that Change Theory in Education established that educational change is unlikely to succeed when teachers are insufficiently prepared for new curricular demands (Fullan, 1982). The resulting effect can undermine teacher confidence, restrict pedagogical creativity, and reduce the overall effectiveness of interdisciplinary instruction. This limits students' ability to comprehend the interconnections between climate change and sustainable development.

Identifying these training needs has profound significance for targeted professional development and support. In this light, professional development programs focused on curriculum design are critical for enhancing teachers' competencies (Mpuangnan, 2024). Such programs can further foster collaborative planning and facilitate the systematic integration of sustainability principles across subjects. It is essential to strengthening teachers' curriculum design and integration skills for the sustainable and effective implementation of interdisciplinary climate change education.

Attitudinal and Value-Based Preparation

Attitudinal and value-based development is critical for the effective implementation of an interdisciplinary climate change curriculum. This focuses on the cultivation of mindsets, attitudes, and ethical values essential for effective teaching and meaningful learning of climate change education (Mpuangnan, 2025). When teachers have comprehensive knowledge in the scientific, ethical, social, and environmental dimensions of climate change, they create inclusive and meaningful learning environments.

Moreover, teacher's knowledge on climate related concepts helps in making lessons relevant to learners encouraging them to demonstrate responsible attitudes and behaviours towards environmental stewardship. For instance, contextualizing events such as the 2017 "Day Zero" water crisis in Cape Town, droughts in the Eastern Cape (Gugenishvili, 2025), gives them concrete, locally relevant insights. Also, teachers who internalize principles of sustainability can serve as models for pro-environmental behaviour (Demirtaş et al., 2025). For instance, a Life Orientation teacher organizing a community clean-up or a Geography teacher guiding learners in local biodiversity mapping exemplifies applied climate action. This approach helps them understand the real-world impacts of climate change in their own communities.

Furthermore, learners show higher engagement in pro-environmental behaviours when they understand the reasoning behind climate actions (Ali et al., 2025). Consequently, many contribute to the meaningful integration of climate change concepts across multiple disciplines. Over time, this approach enhances curriculum coherence and promotes critical thinking about both local and global environmental challenges. According to Change Theory in Education, educational reforms are unlikely to succeed if teachers are unprepared to act as change agents. Low motivation and misaligned values can limit the adoption of innovative pedagogical practices (Cai et al., 2026). This reduces learner engagement and hinders the interdisciplinary integration essential for effective climate education.

Digital and Technology Skills

Digital and technology skills is another critical training needs teachers require in facilitating the implementation of an interdisciplinary climate change curriculum. Proficiency in digital tools, online resources, and educational software enables the design of interactive lessons (Adera, 2025). Teachers' knowledge in education software such as Canvas, Schoology, Moodle and Google Doc (Bondre & Yadav, 2025), is crucial for integration of climate change content across multiple subjects. Change Theory in Education identifies teachers as main agents of change in the learning process. Improving their technological competence supports the effective use of innovative teaching practices. In addition, the use of simulations, virtual experiments, and data-driven activities enables students to engage critically with complex climate concepts and fosters deeper understanding.

However, limited proficiency in digital and technological skills can constrain effective curriculum implementation. Inexperience teachers regarding technology use often rely primarily on traditional lecture-based approaches (Kerich & Kimwarey, 2026). Meanwhile the lecture approach can reduce student engagement (Cherukuri, 2025), and limit interdisciplinary integration. Deficits in digital skills may inhibit pedagogical innovation (Stevens, 2025), hinder learner-centered learning and compromise the overall effectiveness of climate change education. Therefore, ongoing technical support, and opportunities for practical application of digital tools are essential. Improving teachers' technology skills boosts their confidence in the classroom. It also encourages them to use creative teaching methods that connect multiple subjects.

Assessment Competence

Assessment competence constitutes a critical factor in the effective implementation of an interdisciplinary climate change curriculum. The Curriculum and Assessment Policy Statement (CAPS) South African for instance, emphasizes continuous assessment (Chinengundu, 2025). This requires teachers to demonstrate proficiency in both formative and summative evaluation practices. Teachers who possess strong assessment literacy are better positioned to facilitate meaningful learning experiences (Mpuangnan, 2024). Their knowledge in formative strategies must include structured discussions, project-based learning, and reflective journals. Acquiring these competencies enable teachers to monitor learners' conceptual understanding of climate-related issues. This is in line with Fullan's Change Theory in Education

(1982), which identifies teachers as primary agents of reform, the enhancement of assessment competence supports the adoption of innovative and collaborative pedagogical approaches.

The implications of advanced assessment competence become particularly evident when evaluation tasks are aligned with authentic environmental challenges. For example, teachers should use community investigations, such as studying local water shortages or examining waste management systems, as formative assessments. These community investigations help develop learners' critical thinking, problem-solving (Karaoglan-yılmaz et al., 2023), and teamwork skills. In the summative assessments, teachers can focus on learner portfolios. These assessments give learners an opportunity to combine knowledge from different subjects. Such assessment procedure also helps learners show a deeper understanding of sustainability concepts. In this context, Change Theory in Education can be referred. Fullan (1982) justifies that educational change is more successful when teachers have the right skills and confidence to use new teaching methods. Eventually, the new teaching methods seeks to improve learner engagement leading to better academic performance.

It can be established that teacher training is crucial for the effective implementation of an interdisciplinary climate change curriculum, influencing both positive and negative outcomes. Strong content knowledge in climate science enables teachers to convey complex environmental concepts and link them to sustainability, while limited knowledge may lead to superficial instruction. Interdisciplinary pedagogical skills facilitate integration across subjects and promote collaborative learning, whereas insufficient skills can result in fragmented lessons. Competence in curriculum design, attitudinal and value-based preparation, digital and technology skills, and assessment literacy further enhances instructional quality and student engagement. Change Theory in Education states that teachers are main drivers of change in schools. Strengthening teachers' skills helps ensure educational reforms are successfully adopted, implemented, and sustained

CHALLENGES AFFECTING TEACHER PREPAREDNESS

Lack of subject expertise is a significant barrier to preparedness for implementing an interdisciplinary climate change curriculum. This is because the topic requires the integration of scientific, social, and ethical knowledge, which can be difficult to achieve without adequate content understanding. Qiu et al. (2022) found that limited content knowledge can reduce confidence, restrict innovative teaching practices, and lead to fragmented curriculum. Additionally, evidence from South Africa indicates gaps in teachers' understanding of climate processes (Thenga, 2025). This implies that many teachers have challenges in demonstrating deeper scientific knowledge especially concerning climate change instruction. Over time, lack of subject expertise contributes to the struggle for effective integration of climate change education into the national curriculum in South Africa (Mavuso et al., 2022). These limitations reduce opportunities for interdisciplinary learning, critical thinking, and real-world problem-solving. Therefore, there is a need for targeted professional development to strengthen subject expertise and support meaningful curriculum reform.

Moreover, limited resources and instructional materials pose a significant challenge to teacher preparedness for the implementation of an interdisciplinary climate change curriculum. Effective climate education requires contextually relevant materials that support connections across disciplines Henninger & Christmann, 2023). For example, access to updated textbooks, digital technologies, field-based learning opportunities are critical for effective teaching and learning. However, many rural schools especially in South Africa face shortages of basic resources and infrastructure (Omojemite, 2025; Lombo & Subban, 2024). In the absence of adequate materials, climate change education may become overly theoretical and less engaging. This is because teachers are restricted from using creative pedagogical strategies and reinforce reliance on traditional teaching approaches. Providing adequate materials and institutional support is essential for improving teacher preparedness. It also supports the effective implementation of interdisciplinary climate change education.

Additionally, time constraints constitute a significant barrier to teacher preparedness for implementing an interdisciplinary climate change curriculum. The extensive scope of climate change education content often required sufficient time for implementation. This is often a challenge in many secondary schools. In South African for instance, curriculum overload constrains the allocation of time for in-depth discussions and learner-centred instructional activities (Mhlongo & Chaane, 2025). This limitation further reduces the opportunity to employ innovative pedagogical strategies that facilitate comprehensive student understanding. Awayehu (2025) found that insufficient time impedes the integration of environmental content across multiple subjects. This aligns with the Change Theory in Education (Fullan, 1982), that effective educational reform necessitates adequate time to facilitate planning and the implementation of innovative instructional strategies. This means insufficient time seeks impedes teachers' ability to adopt and apply pedagogical innovations. Addressing the time factor is critical for teacher preparedness and facilitating the successful implementation of interdisciplinary climate change education.

Furthermore, insufficient professional development limits teachers' preparedness to implement an interdisciplinary climate change curriculum. Inadequate professional development constrains teachers' content knowledge, instructional planning, and capacity to integrate concepts across multiple disciplines (Popova et al., 2022). In South Africa, Mavuso et al., (2022) reported that many teachers had minimal training in climate change education. This results in fragmented lessons which superficially produces learning outcomes. Mumcu et al., (2022) justify this asserting that without adequate professional development, teachers are unable to adopt innovative interdisciplinary approaches. Targeted training is therefore essential to enhance teacher preparedness and facilitate meaningful and effective climate change instruction.

Resistance to pedagogical change represents a significant barrier to teacher preparedness for implementing an interdisciplinary climate change curriculum. Teachers who are familiar with traditional, lecture-based instruction often hesitate to adopt innovative approaches such as project-based learning, experiential activities, and interdisciplinary learning (Arrasyid et al., 2025). In Ghana for instance, studies indicated that some teachers resisted the transition from content-focused instruction to learner-centred methodologies (Buabeng & Amo-Darko, 2025). This resulted in inconsistent adoption and reduced instructional effectiveness. Therefore, it is incumbent for teachers' educators to focus on targeted professional development, mentoring, and institutional support to improve teachers' capacity. When this is done, teachers will be equipped to adopt innovative teaching practices and enhance the effective delivery of climate change education.

It can be deduced from the foregoing paragraphs that implementing an interdisciplinary climate change curriculum presents significant challenges for teachers. This is because multiple factors can impede effective lesson delivery. Limited subject knowledge constrains teachers' ability to integrate concepts from science, social studies, and ethics. This often resulted in fragmented instruction and reduced instructional confidence. Inadequate access to resources, including textbooks, digital tools, and experiential learning opportunities, further limits the use of innovative and student-centred pedagogical approaches. Time constraints and excessive workloads limit teachers' capacity to implement learner-centred instructional activities. Insufficient professional development further constrains their content knowledge, lesson planning abilities, and instructional confidence. Additionally, resistance to pedagogical change can hinder the adoption of project-based and interdisciplinary teaching strategies. Change Theory in Education posits that successful educational reform depends on teachers functioning as informed agents of change. However, these challenges impede their capacity to implement and sustain innovative practices. Targeted professional development, sufficient instructional resources, and institutional support are essential for improving teacher preparedness. These measures also facilitate the effective implementation of an interdisciplinary climate change curriculum.

CAPACITY BUILDING STRATEGIES

Specialized training workshops

Specialized training workshops constitute an effective strategy for building teachers' capacity in implementing an interdisciplinary climate change curriculum. Workshops enhance teachers' content knowledge, pedagogical skills, and instructional confidence (Rivai & Rahmawati, 2025). Specialized training workshops offer targeted professional development that supports the integration of concepts across science, social studies, ethics, and related disciplines. It is for this reason that the Ghana Education Service (Bashiru, 2026), has mandated basic school, teachers to produce portfolio of evidence for participation in profession development programme for promotion. This intervention has impacted on the students producing expected learning outcomes at the basic level.

Moreover, teachers who are exposed to training programmes sharping their skills to be effective (dos Santos, 2025). Such teachers become conversant with many teaching approaches including innovative instructional approaches like project-based learning and experiential activities. In Finland for instance, climate education workshops equip teachers with strategies for interdisciplinary teaching, collaborative lesson planning, and practical classroom implementation (Egbele, 2025). As a result, there have been improvement in curriculum coherency student engagement over time. This is consistent with Dai and Ran (2026) arguing that specialized workshops enable teachers to act as informed change agents. Therefore, it is imperative to recognise Specialized training workshops as a transformational tool if teacher capacity must improve.

Access to teaching resources

Access and usability of teaching resources represent a critical capacity-building strategy. This denotes teachers' ability to recognise the available teaching resources and be able to use them to facilitate teaching and learning. Li et al. (2021) opines that teachers' knowledge in teaching resources on climate change education often motivate them to implement interdisciplinary curriculum successfully. This includes all forms of teaching resources including digital aids, visual aids, and audio-visual aids (Mamadiyorov & Rashidova, 2025). In the New Zealand context, the Ministry of Education's Environmental Education for Sustainability (EES) initiative offers comprehensive resources. These include lesson plans, digital platforms, and guidelines for community engagement (Gundersen, 2025). These resources are available in schools and accessible to both teachers in facilitating the integration of climate science across curricula. As a result, teachers deliver climate change lessons in an interdisciplinary manner, fostering students' understanding of its multifaceted nature.

This approach aligns with Fullan's Change Theory, which posits that sustainable educational reform requires the development of teacher capacity through access to knowledge, collaborative practices, and practical resources. The EES initiative provides teachers with resources that increase their confidence and teaching skills. This support helps them deliver an effective interdisciplinary climate change curriculum and improve environmental education across the system.

Collaborative learning communities

Collaborative learning communities serve as a team effort for teachers to grow together. When teachers engage collectively, they have opportunity to share expertise and co-develop instructional strategies. This is particularly critical for addressing complex, interdisciplinary topics such as climate change. This collaborative capacity building approach is prominent in India where The National Institute of Educational Planning and Administration (NIEPA) has facilitated networks for teachers (Rawal, 2025). Teachers come from science, geography, and social studies, collaborate to design climate change modules that integrate environmental, social, and policy dimensions. Therefore, interdisciplinary curriculum implementation in Indian context, has manifested over the years. Taking a lesson from this approach, it is critical to encourage collaborative learning communities for effective capacity building of teachers to implement Interdisciplinary climate change curriculum in schools.

Moreover, sustainable collaborative learning is critical for effective teacher capacity building. Ongoing collaboration promotes collective problem solving and encourages innovation in teaching (Qudrat-Ullah, 2025; Khan, 2024; Cojorn, 2024). Sustained collaborative learning contributes to effective team practices and improves teachers' capacity and learning outcomes.

Ongoing mentorship and supports

Ongoing mentorship and support are critical factors in the effective implementation of an interdisciplinary climate change curriculum. Mentorship and support are important for both experienced and inexperienced teachers (Keiler, et al., 2023). While the experienced teachers update their skills with mentorship, the inexperienced teachers get introduced to new skills. This capacity building strategy helps teachers to navigate complex content, adopt innovative pedagogical approaches, and engage in collaborative practices across subject areas.

Mentorship facilitates teachers' professional development and progress (Bradbury & Fitzgerald, 2025). Education institutions have invested in the mentorship programmes to equip their employees with knowledge and skills. The lack of mentorship constitutes lack of professional development (Saad et al., 2026). This finding suggests that mentorship and support for teachers can lead to effective interdisciplinary teaching.

IMPLICATIONS FOR EDUCATIONAL POLICY

1. Teacher training in climate science should focus on specialized programs that enhance teachers' subject knowledge and build their confidence in delivering engaging climate change lessons.
2. Interdisciplinary pedagogical skills can be developed by facilitating professional workshops. This can help teachers integrate concepts from multiple disciplines.
3. Curriculum design and integration require policy support for frameworks that enable teachers to embed climate change across different subjects.
4. Attitudinal and value-based preparation can be addressed by incorporating sustainability principles, climate ethics, and environmental responsibility into teacher education.
5. Digital and technological competence can be enhanced through access to interactive tools, online platforms, and training in innovative teaching strategies. This can help to engage students in climate change education.
6. Assessment competence can be strengthened by guiding teachers to design evaluations that capture both subject-specific knowledge and the ability to apply interdisciplinary thinking to climate-related problems.
7. Future research should examine the extent to which teachers design an effective interdisciplinary assessment that measure both students' climate change knowledge and their capacity for critical, cross-disciplinary thinking.

CONCLUSION

Climate change education has become a critical component of secondary school curricula. This requires teachers to integrate knowledge from science, social studies, economics, technology, and ethics in a coherent and pedagogically meaningful manner. In this light, it was found that teachers' comprehensive content knowledge, interdisciplinary pedagogical skills, curriculum design competence, digital literacy, and robust assessment strategies are critical. In the absence of these capacities, instructional delivery can become fragmented. Learner engagement may decline, and opportunities for developing critical thinking and pro-environmental behaviours may be diminished. It was further established that targeted professional development through specialized workshops, access to instructional resources, collaborative learning networks, and ongoing mentorship can enhance teacher efficacy and preparedness. This is particularly important because teachers are pivotal agents of educational reform. Therefore, their capacity-building is imperative for the implementation and sustainability of interdisciplinary climate education.

REFERENCES

Adera, N. (2025). Innovative learning spaces and blended learning: Quest for 21st century competency teaching and learning approaches. In *Creating dynamic space in higher education: Modern shifts in policy, competencies, and governance* (pp. 139–174).

Ali, M. I., Malik, A., & Latief, M. M. (2025). Developing pro-environmental attitudes through integrated environmental education: Insights from a high school in Indonesia. *International Journal of Sustainable Development and Planning*, *20*(6).

Alimehmeti, G., Ndoka, E., & Paletta, A. (2026). Cultivating green pioneers: Examining the antecedents of sustainable entrepreneurial intent in higher education. *International Journal of Sustainability in Higher Education*, *27*(2), 442–460. DOI: 10.1108/IJSHE-03-2024-0206

Almazroa, H. (2024). Bridging the knowledge–practice gap: Assessing climate change literacy among science teachers. *Sustainability (Basel)*, *16*(20), 9088. DOI: 10.3390/su16209088

Anand, A., & Singh, P. (2025). Interdisciplinary studies: Issues, perspectives, and challenges from National Education Program 2020. In *Integrating personalized learning methods into STEAM education* (pp. 233–260).

Arrasyid, A. K., Herpratiwi, H., Sinaga, R. M., & Fitri, D. M. (2025). Transformation of traditional learning models in elementary schools in Lampung Province: An adaptation approach. *Jurnal Pendidikan Progresif*, *15*(2), 1310–1327. DOI: 10.23960/jpp.v15i2.pp1310-1327

Awayehu Gugssa, M. (2025). Barriers to environmental education in Ethiopia: Do they differ from a global analysis? *International Research in Geographical and Environmental Education*, *34*(2), 156–173. DOI: 10.1080/10382046.2024.2352285

Bashiru, M. (2026). Shaping education in Ghana: A historical and policy analysis of key reforms. In *Education policies and practices in the Global South, Volume One: Basic Education* (pp. 93–123). Springer Nature Singapore. DOI: 10.1007/978-981-95-3268-1_7

Ben Zvi Assaraf, O., Dawson, V., Eilam, E., Gokpinar, T., Goldman, D., Naugauker, N., Eka Putri, G. A. P., Subiantoro, A. W., Tolppanen, S., White, P., Widdop Quinton, H., & Dillon, J. (2025). Climate change education implementation: The voices of policymakers, professional development providers, and teachers in five countries. *International Journal of Science Education*, *47*(2), 191–213. DOI: 10.1080/09500693.2024.2314572

Bondre, S. V., & Yadav, U. (2025). Cloud architecture for education technology. In *Establishing AI-specific cloud computing infrastructure* (pp. 103–122). IGI Global Scientific Publishing. DOI: 10.4018/979-8-3693-9694-0.ch004

Botha, I., & Klopper, D. (2025). Navigating change: A journey through curriculum review. *The South African Geographical Journal*, •••, 1–16.

Bradbury, O., & Fitzgerald, A. (2025). Experiences of mentor teachers in supporting pre-service teachers' professional growth and development using conversation and professional standards. *Educational Process: International Journal*, *14*(1), e2025005. DOI: 10.22521/edupij.2025.14.5

Buabeng, I., & Amo-Darko, B. (2025). Curriculum reforms without foundation: The effects of inadequate preparation in curriculum reforms on Ghanaian teachers and the education system. *Curriculum Perspectives*, *45*(2), 1–15. DOI: 10.1007/s41297-025-00309-7

Cai, H., Sun, J., Sun, D., Han, B., & Wong, L. H. (2026). Examining teacher change in reflective practice: Insights for professional development on adopting representational scaffolding in the inquiry-based learning. *Instructional Science*, *54*(1), 20. DOI: 10.1007/s11251-025-09774-z

Chamo, N., & Broza, O. (2025). Bridging disciplines: Exploring interdisciplinary curriculum development in STEM teacher education. *Education Sciences*, *15*(8), 1064. DOI: 10.3390/educsci15081064

Chen, M. J., Guo, C. Y., Tseng, L. T., Chiu, M. Y., & Weng, H. J. (2025). Understanding teachers' interdisciplinary role perceptions: Scale development and importance–performance analysis. *Psychology in the Schools*, *62*(9), 3128–3143. DOI: 10.1002/pits.23529

Cherukuri, J. (2025). Beyond lectures: Innovative approaches to promote student engagement through scenario-based learning (SBL). *Journal of Engineering Education Transformations*, 570–575.

Chinengundu, T. (2025). Analysis of the South African Construction Technology Curriculum and Assessment Policy Statement Vocational Didactics. *TVET@ Asia*, (24), 1-16.

Cojorn, K. (2024). A collaborative professional development and its impact on teachers' ability to foster higher order thinking. [EduLearn]. *Journal of Education and Learning*, *18*(2), 561–569.

Custodio-Ferrando, A., & Cabero-Fayos, I. (2025). Transdisciplinarity in education: A narrative review of its didactic validity and contributions to sustainable learning. *Sustainability (Basel), 17*(22), 10320. DOI: 10.3390/su172210320

Dai, X., & Ran, Y. (2026). The predicting role of continuing professional development (CPD) in Chinese EFL teachers' agency and pedagogical effectiveness. *European Journal of Education, 61*(1), e70407. DOI: 10.1111/ejed.70407

Demirtaş, Ş., Karasu, N., Afacan, K., Aykut, Ç., Karahan, M. B., & Sert, C. (2025). Integrating pro-environmental behaviors into school-wide positive behavioral interventions and supports for creating green schools. *Improving Schools, 27*(1), 3–19. DOI: 10.1177/13654802241281185

dos Santos, D. M. B. (2025). Teacher training course for a pedagogy of meaning in life: Effects on professional development. *Social Sciences & Humanities Open, 12*, 101682. DOI: 10.1016/j.ssaho.2025.101682

Egbele, N. (2025). Understanding teacher educators' perceptions on sustainability education: A case study from the University of Eastern Finland Teacher Training School, Joensuu (Master's thesis, Itä-Suomen yliopisto).

Emam, M. M., & Al-Mahdy, Y. F. H. (2026). Exploring the role of school climate and self-efficacy in fostering teacher commitment within inclusive education settings. *Psychology in the Schools, 63*(2), 301–313. DOI: 10.1002/pits.70091

Fullan, M. (1982). *The meaning of educational change*. Teachers College Press.

Green, N. (2025). *Climate change mitigation and adaptation strategies in African UNESCO biosphere reserves*. BoD–Books on Demand.

Guimerà-Ballesta, G., Calafell-Subirà, G., Jiménez-Valverde, G., & Esparza-Pagès, M. (2025). Climate change education in secondary schools: Gaps, challenges and transformative pathways. *Encyclopaedia, 6*(1), 8. DOI: 10.3390/encyclopedia6010008

Hartmann, D., Wang, S. M., Pohlmann, L., & Berendt, B. (2025). A systematic review of echo chamber research: Comparative analysis of conceptualizations, operationalizations, and varying outcomes. *Journal of Computational Social Science, 8*(2), 52. DOI: 10.1007/s42001-025-00381-z

Heikkilä, M., Häkkinen, M., Vidbäck, A., Mikkilä-Erdmann, M., & Sääksjärvi, I. E. (2025). Student teachers' conceptual understanding of biodiversity loss in a multiple-source learning environment. *Environmental Education Research, 31*(7), 1356–1370. DOI: 10.1080/13504622.2024.2436110

Henninger, S., & Christmann, D. (2023). Teaching about climate-efficient buildings in the context of geographic education for sustainability. *Sustainability (Basel), 15*(12), 9660. DOI: 10.3390/su15129660

Karaoglan-Yılmaz, F. G., Ustun, A. B., Zhang, K., & Yılmaz, R. (2023). Metacognitive awareness, reflective thinking, problem solving, and community of inquiry as predictors of academic self-efficacy in blended learning. *Turkish Online Journal of Distance Education, 24*(1), 20–36. DOI: 10.17718/tojde.989874

Keiler, L. S., Diotti, R., & Hudon, K. (2023). Supporting teachers as they support each other: Lessons concerning mentor teacher feedback to teacher mentees. *Professional Development in Education, 49*(2), 225–242. DOI: 10.1080/19415257.2020.1839781

Kerich, M., & Kimwarey, M. C. (2026). Pedagogical Practices Employed by Teachers in Developing Learner Communication and Collaborative Competencies in Junior Secondary Schools in Uasin Gishu County, Kenya. *Africa Journal of Technical and Vocational Education and Training, 11*(1), 50–61. DOI: 10.69641/afritvet.2026.111201

Khan, N. (2024). Collective intelligence in education: Teacher collaboration and professional development.

Kotsis, K. T. (2026). Toward an integrated model of science teacher education in Greece. *European Journal of Contemporary Education and E-Learning, 4*(1), 45–58. DOI: 10.59324/ejceel.2026.4(1).04

Lam, M. H. A. (2025). From defining to enacting interdisciplinary education: Curriculum integration strategies among Hong Kong undergraduate interdisciplinary programmes. *Studies in Higher Education, 50*(12), 2762–2781. DOI: 10.1080/03075079.2024.2431582

Lett, E., Asabor, E., Beltrán, S., Cannon, A. M., & Arah, O. A. (2022). Conceptualizing, contextualizing, and operationalizing race in quantitative health sciences research. *Annals of Family Medicine, 20*(2), 157–163. DOI: 10.1370/afm.2792 PMID: 35045967

Li, C. J., Monroe, M. C., Oxarart, A., & Ritchie, T. (2021). Building teachers' self-efficacy in teaching about climate change through educative curriculum and professional development. *Applied Environmental Education and Communication, 20*(1), 34–48. DOI: 10.1080/1533015X.2019.1617806

Li, X., & Hu, W. (2026). Interdisciplinary collaborative teaching in vascular surgery training: A randomized trial of radiologist-surgeon partnership in China. *PLoS One, 21*(1), e0341266. DOI: 10.1371/journal.pone.0341266 PMID: 41592009

Lombo, N., & Subban, M. (2024). Physical infrastructure challenges in rural schools: Reflections to promote quality education. *Administratio Publica*, *32*(1), 69–101.

Mamadiyorov, J., & Rashidova, M. (2025). Teacher's use of multimedia resources technology in teaching mathematics lesson in primary grades. *Multidisciplinary Journal of Science and Technology*, *5*(1), 78–81.

Marschall, G. (2022). The role of teacher identity in teacher self-efficacy development: The case of Katie. *Journal of Mathematics Teacher Education*, *25*(6), 725–747. DOI: 10.1007/s10857-021-09515-2

Mavuso, M. P., Olawumi, K. B., Khalo, X., Kafu-Quvane, B., & Mzilikazi, B. (2022). Implementation of teacher capacitation programs to integrate climate change education. *International Journal of Learning. Teaching and Educational Research*, *21*(11), 73–86.

Melesse, D., Menkir, S., Yemata, G., & Seifu, A. (2025). Effect of context-based instructional approach on students' science process skills acquisition in environmental concepts. *Education Inquiry*, 1–17.

Mhlongo, A. R., & Chaane, T. M. (2025). Challenges experienced by Gauteng Foundation Phase departmental heads in managing teaching and learning. *South African Journal of Childhood Education*, *15*(1), 1590. DOI: 10.4102/sajce.v15i1.1590

Mpuangnan, K. N. (2024). Teacher preparedness and professional development needs for successful technology integration in teacher education. *Cogent Education*, *11*(1), 2408837. DOI: 10.1080/2331186X.2024.2408837

Mpuangnan, K. N. (2024). Determining skills of basic schoolteachers in test item construction. *Review of Education*, *12*(2), e3478. DOI: 10.1002/rev3.3478

Mpuangnan, K. N. (2025). Climate change or attitude change? Traditional leaders and school management at the war front. *Critical Insights in Climate Change*, *1*(1), 2557067. DOI: 10.1080/29931495.2025.2557067

Mpuangnan, K. N. (2025). Embracing diversity in multi-racial classrooms through pedagogical strategies. *Journal of Intercultural Communication*, *25*(2), 202–217. DOI: 10.36923/jicc.v25i2.1075

Mumcu, F., Atman Uslu, N., & Yildiz, B. (2022). Investigating teachers' expectations from a professional development program for integrated STEM education. *Journal of Pedagogical Research*, *6*(2), 44–60. DOI: 10.33902/JPR.202213543

Muzakki, H., Na'im, N., & Qomar, M. (2026). The management of innovation initiation in Islamic education and its implications for madrasah competitiveness. *International Journal of Education Management and Religion*, *3*(2), 566–583.

Nassor, H. A., Ephrem, A., & Mbaula, W. (2026). Teachers' pedagogical competencies on the utilisation of instructional materials for teaching history. *Educational Dimension*, *14*, 51–71.

Newsome, D., Newsome, K. B., & Miller, S. A. (2023). Teaching, learning, and climate change: Anticipated impacts and mitigation strategies for educators. *Behavior and Social Issues*, *32*(2), 494–516. DOI: 10.1007/s42822-023-00129-2 PMID: 38625183

Omojemite, M. D. (2025). Teacher preparedness for transformative education in the intermediate phase. *Interdisciplinary Journal of Education Research*, *7*(1), a22. DOI: 10.38140/ijer-2025.vol7.1.22

Papademetriou, C., Anastasiadou, S., Belias, D., & Ragazou, K. (2025). Integrating sustainability into human resource management: Building a greener workforce for the future. *Sustainability (Basel)*, *17*(3), 1113. DOI: 10.3390/su17031113

Popova, A., Evans, D. K., Breeding, M. E., & Arancibia, V. (2022). Teacher professional development around the world: The gap between evidence and practice. *The World Bank Research Observer*, *37*(1), 107–136. DOI: 10.1093/wbro/lkab006

Pradhan, K. (2023). Emotional maturity assessment in higher secondary school students: A comprehensive study. *European International Journal of Philological Sciences*, *3*(09), 06–09.

Priya, P. D. (2025). Exploring the role of multidisciplinary learning in developing critical thinking skills. *Innovative Journal for Advances in Education, Science. Commerce & Multidisciplinary Learning*, *1*(1), 32–48.

Qiu, C. A., He, H. X., Chen, G. L., & Xiong, M. X. (2022). Pre-service teachers' perceptions of technological pedagogical content knowledge in mainland China. *Education and Information Technologies*, *27*(5), 6367–6391. DOI: 10.1007/s10639-022-10888-x PMID: 35035257

Rawal, D. M. (2025). Operationalising cluster school complex in accordance with New Education Policy, 2020. *IASSI Quarterly*, *44*(1).

Rivai, F. A., & Rahmawati, N. (2025). Workshop, assistance, and capacity building in the development of teaching materials based on 21st-century learning. *Al Ghazali: Jurnal Pendidikan Dan Pemikiran Islam*, *5*(2), 247–262.

Rodríguez, N. N., & Swalwell, K. (2025). *Social studies for a better world: A guide for elementary educators*. Taylor & Francis. DOI: 10.4324/9781003468462

Romero Ariza, M., Quesada Armenteros, A., & Estepa Castro, A. (2024). Promoting critical thinking through mathematics and science teacher education. *European Journal of Teacher Education*, *47*(1), 41–59. DOI: 10.1080/02619768.2021.1961736

Saad, W. Z., Alias, N. A., Chong, C. M., & Sabri, S. (2026). Catalysing sustainability in higher education. In *Digital transformation and sustainability in higher education* (pp. 1–26). IGI Global Scientific Publishing.

Stevens, C. (2025). Teachers and teaching: Pedagogy, digital skills and professional development. *Open Learning*, *40*(1), 1–3. DOI: 10.1080/02680513.2024.2436665

Tang, K. (2025). Teacher conceptions of climate change and their role in climate change education. *Discover Sustainability*, *6*(1), 486. DOI: 10.1007/s43621-025-01334-2

UNESCO. (2025). UNESCO Launches New Intergovernmental Hydrological Programme Water Information Network System (IHP-WINS) Platform." https://unesco.org.uk/news/unesco-launches-new-ihp-wins-platform-towards-improved-water-data-sharing

Vettriselvan, R., & Ramya, R. (2025). Sustainable curriculum design and development: A comprehensive approach. In *Smart education and sustainable learning environments in smart cities* (pp. 471–486). IGI Global Scientific Publishing.

Wang, F., Wong, W. K., Wang, Z., Albasher, G., Alsultan, N., & Fatemah, A. (2023). Emerging pathways to sustainable economic development. *Resources Policy*, *85*, 103747. DOI: 10.1016/j.resourpol.2023.103747

Chapter 6
A Reflection on Preparing New Creative Arts Teachers:
Climate Change and Curriculum Redevelopment

Thelma Kathleen Buchholz Mort
http://orcid.org/0000-0002-8775-0618
University of South Africa, South Africa

ABSTRACT

Currently in South African primary schools, Creative Arts forms part of the Life Skills curriculum (on the current CAPS curriculum). This chapter explores how climate change educational learning opportunities may occur in this curriculum, and the forms these opportunities take or might take. The chapter is also a lecturer's reflection on her practice in teaching student teachers about how visual arts in the Creative Arts education has value in educating learners on climate change in the following ways: through teaching about local artists whose work reflects climate change ideas, through materiality: in use of materials, through stewardship of materials and the environment, and through designing projects which directly address climate change, on the curriculum. These three areas are key to how future teachers might be prepared to teach this curriculum and empowered to enact climate change in future classrooms. This chapter is a reflection about incorporating this knowledge and teaching in this way, my own practice as a teacher educator at a distance university in South Africa.

DOI: 10.4018/979-8-3373-6975-4.ch006

INTRODUCTION

South Africa, like many parts of the world, is degraded environmentally. This is easily discernible, from individual reports, casual conversation, as well as news, academic articles and environmental impact studies. South Africa is a major emitter of greenhouse gases (GHGs) which account for 65% of all Africa's emissions (Struckhoff et al. 2013). This is of course due to its fossil fuel-based economy. There is also significant air pollution caused by mining, for example, in South Africa mercury emissions are caused by gold and coal mining (Fayiga et al. 2018). Due to most shops and supermarkets still providing customers with plastic bags, plastic pollution is a problem. Much plastic goes into the sea and rivers, and due to infrastructure challenges, "according to The State of Waste Report4, 34% of South Africans do not have access to any waste management or removal services" (Hanekom, 2020, p.1). Against this backdrop, that Reddy writes

> *Climate change is taking on increasing importance as we begin to feel the physical effects of the climatic changes in our daily lives. Some of the effects include increased frequency of extreme weather events such as cyclones and tornados, more frequent cycles of floods and droughts, reduced agricultural yields, quicker melting of the polar ice caps and the extinction of plant and animal species (Reddy 2021, p.3).*

In the light of this universities have a crucial role in creating a culture for societal sustainability and democracy (Mncube & Olawale, 2021), due to "escalating climate change challenges and environmental degradation, enhancing eco-literacy through educational frameworks has never been more critical. Educators are pivotal in cultivating awareness and behaviours that foster sustainability among future generations" (Kouam, 2025, p.38).

The role of education is clearly defined as extremely important for driving the United Nations' 2030 Agenda for Sustainable Development, as it "identifies education (Goal 4) as both an end and a key enabler of all other Sustainable Development Goals (United Nations, 2015)" (Egbele 2025, p. 8) and due to the likely possibility of teachers influencing and impacting younger generations, teacher education has been seen as occupying "a strategic position in advancing sustainability" (Egbele, 2025, p.8)

LITERATURE REVIEW

Environmental education is related to many other fields and underpinned by many values. One of the key goals of environmental education is to create people who are Eco-literate. Eco literate people are those who: "have the ability to understand, analyze, and actively engage with environmental issues" (Ginju, 2020)." These skills are linked to making informed choices and becoming "global citizens" (Ginju, 2020). Therefore environmental education is linked to citizenship education. Fundamentally environmental education is driven by and linked to values.

Values and environmental education

Some teacher educators describe environmental education as developing "knowledge for common good" (Mandikonza & Lotz-Sisitka, 2016, p. 110. A common good is defined as something, like a public good, which is good for everyone, and therefore reflects a democratic principle. Furthermore, the idea of a common good reflects a return not just to subject knowledge but to values-based education. This idea is strongly supported by Msezane (2020).

The values underpinning environmental education have been regarded as important and linked also to citizenship education or preparing learners and students to be citizens aware of their roles of responsible earth-stewardship. The focus on values also leads to a strong overlap between Environmental Sustainable Education and Character Education as noted by Jordan (2023). The importance of values being guiding principles was recognised early by Scott and Oulton who defined 'environmental values education' linking it clearly to sustainability:

We see the values which individuals hold as being those actions, ideas and ideals which are of fundamental importance to them, and which act as guides to how they feel they ought to live their lives, interacting with other people and with other species. In this sense, values education can be seen as the systematic and planned attempts by teachers to explore such issues with learners—both in the context of the formal and informal curriculum and in the ways that the school as an organisation conducts itself, both internally and in its relationships with the wider community (Scott & Oulton, 1998, p. 211).

Values education feature prominently in discussions of the environment and sustainable development:

"Values featured heavily throughout the UNESCO 2006 Framework for the DESD International Implementation Scheme, stating for example Education for Sustainable Development (ESD) 'is fundamentally about values, with respect at the centre: respect for others, including those of present and future generations, for difference and diversity, for the environment, for the resources of the planet we inhabit' (UNESCO, 2006, in Jordan, 2023, p. 2).

Eco-pedagogy relates to an ecologically driven set of teaching skills and orientation. Similarly, Egbele (2025) also notes that eco-pedagogy is holistic and value-driven.

Values, Sustainability and Social Justice

With the emergence of environment and sustainability education (ESE) in local teacher education contexts (Mandikonza & Lotz-Sisitka, 2016) it must be noted that there is a broader concern relating to social justice. Berkowitz writes of 'the centrality of CE for creating and sustaining a just world' (2017, p. 83). Embedded in the Sustainable Development agenda of 2030 are multiple Sustainable Development Goals which are interconnected in that if achieved, they will "eradicate poverty, safeguard the planet, and promote prosperity for all. When effectively implemented, it holds considerable potential to advance global sustainability" (Egbele, 2025, p.15). This shows the potential impact and social justice ramifications of sustainability.

Developing future Teachers to teach with Environmental sensitivity

"Environmental education has in many ways been a response to the perceived environmental crisis on local and global levels for almost forty years" (Reddy, 2021, p.2). There have been many calls to environmental education in teacher education (Khan 2013), in fact the South African 1995 White Paper on Education and Training directs the inclusion of environmental issues in formal education (Ruthanam, Reddy & Pillay, 2021).

It has also been found that teacher education programmes significantly shape teachers' attitudes (Egbele, 2025, p. 21). Education is therefore an obvious vehicle to create awareness of and drive environmental issues and sustainability, and a key area of importance is teacher education. "When teachers internalize sustainability as a professional value and pedagogical principle, they become agents of change capable of shaping future generations' understanding of sustainable living" (Egbele, 2025, p. 9).

Zelenskil & Nisbet (2014, p. 4) found that a connection to nature "strongly predicts sustainable attitudes and behaviours, and this relationship holds across many assessment tools".

But what does that mean in South Africa where many have grown up in environmentally degraded environments? Kristjansson writes of the role situationism plays in character development and behaviour: "personality psychologists seem nowadays to agree that behaviour is a function both of character and situations. In other words, since situations are the arena where character plays out, it would be a conceptual, no less than empirical, error to maintain that character is completely situation independent" (2013, p. 273). Kristjansson therefore argues for education which has real hands-on learning opportunities and situations.

To this end Ndebele, et al. (2024) argue for a reconceptualization of initial teacher education in South Africa in order to create an imaginary of transformative and sustainable alternatives (2024, p. 1). This has been echoed by others who see that it must be endorsed across the university. For example, Ikhsan et al. suggest that Environmental Education be implemented comprehensively across through faculties, in policies, curriculum, teaching methods, and universities' campus cultures. In this way students' ecological literacy will be developed and scaffolded. Jannah et al. (2013) suggest that universities adopt specialised Environmental Education tool kits to improve environmental knowledge, literacy and behaviour. Social studies teacher education students perceive Environmental Education as beneficial for developing cognitive, affective, and behavioural dimensions of ecological literacy (Durmuş & Kinaci, 2021).

However, the perspectives of teachers responsible for delivering this environmental education on initial teacher education courses have largely been overlooked. Eco-literacy can be categorised into three levels: nominal, functional, and operational, reflecting increasing environmental awareness and action (Ginju, 2020). Due to the complexity of Eco-literacy, or the multi-layeredness of environmental living and thinking, research which has focused on specific teaching strategies or interventions has done so without necessarily giving attention to "how teachers engage with green pedagogies in their daily practices. Despite the increasing recognition of the importance of eco-literacy, there is limited research on how teachers perceive and implement green pedagogies in the classroom." (Kouam, 2025, p.38)

A Curriculum as Sensitization

The preparation of future teachers involves the marriage of theoretical knowledge and practical knowledge, which together inculcate professional teaching ability, a sense of adaptability, and a commitment to excellent teaching (Darling-Hammond et al., 2017). However, a teacher education curriculum can also sensitize students to other issues, such as environmental issues. As a precursor to my involvement with the academics' group, the Climate Change in Education group, I was aware of how several environmental issues were already present on the current curriculum. Kouam's study (2025) of international changes made to TE found that five themes emerged to infuse ES into ITE curricula: "Thematic analysis revealed five key areas: definitions and understanding of green pedagogies, practical implementation strategies, fostering active citizenship, emotional impacts of eco-education, and perceived effectiveness of these approaches"(Kouam 2025, p. 37).

Reorientation from Science to All Subjects

"Faculties of education are considered central to efforts to reorient education for the values of sustainability" (Howard, 2023, p. 1). The UN Guidelines asks Teacher Education to use an "approach to fostering the values of sustainability that is locally relevant and culturally appropriate" (in Howard, 2023, p. 1)The Climate Change in Education group's interest in re-writing and redeveloping curricula to include climate change or climate justice education issues was a way of preparing future teachers to have maximum environmental impact, as they are envisaged as taking such knowledge and applications gleaned from their B. Ed courses to their future classrooms, thereby creating a ripple effect in society. It is for this reason that climate change is extremely important to teacher education. Climate change has typically infused science and biology curricula, as Howard (2023, p. 4) notes: "... throughout all grades environmental education emphasizes technology, trade and resources. Children learn about the natural world and appropriate human- environment relationships as a subset of the science curriculum."

However, environmental education is no longer seen as only belonging in sciences. "rather than being confined to discrete "environmental" lessons, environmental sustainability education is conceived as an integrative and transformative educational agenda" notes Egbele (2025, p. 22). The broadening and integration of environmental, sustainable education has led to it being integrated across all subjects in teacher education.

Inwood & Kennedy (2020, p. 587) in their Canadian study note that it seems like a leap to have art educators involved in climate change pursuits noting that they did not fit the "...profile of a typical faculty member and researcher in environmental and sustainability education, who is often grounded in the disciplines of environmental science or environmental studies".

One of the advantages of environmental education not being solely pocketed in the sciences is that by its incorporation in and across other subjects and knowledge domains, it can be freed of a purely scientific (or scientistic) approach. This is necessary as the reductionist approach of the sciences has been critiqued in environmental education in that it neglects the emotional side of nature. Wendell Berry (2000) says, "We know enough of our history by now to be aware that people exploit what they have merely concluded to be of value, but they defend what they love" (p. 39). Biesta (2010) puts this another way: "They begin to value what is measured rather than measuring what they value." Similarly, Doerr writes: "I needed to find something that would move my students from "I know" to "I care." (2004, pp. 30-31).

This leads us to the question – how can we teach others to love and respect nature (Ruthanam, Reddy & Pillay 2021, p.133), and subsequently adopt processes and actions which would lead to sustainable behaviours? And moreover, how would we do this in South Africa where the access to nature is uneven?

What approach should one take to environmental education? Variously different approaches have been suggested with regard to Character Education:

It has been recognised that environmental education "should be cross-curricular, collaborative, project-based, civic, and promote a sense of citizenship" (Curren & Metzger, 2017; 2019);that the modelling of sustainability by teachers is important (Ferkany, 2021), school approach or climate, role-modelling, experiencn nature, exploring sustainability dilemmas (Jordan & Kristjánsson, 2017), look at the 'deep exploration and articulation of issues pertaining to sustainability' (Curren & Metzger, 2017, p. 178) and promoting learners thinking about how 'to think creatively about how to live flourishing lives in ways consistent with sustainability?' (Curren & Metzger, 2017, p. 68).

Modelling behaviour and actions has been given much attention Howard (2023, p. 10) writes: Modelling the values of sustainability as a way of being in the world allows teacher educators to communicate the values as they are lived in the classroom. Teacher educators who are committed to and practice interdisciplinary course work, support participatory learning, seeks ways to incorporate shared decision making, provide opportunities for teachers to reflect on their own values, and explore issues around local sustainability provide powerful opportunities to foster the values of sustainability".

Materials and modelling sustainability

With regard to materials, a participant in Inwood and Kennedy's 2020 study noted that:

"I really liked that it was something that we could access, like the materials were things you could easily find or source out. So it wasn't something that you could implement only with the right resources, so I was thinking this is something you could totally do in a school" (Inwood & Kennedy, 2020, p. 594)

This is a particularly important consideration is South African schools which are mostly under-resourced. Pedagogically it is also important as resourcefulness and learning to work with different materials are key factors in learning to make art. Therefore these same issues should be key factors in teaching student teachers how to teach art.

Cultural Specificity

Furthermore, Howard highlights the cultural specificity of sustainability as an understanding and a set of practices: "To understand how a community might best develop sustainably, it is essential to bring to light the ontological possibilities that emerge from an understanding of historical traditions, unique cultural narratives, languages and place-based ways of knowing and acting"(Howard 2023, p. 3). This suggests pertinently the roles that might be played by traditional or indigenous knowledge, which is receiving much attention in South Africa currently. Ndebele et al. (2025, p. 4) integrate the idea of appropriate teacher education pedagogies to local contexts - an approach which can be linked to climate justice, and social justice:

> *The efficacy of teacher education programs is further enhanced through the use of pedagogies that are relevant to the local context. By incorporating culturally responsive and contextually appropriate pedagogies, teacher educators can ensure that prospective teachers possess the necessary skills and knowledge to effectively address their students' unique needs and challenges. This entails integrating indigenous knowledge, language, and cultural practices into the curriculum while also fostering critical thinking, problem-solving, and socio-emotional skills that are essential for sustainable development and social justice.*

As Khan notes: "there is a wrong notion that economic development is based only on industrialization" (2013, p. 9) which has often led to the devaluing of traditional cultural practices. It can also lead to individuals feeling that because of the negative consequences of non-green technological industrialisation, they are being punished for unsustainable industrialisation.

Likewise, the "UNESCO document Guidelines and Recommendations for Reorienting Teacher Education to Address Sustainability (Hopkins & McKeown, 2005) state that "education for sustainable development is locally relevant and culturally appropriate, … is based on local needs, perceptions, and conditions but recognizes fulfilling local needs often has global effects and consequences" (p. 16in Howard, 2023, p.3).

Another reason for cultural specificity is the necessity of working with lived experience of a learner, their family, their community in a specific place. When environmental issues may seem too complex, it becomes necessary to connect with 'place'. This as a starting point allows the students to consider the local issues, and their relationship to global issues (Sobel, 1996; Davis, 2018). This also links to local knowledge and an epistemology of working with student experience.

Environmental Concerns and Visual arts

This chapter focuses on Visual arts, and teaching teachers how to teach Visual Arts in the Intermediate Phase.

There is a rich international history of visual artists working with environmental ideas and concerns, beginning from the 1970's with the land artists Michael Heizer, Robert Smithson, and Richard Long in North America and Andy Goldsworthy in

England (to mention a few) (Inwood & Kennedy 2020, p. 587). There are also other artists who express critiques of environmental exploitation and consumption such as the English artist Banksy.

There are also many South African artists working currently working whose work can be seen as containing environmental concerns ranging from showing a love of nature through landscape, a concern with environmental degradation such as the work of Kentridge. This degradation and its related, accompanying and pressing environmental issues, have been depicted and used in visual art (Mashazi et al., 2024).

A concern with plastic pollution such as seen in the work of Pia Truscott and Mbongeni Buthelezi (Santos, 2026) who make "paintings" through heating plastics and applying them to surfaces, to which they adhere. Nobukho Nqaba uses Chinese plastic bags in her installation works (Cheng, 2018). Other African contemporary artists such as Cyrus Kabiru and El Anatsui also make use of waste and found objects (Stanhouse, 2016) in their artworks, making them prolific and inventive recyclers.

But what has been missing from general education is the pedagogical knowledge of how to integrate environmental education ideas (Sikhosana 2022), and this lack is also felt in arts education, where the conceptualization of a critical and pedagogical approach to environmental art education is needed (Inwood& Kennedy, 2020) and especially on initial teacher education programmes. This is particularly needed as "Arts educators around the globe are increasingly concerned with playing their part in helping their students grapple with big issues of our times related to climate change and environmental destruction" (Davis, 2018).

However, another lucky element in the South African education system is the presence of environmental art concerns on the state curriculum. These range from recycling, to collecting things from the environment to use in artworks (such as found objects, natural objects, and sounds to be recorded from the environment, observational work (drawing, painting, sculpting and making: buildings, landscape, people, fauna and flora) and taking care of the environment including tools, art materials, artworks and being mindful of sharing spaces.

Materials as message too

McLuhan famously stated that the "media is the message" (1967) meaning that the materials or media also were part of the content or meaning of an art form or communication. This is an important part of what has now been called materiality. In the Inwood and Kennedy 2020 study one of the students noted the difference between *depicting* an environmental disaster whereas actually *using* the media as part of the message:

not just drawing a picture about, you know, an iceberg, but actually creating something that is more relevant where kids are thinking about how the art itself is impacting the environment to bring awareness.
(2020, p. 595)

How does one renew or infuse Environmental Education or Climate change curricula in a way which does not present the environmental concern as an add-on, and a way which avoids siloing the information, and developing it as a practice? One of the restrictions felt keenly was that in distance learning education teacher training system, it is hard to actually physically model environmental practices for as Wittgenstein said, "What can be shown cannot be said." (Kenny, 2006, p. 37in Howard, 2023, p. 4) .

I had become involved with the Climate Change in Education group as a response to an institutional call for all academic departments and all module-leaders who teach across the teacher education college to commit to infusing their teaching areas with climate change or Environmental sustainability education. The idea was clearly to show the institutional commitment and systemic reform to support ESE 's success at the university (Lozano et al., 2017).

Egbele (2025, p. 22) writes that

Empirical research shows that embedding sustainability within intended learning outcomes and teachers' profiles facilitate curriculum design, pedagogical coherence, and assessment.

This is consistent with the literature on climate change education: as being contextualized in many ways, not merely confined to the sciences, but that "young people learn about climate change in relation to the multiple viewpoints of parents, teachers, media, communities, environments and institutions, among many other influencing factors" (Rousell & Cutter -Mackenzie-Knowles, 2020). It is therefore necessary to see climate change education as something which should occur across the curriculum but also that it can occur in different ways, educationally. Some of these ways are through deliberate curricular adaptation, lectures, conversations at the end of class, assignments, peer interaction, shared descriptions of personal and community experience and prompts in discussion forums. As a lecturer, it is necessary to bear all these different learning dimensions in mind as I attempt to bring climate change education into the primary teacher education curriculum.

THEORETICAL FRAMEWORK

This qualitative, interpretive study is informed by multiple ideas, including Mezirow's 1991 Transformative Learning Theory (Mezirow, 1991) which "emphasizes the process of critical reflection as a means to foster personal and social change"

(Kouam, 2025, p.39). Transformative Learning Theory is related to Critical Realism and akin to Critical pedagogy. There is a relationship between this and the theoretical frameworks of the Ecological Systems Thinking (Capra & Luisi, 2014) which examines the relationships and interconnectedness between environment, culture, society and economics, and also Eco-pedagogy which emphasises critical thinking rooted in ethical and engagement (Kahn, 2010), which like Critical pedagogy, is rooted in the liberation politics of Paolo Freire, and seeks to create change in the student and in society as a result.

Schon's ideas of the reflective educational practitioner are relevant here as I reflect on my own process of introducing environmental sustainability ideas into my teaching of a module preparing creative arts future teachers.

This paper is a reflection on my own ideas and practice as I, influenced by my colleagues in the Climate Change in Education Group, initiate what I feel to be very pressing environmental issues, in a part of a subject which I love, and regard as very important for child development, human potential, education and social change in South Africa – the visual arts component of Creative Arts education. The paper is therefore informed by Schön's notions of reflection and its importance in teacher education, as well as Dewey's original definition of it as the ''active, persistent careful consideration of any belief or supposed form of knowledge in the light of the grounds that support it and the further conclusions to which it tends'' (Dewey, 1933, p.6).

The paper is also informed by readings and knowledge seeking about climate change justice agenda. Here it is necessary to refer back to the UN Sustainable Development Goals as guiding practice. "The UN Assembly (2018) ratified 17 Sustainable Development Goals (SDGs). Goal 4 emphasises education to promote sustainable living, with inclusive education (Goal 4.7) fostering gender equality, human rights, peace, and cultural diversity. SDG Goals 2, 6, 7, and 12 empower students to address hunger, water, sanitation, energy, and consumption. Efforts include sustainable agriculture, waste reduction, clean water access, and promoting renewable energy for a sustainable future." (Gesthuizen, et al., 2025, p. 27).

Egbele notes that of the United Nations' 2030 Agenda for Sustainable Development, it is Sustainable Development Goal 4) which concerns itself with education, which is (seen as) "…both an end and a key enabler of all other Sustainable Development Goals…" Egbele, 2025, p. 8)

Because this (environmental) knowledge is not my original field, it is relatively new knowledge which needs to be thought about and considered in relation to other knowledge, in this case Creative Arts teaching, with a focus of visual arts. This is therefore very appropriate to reflective practice as, according to Schön (1983, p. 50) when "there is some puzzling or troubling or interesting phenomenon" with which an individual is grappling she "also reflects on the understandings which have been implicit..., understandings which ...[are] surfaced, criticized, restructured, and embodied in further action" (Schön,1983, p. 50). The complexity of this process is echoed both in the complexity creative arts education and of environmental education for sustainability. As Nelson writes, "sustainability education is multidimensional, reflecting the interconnected nature of sustainability challenges" (Egbele, 2025, p. 14).

This chapter is therefore a reflection of a process of making sense of climate change education's application in the creative arts, both from a perspective of preparing future teachers for teaching for disciplinary knowledge and its application, as well as for a future in which climate change has increasingly urgent interdisciplinary educational applications and currency. This process of reflection also led to action, specifically around redesigning curriculum and assignments in order to introduce climate change issues. This process was transformative. As Khan notes (2013, p. 8): "The sustainable development framework entails thinking far into the future and taking into account how our present actions might affect our ability to live a wholesome and fulfilling life."

When considering how transformation could occur in a virtual education setting, the thoroughness of Gesthuizen et al.'s work (2025, p. 27) was informative: "transformative learning as a process often triggered by disorienting dilemmas, which can be a personal or social crisis".

In the findings section I will discuss my considerations when introducing environmental education ideas into Creative arts teaching module.

STATEMENT OF THE PROBLEM

This is a reflective self-study, but the problem really underpinning this study is,

What pedagogical approaches can one use to infuse environmental education into a Creative Arts teaching module for future primary school teachers?

METHODOLOGY

This chapter is an interpretive qualitative study. It is primarily focused on my reflection as a teacher educator bringing environmental education into my subject area, creative arts teaching, where student teachers are prepared to teach Creative Arts education in the primary school years. To structure this reflection, I first look at the CAPS curriculum, and see how environmental education concerns are mandated therein. I then consider ways I have incorporated environmental education concerns into the curriculum.

FINDINGS

As part of this process, it was necessary to look at two existing curricula – the CAPS curriculum (which is the curriculum currently taught in South African schools) and the curriculum I have designed to prepare future teachers to teach the subject of Creative Arts. While school curricula do change (and I do not slavishly prepare my student teachers to teach it, knowing that education is a highly portable qualification and they may well be working abroad) I realize that in order for student teachers to feel confident and competent they do need to have had a university teacher education curriculum which references the current school curriculum. This chapter therefore begins with a discussion of the CAPS curriculum as opening up learning areas which are related to environmental concerns. It is to be noted that much of the learning in the arts is physical, through making, experiences and using materials. These 'topics' on the CAPS curriculum are discussed. The chapter then goes on to describe ways in which I began to think about introducing environmental issues in a more robust way, and how a focus on local Southern African artists and culture could coincidentally serve that purpose. In this way I combined my understanding of art history, materiality in Southern African artmaking and teaching and learning in the classroom, and assessment, with an environmental education agenda.

CAPS Curriculum

The current Curriculum and Assessment Policy Statement (CAPS) was introduced in 2011 by the Department of Basic Education, and seeks to integrate environmental concerns (Bopape, Mudau, & Msezane, 2021) in all subjects from Grade R to Grade 12, guided by two principles. The first principle relates to "promoting human rights, inclusivity, and environmental and social justice by encompassing practices aligned with social and ecological justice and human rights as defined in the South African Constitution" and the second is about "valuing indigenous knowledge systems and recognizing the historical and cultural significance of indigenous knowledge systems" (Mngomezulu & Sikhosana, 2025, p. 380).

Some of these environmental issues touched on in the curriculum are reusing, recycling, and being inventive with materials, as well as the fact that many artefacts are made from local materials, in Southern African cultures. However, despite these areas being present on the curriculum, there was no explicit instruction about the environment or working environmentally given for teachers.

It is necessary here to described how the creative arts functions as a subject in South African primary schools. It is nested within the subject Life Skills. There are three parts to the Life Skills curriculum, namely Personal and Social Well-being, Physical education, and Creative Arts. In the curriculum, Personal and Social Well-being is described as "the study of the self in relation to the environment and society. The study area provides opportunities for learners to practise life skills required to make informed choices regarding personal lifestyle, health and social well-being. It provides learners with skills to relate positively with and contribute to family, community and society. Learners are equipped with skills that will assist them to deal with challenging situations positively and recognise, develop and communicate their abilities, interests and skills with confidence. They learn values such as respect for the rights of others and tolerance for cultural and religious diversity in order to build a democratic society. In the Life Skills curriculum for Grades 4 to 6, Personal and Social Well-being is expressed as a study area containing three topics. The three topics are: 1) Development of the self (2) Health and environmental responsibility, (3) Social responsibility" (CAPS curriculum, 2011).

An example of a part of the Personal and Social Well-being part of the curriculum is shown below in Figure 1:

Figure 1. Personal and Social Well-being topics from the Life Skills curriculum

Topics	Grade 4	Grade 5	Grade 6
1. Development of the self	• Personal strengths • Respect for own and others' bodies • Emotions: understanding a range of emotions • Dealing with conflict • Personal experience of working in a group • Bullying: appropriate responses to bullying • Reading for enjoyment	• Positive self-concept formation • Receiving and giving feedback • Coping with emotions • Relationships with peers, older people and strangers • Reading skills: reading with understanding and using a dictionary	• Positive self-esteem: body image • Abilities, interests and potential • Peer pressure • Problem solving skills in conflict situations • Self management skills • Bullying: getting out of the bullying habit • Reading skills: reading with understanding and fluency
2. Health and environmental responsibility	• Dangers in and around water: home and public swimming pools, rivers and dams • Traffic rules relevant to road users: pedestrians and cyclists • Personal and household hygiene • Dietary habits of children • Healthy environment and personal health: home, school and community • HIV and AIDS education: basic facts	• Safety measures at home and the environment • Water as an important basic need • Healthy eating for children • Local environmental health problems • HIV and AIDS education: dealing with stigma • Substance abuse	• Basic first aid in different situations • Food hygiene • Communicable diseases • HIV and AIDS education: myths and realities

Here it can be noted that large learning area "Health and environmental responsibility" which deals with topics like "Water as an important basic need • Healthy eating for children • Local environmental health problems."

The Sport section of the CAPS curriculum is very short. It gives ideas for physical exercises, movements and games that can be played. The focus is on developing co-ordination, enjoying movement and sport.

Before I leave the general subject Life Skills, with its other areas of Personal and Social Well-being and Sport, it is necessary to consider the roles of the teacher who is envisaged as delivering this curriculum. Msezane (2022, p. 746) gives attention to this:

"The South African DBE establishes norms and standards of education that educators in all fields are required to incorporate into the learning program in support of the five primary areas of duty for educators.

The DBE (2014) lists the following roles:

- In this capacity, the educator mediates learning while being sensitive to the various requirements of learners. The teacher should show a thorough understanding of the subject matter and numerous principles.
- Interpreter and designer of educational programmes and materials: In this position, the teacher is responsible for explaining environmental concepts to students and creating educational programs that are tailored to their requirements.
- Learning subject expert: The teacher is expected to possess a solid foundation in the subject-specific knowledge, abilities, values, principles, methods, and processes.

The teacher should be aware of the many methods for disseminating Environmental Education in ways that are suitable for the students and the setting. To make sure that teachers are well-equipped to integrate knowledge and experience in the teaching of Environmental Sustainable Development in schools, the roles listed above are crucial."

While it is envisaged that the teacher should have take on these roles and have this expertise, what pedagogies s/he will use are not delineated or integrated into the curriculum.

The Creative Arts part of the Life Skills curriculum, it can be noted that the Creative Arts part has four areas: dance, drama, visual arts and music. In this chapter I am going to focus on the visual arts, as it in that regard that environmental education becomes explicit, and that is my subject area. However, it is worth noting, as has been shown above, that it is also embedded in the Personal and Social well-being part of the curriculum.

As described earlier, the visual arts component is the area where environmental concerns and activities are most obviously lodged. It is where four ways of sensitising learners to environmental concerns can be found. These are: (1) recycling; (2)

collecting things from the environment to use in artworks (such as found objects, natural objects, and sounds to be recorded from the environment, (3) observational work (drawing, painting, sculpting and making: buildings, landscape, people, fauna and flora) and (4) stewardship or taking care of the environment including tools, art materials, artworks and being mindful of sharing spaces.

Various examples of the visual art curriculum can be shown in Figure 2: Topic 1 continued: Grades 4,5 and 6 Visual Art below, with 'topics' to explore. These will be discussed in terms of the environmental learning opportunities they afford.

Figure 2. Topic 1 continued: Grades 4, 5 and 6 Visual Art

Grade 5	Grade 6
• Introduce local environment and deepen awareness of art elements and design principles in visual stimuli	• Introduce national environment and extend range and observation of art elements and design principles: contrast, proportion, emphasis, balance and unity
• Apply and identify in own work	• Apply, identify and personally interpret in own work
• Extend themes to include local environment; develop techniques	• Extend to include national environment; further develop use of media and techniques

Figure 3. Topic 1 continued, Visual Art, Create in 3-D

3. Create in 3D	• Themes interpreting the personal and social world using 3-dimensional techniques that encourage manipulation of media and awareness of shape in space: 'behind, in front of, next to, above, underneath', etc.	• Extend to include local environment; develop techniques; deepen awareness of personal use of space	• Extend to include national environment; further develop use of media and techniques; properties of 3-dimensional work (form/volume, deep and shallow space)
	• Art elements: texture, shape/form and colour	• Increase conscious use of all art elements	• Deepen and extend conscious use of all art elements
	• Design principles: contrast, proportion	• Introduce emphasis and further develop use of design principles	• Introduce conscious use of balance, and further develop use of design principles
	• Skills and techniques for 3D work	• Develop techniques	• Develop techniques
	• Use of tools: safety, consideration of others, shared resources	• Care of tools	• Responsibility for shared space and tidiness
	• Concern for the environment: use of recyclable materials	• Awareness and value of recycling	• Intentional choice of recyclable materials

What can be noted in each of the 3-D learning areas are environmental concerns. They appear in three ways: as resource management ("care of tools"), as spatial and community responsibility ("responsibility for shared space and tidiness") and then recycling. The curriculum is a spiral one, with the same concerns being touched on in greater depth each successive year. In Grade 4 it is enough for learners to show "concern for the environment: use of recyclable materials", but in Grade 5 this has to built on to show both "awareness and value of recycling" and by Grade 6, there is the "intentional choice of recyclable materials". This shows a depth and a progression.

In Grade 4 Visual Art, the use of recycled materials is explicit in Term 2's Topic 2:

> *"Topic 2: Create in 3D, mobiles or stabiles (Suggested time: 2 hours) Recyclable materials: cardboard/ paper off-cuts, beads, sequins, ribbon, natural objects, various other suitable materials, cotton, wire for hanging, wood glue, etc" (Department of Basic Education Life Skills, 2011, p.45)*

Again, the use of recycled materials occur in Term 4:

Topic 2: Create in 3D, a kite/dream catcher/ bird feeder (Suggested time: 2 hours) Recyclable materials: cardboard/ paper off-cuts, beads, sequins, ribbon, cotton, string, natural objects and any other suitable materials, wood glue, etc (Department of Basic Education, 2011, p. 46).

Figure 4. Grade 5 Visual Art topics: Self and others in local environment (Department of Basic Education, 2011, p. 48)

Topic 1: Create in 2D, images of self and others in local environment	2 hours	Materials: 2/3B pencils, charcoal, coloured inks, oil pastels, tempera paint
• Drawing and/or colour media: exploring a variety of media and techniques • Art elements: use of complementary colour in own images of self and others in local environment • Design principles: emphasis (focal point) used in own images of self and others in local environment		
Topic 2: Create in 3D, self and others in local environment	2 hours	Earthenware clay
• Skills and techniques: earthenware clay • Art elements: reinforce texture, shape/form in own models of human figure • Design principles: introduce emphasis in own models of human figure • Spatial awareness: reinforce conscious awareness of working in space, e.g. front, back and sides of model to be completed • Appropriate use of tools		

In the excerpt from the curriculum (Topic 2) shown in the Figure 4 above, the learners' attention is drawn to the human figure in the environment. This requires skills of observation, understanding of proportion and perspective. When it comes to the use of earthenware clay in the "create in 3D" part, there is nothing to stop the teacher, if she were in a rural area, taking the learners to a riverbed to make the figures there (as has been done for time immemorial in places like rural KwaZulu Natal), nor to stop the teacher collecting clay from the river or the local area. By making the figures out of clay, and placing them in a setting also made out of clay, and using clay from the local environment, the environment repeats as a topic, a setting and a material, throughout the project.

While this chapter is focused on the Visual Arts, it must be noted that in the music section similar creative environmental possibilities are afforded such as making "Instruments using found objects (e.g. stones, cans, seeds, pipes, bottles etc.)" (Department of Basic Education, 2011, p. 34).

The environment is also used when learners make "Sound pictures based on themes (a thunderstorm, a train journey) using voice, body and found or made instruments (considering pitch, dynamics, tempo and rhythms). This is a way that people would be asked to pay very close attention to their environment in terms of sounds, and sensitize learners to their environment from another sense, another perspective. These are also environmental educational experiences involving the arts. Finley (2003, p. 293) describes environmental art experiences as "useable and responsive' research that can lead to action in a community as well as spark 'visionary critical discourses that examine how things are but also imagine how they could be otherwise.'" This kind of synthesis of learning in multiple ways, is described by Egbele who writes that "the Art lesson demonstrated how environmental awareness and cultural reflection can coexist in creative learning" (Egbele, 2025, p. 68).

Reflection

I discuss my reflection on the learning which occurred.

On reflection, I looked at the types of knowledge that future arts teachers would need. This is shown in the below diagram: Figure 5. Knowlwdge for teacher (author-generated).

Figure 5. Knowledge for teachers (author-generated)

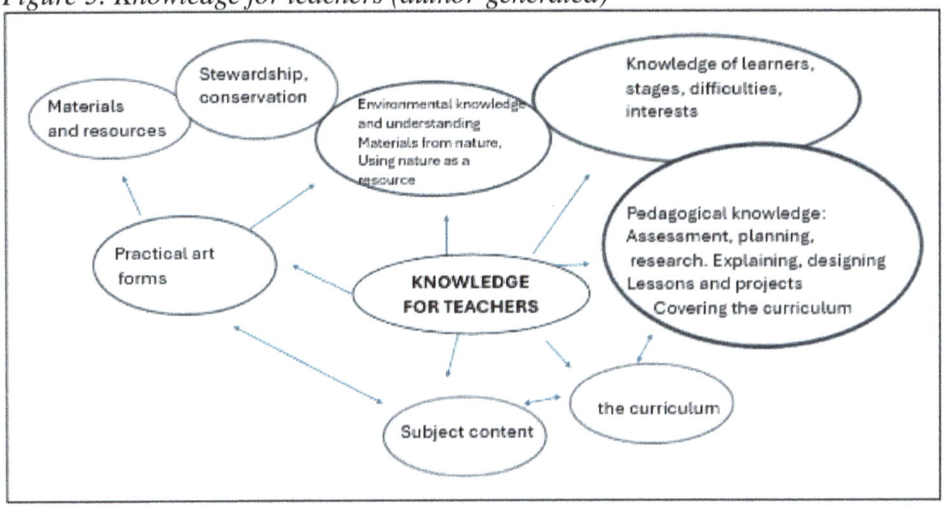

This gave me information about areas of learning and prompted me to as to how I could infuse environmental concerns into teacher education.

The three ways I considered that I could use environmental educational issues in relation to the subject creative arts, with a focus of visual arts education, were:

1. Through teaching content. This involved introducing local artists who have an environmental message, or used materials in an environmentally pointed and innovative way.

To go into some detail about this: artists such as Cyrus Karibu, El Anatsui, Willie Bester, Mbongeni Buthelezi, Nicholas Hlobo use recycled materials in very innovative ways. An artist like Buthelezi is explicit in his use of plastic as a response to plastic pollution in South Africa. Then there are the land artists like Strijdom van der Merwe, who work with natural materials. Other artists who work in more traditional ways, depicting nature through landscapes, and still other artists like Kentridge who depict environmental degradation through depicting human interference in the landscape, such as the mine dumps around Johannesburg. Then other artists such as Peter Clarke who also show figures in the landscape, often struggling in hostile landscapes due to forced removals. There are also the 'urban landscape' artists like Sekoto, David Goldblatt, Siyabonga Sikosana, and George Pemba who depict urban spaces and the way people live in them, often bringing political and social justice ideas into the environmental degradation which they depict. Rural artists like Esther Mahlangu use traditional materials, and art forms, but take them into the future.

Without wanting to over-simplify, I think many visual artists are working in environmentally interesting or pertinent ways, and/or are making statements about the environment. From this kind of content, which I introduce to my students to give them a foothold in the subject, and for enrichment, it is possible to teach about many things. Most of my students have not had much access to creative arts education, and when they are introduced to local artists, it is giving them access to their heritage. It is also showing them ways they can work with artists' ideas, and bring in some art history into their lessons. For example, when they are working with the topic of adornment, which is on the curriculum, they could begin with Cyrus Kabiru, and how he made mask-like face decorations (which he called C-Stunners) from scrap.

2. Through the idea of stewardship. This entails discussing knowledge of materials and how materiality is embedded in the arts and also involves teaching student teachers how we can clean, preserve, make, store, source, reuse, recycle and find materials. This involves looking after what one has bought (such as brushes) and thus extend the life of such materials, but also how we can re-use materials (magazines, old books, newspaper, fabric, cans, plastic, cardboard, buttons, ribbons, pins – virtually anything). When it comes to making materials, it is possible to make one's own charcoal, to experiment with drawing with inks made from plant/vegetable dyes, and tea, to make glue from paste made with flour and water, to make playdough. The important part of this is that future teachers learn to be innovative and resourceful, and to model this to learners, to show them that you can make materials, or find materials in nature, and that visual art can be made out of anything. The preservation, correct cleaning, storage and management of materials will also help future teachers considerably with budgeting. In this the teacher is a prime resource too, as she must model this. In terms of resource constrained schools, which South African schools frequently are (Ruthanam, Reddy & Pillay, 2021, p. 136), this is an important educational approach which is pragmatic and pedagogically intelligent.
3. Through the concept of materiality, which is defined as "the fact of being a physical substance or object, or something that is a physical substance or object" (Cambridge Dictionary, online). Materiality also has a more artistic meaning which concerns the ability to handle, know and explore materials is linked to

the point above. The use of innovation in handling materials is a kind of resourcefulness which is encouraged in both art education and environmental or sustainable education. An example of innovation in handling materials would be the full exploration of a material, for example, plastic or cardboard. As part of resource constraint, a fuller exploration of materials is needed, but this is paradoxically, an educationally enriching possibility. Materiality also relates to the qualities of a material. In visual arts education one of the prime learning areas is about what a material can do, and what its potential uses are. In terms of using materials innovatively, often materials are joined or used together, like in a collage or montage, but an important thing for artists or learners to know, is what materials can be joined. There is also, to do with materials, as Asin notes "Another area of concern—as a motivation for exploring sustainability within the field of art education—highlighted in the data is the issue of toxicity." (Asin, 2025, p. 108). Some materials are toxic and it is our duty and part of knowledge of materials and stewardship, that we handle and store these appropriately and teach learners about this, and their dangers.

When it comes to resources teachers have are the environment itself, which they can use as a site for artwork, as a place here resources can be found (from litter to found objects to natural materials), the curriculum, and the human resources (including community, local artists, the learners or children themselves) and of course, the teacher. These are shown in Figure 6 below.

Figure 6. Some types of learning resources available for teachers (author-generated)

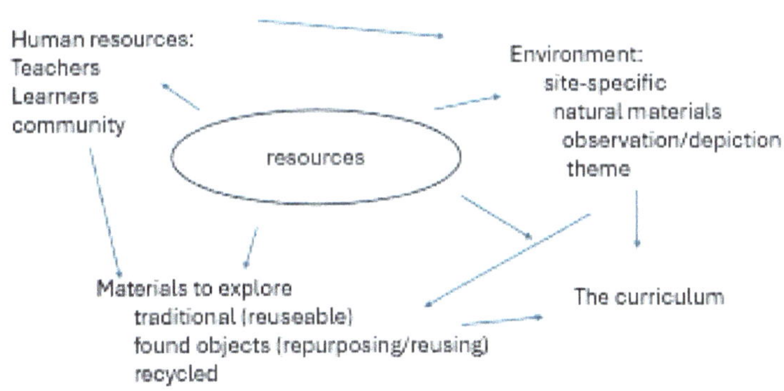

4. Through setting assignments in which student teachers had to reflexively use their knowledge of environmental issues, as well as possibly 1 and 2 above, to develop environmental projects for their (future) learners.

While considering the above points, this understanding is being filtered through three other sources: the CAPS curriculum, environmental readings and the Sustainable Development goals, as they relate to education and the environment. This involves deep reflection on my part as a lecturer and it is this which I share primarily, including my problems and concerns.

Epiphanies

Gesthuizen et al. describe the importance of an epiphany or a wake-up moment in education.

The authors discuss this epiphany as such:

> *"a historical case study of three environmental scientists (Aldo Leopold, Rachel Carson and David Suzuki) conducted by Walter, (2013) noted that each underwent a transformation driven by critical dilemmas, which ignited their passion and drove towards a greater ecological, environmental consciousness and global awareness. These dilemmas were often emotional experiences with nature, family relationships and natural objects such as dead wolves, birds, and trees. Replicating this epiphany and real-world experiences inside a classroom poses a challenge for teachers. Nevertheless, environmental educators have devised methods to introduce contexts and situations that capture students' attention and require their deep focus. Teachers can instil a sense of purpose and urgency towards a sustainable future by distinguishing between education for change and education in change (Sterling, 2001)" (2025, p. 28).*

Akey describes engagement as key to successful learning, and indeed the start of self-directed learning: "Engagement in schoolwork involves both behaviours (such as persistence, effort, attention) and attitudes (such as motivation, positive learning values, enthusiasm, interest, pride in success)" (Akey, 2006, p.6). However in order to activate engagement teachers or teacher educators need to create hooks or learning catalysts.

Davis (2018, p. 1) describes the necessity of feelings and knowledge, and also key activities as catalysts:

> *for students to be mobilised to care about and act for the environment requires both knowledge but also emotional commitment and the means for expression and communication. Findings drawn from student reflective responses and focus group interviews identified various triggers for student engagement. These included excursions and outdoor activities framed by arts-based processes.*

I realised also, like Walter (2023) that when introducing environmental issues into the curriculum, I needed to prepare something which while being online, would grab the students' attention and be quite shocking and provocative. To this end I prepared a slideshow with shocking images and statistics.

The next part of this was to relate the climate changes to them, in other words, to bring it home. This meant asking them questions about what the changes were that they saw where they lived.

When giving the student teachers ideas about ways of teaching the arts, I brought in seasons and the kind of natural changes one experiences with different seasons. This is following on from the idea that environmental education should involve "a sense of wonder" (Kahn, 2010, p. 12), and also is to inspire appreciation of nature, and therefore to get emotionally closer to it. I am personally influenced by years spent in the Rudolph Steiner or Waldorf system, both as a learner, and as a child. Jordan (2023) describes this type of education as experiential, and one which places an importance on developing "feelings for nature" (Carson, 1998) as opposed to merely knowing facts about nature. When learning about narure, it is important to go out into nature. Similarly, "outdoor and craft-based learning further demonstrated the role of Eco-pedagogy in nurturing emotional and ethical connections to nature. Students were not only learning about sustainability but also experiencing it through practical, creative, and reflective engagement"(Egbele, 2025, p. 74).

Activities

Tilbury (1995) writes that "Education for Sustainability (ESD) in schools benefits from the inclusion of active learning strategies". Art-making offers many such active experiential learning opportunities.

Davis notes four ways of ensuring engagement:

- Connecting to the idea of the work through personal background and interest
- Out of classroom experiences framed by arts processes and a sense of agency
- Interactions with domain experts
- Arts-based action leading to committed engagement (2018, p.14).

Reddy (2021) asks about how we can ensure deep learning occurs, and how can we ensure that student teachers take this on in their own future work contexts? Gesthuizen et al. think inspiration is an important factor in education, and go some way in answering this question: "By incorporating ecological sustainability and social justice dilemmas into classroom activities and discussions, teachers can inspire students to become reflective agents of change" (Gesthuizen et al. 2025, p. 28). Effective environmental education needs to have activities which enable students to mesh conceptual understanding with actions in order to and interact with climate change in real, local ways (Monroe et al. 2019; Kok, et al., 2025). To this end I designed assignments which would force the students to engage with environmental/sustainability issues.

By giving assessments which compel students to get to grips with environmental concerns, students have to use their knowledge in creative ways, and become reflexive. In assessments, environmental issues are addressed leaving the student teachers with a project whereby they would be forging in their schools and communities.

Below are two examples of this. These two questions are two options out of three for the module. Therefore, it is not compulsory that the students engage with these assignments. See Figure 7. Assignments on the curriculum for one of my modules, 2026.

Figure 7. Assignments on the curriculum for one of my modules, 2026

2A. Greening (Skills assessed: research/public event planning/ project planning, integration of ideas) (30 marks)

Read the texts on climate change and how they relate to the Sustainable Development Goals. Design a project that focuses on one environmental problem, with which you/your family/young people are affected by.

Design a project which has the following components:

i. Background to the problem
ii. A physical intervention/interdisciplinary public event to address the problem (What is it? What does it entail? Who is involved?)
iii. A related practical art /drama/ music or dance project which addresses the issue. Include concepts which work on multiple levels (physical, interpretative, emotional, narrative) and materials needed.

Or 2B. A save-the-species event. (30 Marks)

Research on a species which is threatened. Design a project that focuses on one problem, with which you/your family/young people you know are affected by.

Design a project which has the following components:

i. Background to the problem plus how the species is shown in an art form (a song/artwork/play/dance)
ii. A public event to highlight the problem which uses the arts
iii. A related practical art, drama, music or dance project which addresses the issue.

These projects are both about researching environmental issues and designing an individual project or public/community events to amplify them and using the arts to do this. By designing assignments like this I hope the students will engage with environmental ideas, and think of interesting and original ways they can use art forms in this regard. This is an interdisciplinary approach, bringing together environmental issues, art forms and project design. With the contextualisation of where, when and why these elements come together, each case or project /assignment has individual coherence, its own contextual framework to retain its meaning and coherence (Chang et al., 2019). This also calls for a fairly flexible marking memo or rubric. By designing assessments where I asked students to design an environmental art event I could measure student learning and understanding (Kouam, 2025, p. 42).

This year my intention is to begin an online art-making community for future teachers. The idea is two-fold – to make sure that future creative art teachers are comfortable with art- making in all its messiness and materiality, but also to do some explicitly environmental education projects. These can be used for research and studies to explore what can be learned from the process of creating environmental art … with preservice teachers and how this learning impacts on preservice teachers' own teaching experiences, attitudes and behaviours in relation to environmental sustainability" (Inwood & Kennedy, 2020, p. 586). This will be a way of building on a process which has begun in my teaching, and also a way of checking students' understanding of environmental issues, confidence, and problem-solving abilities.

I would like to conclude the discussion section with a few remarks relating to the need to transform initial teacher education, and what environmental education and creative arts can offer with regard to this.

Firstly, I would like to discuss the need for transformation. Reddy (2021) raises in a critique of teacher education, that it follows an industrialised model, with courses which seem to separate knowledge from context, and based on a "compartmentalised view of reality" (2021, p. 168) which seems to deny the complexity of today's schools. I think into this very difficulty of the siloed quality of teacher education, that Environmental Education concerns offer a very real opportunity of working in an inter-disciplinary or cross-disciplinary way, which is intelligent, creative and productive.

Furthermore, the creative arts, especially the visual arts which use materials innovatively, in which divergent solutions are prized, which encourages interdisciplinary approaches or 'deep thinking', which values making connections and problematizing, and which are deeply interpretive, but where learning is also often hands-on or practical, lends itself to using and expressing environmental education concerns.

The next issue, I would like to discuss, which is really a follow-on from what Reddy (2021) writes about above, when he describes ITE as being outdated, compartmentalized and out of touch with the context of our schools. Ndebele et al. (2024, p. 7) call for "Transformative education requires a shift in paradigms towards student-centred, inclusive, and culturally responsive teaching practices." I would like to argue that this is what environmental education is. It brings everyone into the discussion because it affects everyone, so it is inclusive, it is student-centred. Moreover, it affects everyone's discrete cultural practice and understandings because this is what we bring with us into our living and our approaches to it. Therefore, environmental education concerns can very beautifully bring us into a transformative focus. That environmental education and transformative education are linked is clear (Govender, Maphalala & Mncube, 2025) who write that initial teacher education needs to change to include social-emotional learning and interdisciplinary collaboration.

Lastly, I would like to amplify what Inwood and Kennedy (2020, p. 588) write about the necessity of the conceptualization of a critical and pedagogical approach to environmental art education. While the visual arts are a fertile ground for teaching environmental educational concerns, with many teaching possibilities, there is a lack of pedagogic groundwork. While it is envisaged that the teacher should take on the role of an environmental educator and have this expertise, what pedagogies s/he will use are not delineated, integrated into the curriculum, or in any way apparent. It is therefore essential that teacher educators work to find appropriate pedagogies.

This is echoed by Blose (2025) who writes that there is a need for more research on how teachers are using environmental education knowledge and ideas in their classrooms. There is a need for art teachers to have more knowledge about sustainability (Asin, 2025). Teachers are change agents and in a similar vein, there is also a need to follow the student teachers who are currently having their ITE curricula infused with EE into their classrooms, in a longitudinal study, to see how they are using this knowledge in their classroom. There is a need to study learners' responses, and their inventiveness, in the face of the environmental knowledge they acquire.

It is into this gap in pedagogical approaches into which I write about my own efforts to introduce environmental education into the creative arts curriculum, and into which this reflection is written.

CONCLUSION

This chapter is not a how-to guide, but a reflection on my own practice as a lecturer in teacher education of beginning a process of teaching environmental concerns in a more explicit way. This process entailed going through both the CAPS and university course curriculum carefully, looking for opportunities both to amplify South African heritage and innovation but also ways of creating artwork which could attend to environmental concerns. This process also involved reflecting on how to do this with both sensitivity and urgency. Underscoring all of this is an ethical dimension of teacher education, in a country where climate change is increasingly apparent and affects those communities who are most vulnerable. It is important that we think in an ecological way which sees all living elements of society in inter-relation, as Howard (2023) writes:

> *"The education of the young, and by default the education of the teachers responsible for our children's learning, must reflect a different consciousness than one that sees the destruction of cultures, languages and the earth as inevitable by-products of "progress". We must educate so we see ourselves as part of the web of life, not simply isolated self-maximizing individuals" (Howard 2023, p.11).*

Lecturers and course curricula point student teachers to important issues and understandings. Clearly the largest implication is that if lecturers do not address climate change issues, student teachers will not necessarily be aware of them, and how to address them in the classrooms of the future. I regard myself as being at the beginning of this journey into developing appropriate ways of infusing environmental ideas into my Creative Arts teaching.

However, the process of introducing climate change issues into existing course curricula requires reflection, reflexivity and thoughtfulness. It is also an exciting process, especially in the arts, where often such connections are less obvious. The implication is that climate change action is also linked to sustainable creativity and addressing patterns of making and production in the arts, art education, and design, and the need for this in a developing society is invaluable.

REFERENCES

Akey, T. M. (2006). *School context, student attitudes and behavior, and academic achievement: An exploratory analysis*. MDRC.

Asin, I. D. (2025). *The Role of Sustainability in the Future of Art Education*. Teachers College, Columbia University.

Berkowitz, M. W. (2017). The Centrality of Character Education for Creating and Sustaining a Just World. In Worldwatch Institute (Ed.), EarthEd: State of the World (pp. 83–94). Island Press. DOI: 10.5822/978-1-61091-843-5_7

Berry, W. (2000). *Life is a miracle: An essay against modern superstition*. Counterpoint.

Biesta, G. (2010). 2010 Why 'What Works' Still Won't Work: From Evidence-Based Education to Value-Based Education. *Studies in Philosophy and Education*, *29*(5), 491–503. DOI: 10.1007/s11217-010-9191-x

Blose, P. (2025). Pedagogical Approaches for Teaching Education for Sustainable Development in the Technology Education Curriculum. *Research in Social Sciences and Technology*, *10*(1), 209–232. DOI: 10.46303/ressat.2025.12

Bopape, J., Mudau, A. V., & Msezane, S. B. (2021). Greening the school for sustainable development: A case of Tshwane North District. *Journal for the Education of Gifted Young Scientists*, *9*(2), 161–180. DOI: 10.17478/jegys.901622

Capra, F., & Luisi, P. L. (2014). The systems view of life: A unifying vision. Cambridge university press. The Systems View of Life: A Unifying Vision

Carson, R. (1998). *The Sense of Wonder*. HarperCollins. (Original work published 1965)

Cheng, Y. (2018). "The Bag Is My Home": Recycling "China Bags" in Contemporary African Art. *african arts, 51*(2), 18-31.

Curren, R., & Metzger, E. (2017). *Living well now and in the future: Why sustainability matters*. MIT Press.

Curren, R. R., & Metzger, E. (2019). *Education in the anthropocene. A pragmatic approach*. On Education. *Journal for Research and Debate*, *2*(4).

Darling-Hammond, L., Hyler, M. E., & Gardner, M. (2017). Effective teacher professional development (Research Brief). Learning Policy Institute. (Report PDF:) https://learningpolicyinstitute.org/product/effective-teacher-professional -development report (Report PDF: https://learningpolicyinstitute.org/sites/default/ files/productfiles/Effective_Teacher_Professional_Development_REPORT

Davis, S. (2018) The engagement tree: Arts-based pedagogies for environmental learning. .

Department of Basic Education. (2011). *National Curriculum Statement (NCS). Curriculum and Assessment Policy Statement. Life Skills Intermediate Phase Grades 4-6*. Department of Basic Education.

Dewey, J. (1933). *How we think: A restatement of the relation of reflective thinking to the educative process*. DC. Heath and Company.

Doerr, M. (2004). *Currere and the environmental autobiography: A phenomenological approach to teaching ecology*. Peter Lang.

Durmuş, E., & Kınacı, M. K. (2021). Opinions of social studies teacher education students about the impact of environmental education on ecological literacy. *Review of International Geographical Education Online*, *11*(2), 482–501. DOI: 10.33403/ rigeo.825516

(2019). Education for sustainability – Where do you go from here? InChang, C.-H., Kidman, G., & Wi, A. (Eds.), *Issues in teaching and learning of education for sustainability: Theory into practice* (1st ed., pp. 180–191). Routledge., DOI: 10.4324/9780429450433

Egbele, N. (2025). *Understanding Teacher Educators' Perceptions on Sustainability Education: A Case Study from the University of Eastern Finland Teacher Training School, Joensuu* (Master's thesis, Itä-Suomen yliopisto).

Fayiga, A. O., Ipinmoroti, M. O., & Chirenje, T. (2018). Environmental pollution in Africa. *Environment, Development and Sustainability*, *20*(1), 41–73. DOI: 10.1007/ s10668-016-9894-4

Ferkany, M. (2021). Aristotelian Virtue Education and Education for Sustainable Development: Prospects and Problems. In Kawall, J. (Ed.), *The Virtues of Sustainability*. Oxford University Press.

Gesthuizen, R., Tan, H., & Kidman, G. (2025). Inspired to build understanding: Learning about sustainability and environmental education from within a virtual world. *International Research in Geographical and Environmental Education*, *34*(1), 25–42. DOI: 10.1080/10382046.2024.2348265

Ginju, S. (2020). *Studies and current trends in science of education ecological literacy in the context of institutionalized education*. Psych-Pedagocical Attributes.

Govender, S., Maphalala, M. C., & Mncube, D. W. (Eds.). (2025). *Education and climate change: Discovering strategies for individual and collective action*. AOSIS.

Hanekom, A. (2020). South African initiative to end plastic pollution in the environment. *South African Journal of Science, 116*(5-6), 1–2. DOI: 10.17159/sajs.2020/8095

Hopkins, C., & McKeown, R. (2005). Guidelines and recommendations for reorienting teacher education to address sustainability. Retrieved from: www.unesco.org/education/desd

Howard, P. (2023). Who Will Teach the Teachers?: Reorienting Teacher Education for the Values of Sustainability. In *Teaching sustainability/teaching sustainably* (pp. 149-157). Routledge.

Inwood, H., & Kennedy, A. (2020). Conceptualising art education as environmental activism in preservice teacher education. *International Journal of Art & Design Education, 39*(3), 585–599. DOI: 10.1111/jade.12308

Jordan, K., & Kristjánsson, K. (2017). Sustainability, virtue ethics, and the virtue of harmony with nature. *Environmental Education Research, 23*(9), 1205–1229. DOI: 10.1080/13504622.2016.1157681

Jordan, K. E. (2023). The intersection of environmental and sustainability education, and character education: An instrumental case study. *British Educational Research Journal, 49*(2), 288–313. DOI: 10.1002/berj.3843

Kahn, R. (2010). Critical pedagogy, ecoliteracy, & planetary crisis: The ecopedagogy movement. Critical pedagogy, ecoliteracy, & planetary crisis: The ecopedagogy movement.

Khan, S. H. (2013). Fostering Sustainable Development and Environmental Education Programmes: Role of Teacher Training Institutions and Colleges of Teacher Education. *African Educational Research Journal, 1*(1), 8–17.

Kok, P. J., Govender, S., Kok, L., & Mncube, D. W. (2025). Effective curriculum-aligned climate change education pedagogy: Causal mechanisms in Natural Sciences and Geography. In Govender, S., Maphalala, M. C., & Mncube, D. W. (Eds.), *Education and climate change: Discovering strategies for individual and collective action* (pp. 95–113). AOSIS Books., DOI: 10.4102/aosis.2025.BK491.06

Kouam, A. W. F. (2025). Teachers as environmental educators: Exploring perceptions and practices of green pedagogies in fostering eco-literacy. *Journal of Applied Learning & Teaching, 8*(S1), 37–46.

Kouam, A. W. F. (2025). Teachers as environmental educators: Exploring perceptions and practices of green pedagogies in fostering eco-literacy. *Journal of Applied Learning and Teaching, 8*(Sp. Iss. 1), 37-46.

Kristjansson, K. (2013). Ten Myths About Character, Virtue and Virtue Education – Plus Three Well-Founded Misgivings. *British Journal of Educational Studies, 61*(3), 269–287. Advance online publication. DOI: 10.1080/00071005.2013.778386

Lozano, R., Merrill, M., Sammalisto, K., Ceulemans, K., & Lozano, F. (2017). Connecting competences and pedagogical approaches for sustainable development in higher education: A literature review and framework proposal. *Sustainability (Basel), 9*(10), 1889. DOI: 10.3390/su9101889

Mandikonza, C., & Lotz-Sisitka, H. (2016). Emergence of environment and sustainability education (ESE) in teacher education contexts in Southern Africa: A common good concern. *Educational Research for Social Change, 5*(1), 107–130. DOI: 10.17159/2221-4070/2016/v5i1a7

Mashazi, T., Jacobs, P., Rampedi, I. T., & Modley, L. A. (2024). Conveying Science through Art: The Case of a Polluted Urban Stream in Johannesburg, South Africa. *The International Journal of Interdisciplinary Social and Community Studies, 19*(1), 107–124. DOI: 10.18848/2324-7576/CGP/v19i01/107-124

McLuhan, M. (1967). *The Medium is the Message*. Bantam Books.

Mncube, V., & Olawale, B. (2021). Leadership and management in the context of Covid-19 stringency: are democratic principles a solution? *14th Annual International Conference of Education, Research and Innovation*. November 2021, Spain. DOI: 10.21125/iceri.2021.1973

Monroe, M. C., Plate, R. R., Oxarart, A., Bowers, A., & Chaves, W. A. (2019). Identifying effective climate change education strategies: A systematic review of the research. *Environmental Education Research, 25*(6), 791–812. DOI: 10.1080/13504622.2017.1360842

Msezane, S. (2022). Descriptive analysis of the education for sustainable development programme: Fundisa for change perspective: Fundisa for change perspective. *International Journal of Curriculum and Instruction, 15*(1), 741–761.

Msezane, S. B. (2020). Positioning of environmental education in life sciences (Grade 12). *Ecology. Environmental Conservation, 26*(4), 1450–1458.

Ndebele, C., Marongwe, N., Ncanywa, T., Matope, S., Ginyigazi, Z., Chisango, G., Msindwana, P. N., & Garidzirai, R. (2024). Reconceptualising initial teacher education in South Africa: A quest for transformative and sustainable alternatives. *Interdisciplinary Journal of Education Research*, *6*, 1–19. DOI: 10.38140/ijer-2024.vol6.07

Reddy, C. (2021). Environmental education, social justice and teacher education: Enabling meaningful environmental learning in local contexts. *South African Journal of Higher Education*, *35*(1), 161–177. DOI: 10.20853/35-1-4427

Rousell, D., & Cutter-Mackenzie-Knowles, A. (2020). A systematic review of climate change education: Giving children and young people a 'voice' and a 'hand' in redressing climate change. *Children's Geographies*, *18*(2), 191–208. DOI: 10.1080/14733285.2019.1614532

Ruthanam, M., Reddy, P., & Pillay, D. (2021). Teachers' Choices of Teaching Methods for Environmental Education: A case study of Life Skills teachers at a primary school in South Africa. *Southern African Journal of Environmental Education, 37.*

Santos, F. L. (2026) The Art of Trash: Mbongeni Buthelezi: Painting with Plastic. More Africa. 3 January 2026. The Art of Trash: Mbongeni Buthelezi Painting with Plastic - More Africa

Schön, D. A. (1983). *The reflective practitioner: How professionals think in action.* Basic Books, Inc.

Scott, W., & Oulton, C. (1998). Environmental values education: An exploration of its role in the school curriculum. *Journal of Moral Education*, *27*(2), 209–224. DOI: 10.1080/0305724980270206

Sikhosana, L. (2022). Clarifying the significance of instructional methodologies for environmental education integration. *International Journal of Research in Business and Social Science*, *11*(7), 2147–4478. DOI: 10.20525/ijrbs.v11i7.2016

Sobel, D. (1996). *Beyond Ecophobia: Reclaiming the Heart of Nature Education.* The Orion Society.

Stanhouse, H. L. (2016). *Reframe, Reuse, Recycle: The Found Object in Post-Colonial Africa.* Recontextualized by Contemporary Artists.

Sterling, S. (2001). Sustainable education: Re-visioning learning and change. Schumacher Society/Green Books. https://www.greenbooks.co.uk/sustainable-education

Struckhoff, M., Stroh, E., & Grabner, K. (2013). Effects of mining-associated lead and zinc soil contamination on native floristic quality. *Journal of Environmental Management, 119C*, 20–28. DOI: 10.1016/j.jenvman.2013.01.021 PMID: 23435182

Tilbury, D. (1995). Environmental education for sustainability: Defining the new focus of environmental education in the 1990s. *Environmental Education Research, 1*(2), 195–212. DOI: 10.1080/1350462950010206

United Nations. (2015). Transforming our world: The 2030 Agenda for Sustainable Development (A/RES/70/1).

United Nations Educational, Scientific and Cultural Organization. (2015). *Global Action Programme on Education for Sustainable Development (GAP): Follow-up to the DESD. UNESCO 2015 - UNESCO Digital Library United Nations Educational, Scientific and Cultural Organization. (2017). Education for sustainable development goals: Learning objectives*. UNESCO., DOI: 10.54675/CGBA9153

United Nations Educational, Scientific and Cultural Organization. (2019). Framework for the implementation of Education for Sustainable Development (ESD) beyond 2019. Paris: United Nations Educational, Scientific and Cultural Organization. https://unesdoc.unesco.org/ark:/48223/pf0000370215

United Nations Educational, Scientific and Cultural Organization. (2020). Education for sustainable development: A roadmap (Roadmap for implementing the ESD for 2030 framework). Paris: UNESCO. https://unesdoc.unesco.org/ark:/48223/pf0000374802

United Nations Educational, Scientific and Cultural Organization. (2021). Berlin Declaration on Education for Sustainable Development: Learn for our planet — act for sustainability (UNESCO World Conference on ESD, 17–19 May 2021). Paris: UNESCO. https://unesdoc.unesco.org/ark:/48223/pf0000381229DOI: 10.1891/9780826190123

United Nations Environment Programme. (2025). UNEP education and higher education platforms. UNEP. https://www.unep.org

United Nations General Assembly United Nations. (2015). THE 17 GOALS | Sustainable Development United Nations. (2023). The 17 sustainable development goals.

Vogel, C., Nkrumah, B., Kosciulek, D., Lebea, D., Booth, T., & Brown, M. (2022). 'Empowering youth as change agents for climate change in South Africa': Challenges, caveats and course corrections. *Journal of Youth Studies, 25*(6), 812–832. DOI: 10.1080/13676261.2022.2046259

Walter, P. (2013). Dead wolves, dead birds, and dead trees: Catalysts for transformative learning in the making of scientist-environmentalists. *Adult Education Quarterly, 63*(1), 24–42. DOI: 10.1177/0741713611426348

York, R. A. (2014). *Re-connecting with nature: Transformative environmental education through the arts*. University of Toronto.

Zelenski, J. M., & Nisbet, E. K. (2014). Happiness and feeling connected: The distinct role of nature relatedness. *Environment and Behavior, 46*(1), 3–23. DOI: 10.1177/0013916512451901

Chapter 7 Reflective Professional Development for Climate Change Education: A Framework for Strengthening Teacher Competence and Instructional Practice

Emmanuel Adjei-Boateng
http://orcid.org/0000-0003-1786-8116
University of Ghana, Ghana

Kwaku Darko Amponsah
University of Ghana, Ghana

Phyllis Bernice Kwarteng Donkor
http://orcid.org/0000-0003-3249-3025
University of Ghana, Ghana

ABSTRACT

Climate change is a pressing global challenge, with Africa facing heightened vulnerability due to its dependence on climate-sensitive sectors such as agriculture, water resources, and energy production. Climate Change Education (CCE) plays a critical role in fostering environmental literacy, ethical awareness, and action-oriented learning. Teachers are central to effective CCE implementation; however, evidence suggests that many African educators, despite understanding climate change issues, face limitations in subject knowledge, pedagogy, instructional resources, and continuous professional development. Reflective practice is increasingly recognised as a key mechanism for strengthening teachers' capacities, supporting adaptive

DOI: 10.4018/979-8-3373-6975-4.ch007

teaching, and empowering learners to engage with climate challenges. This chapter highlights the need to integrate experiential, inquiry-based, and action-focused approaches into teacher professional development to promote transformative climate education and sustainable practices in African contexts.

INTRODUCTION AND THEORETICAL POSITIONING

Climate change represents one of the most complex global challenges confronting contemporary societies, with far-reaching environmental, social, economic, and political consequences. Rising global temperatures, shifting precipitation patterns, biodiversity loss, and increased frequency of extreme weather events are already affecting livelihoods, food systems, water security, and public health across many regions of the world (IPCC, 2022). These impacts are pronounced in developing regions, especially across African countries, where large segments of the population depend on climate-sensitive sectors such as agriculture, fisheries, and natural resource-based livelihoods (UNEP, 2021). The scale and urgency of these challenges underscore the need for transformative societal responses that go beyond technological solutions to include educational strategies that foster environmental literacy, climate awareness, and action-oriented citizenship.

Within this context, Climate Change Education (CCE) has emerged as a critical component of sustainability education, equipping learners with the knowledge, skills, values, and competencies necessary to understand climate systems and participate in mitigation and adaptation efforts (Mochizuki & Bryan, 2015; Stevenson et al., 2017). CCE emphasises interdisciplinary learning, inquiry-based problem solving, ethical reasoning, and engagement with local environmental challenges. Rather than focusing solely on the transmission of scientific knowledge, contemporary approaches to CCE prioritise the development of learners' critical thinking, decision-making, and participatory competencies, enabling individuals and communities to respond effectively to environmental challenges (Rousell & Cutter-Mackenzie-Knowles, 2020).

Despite the growing global emphasis on Climate Change Education (CCE), African education systems face unique structural and institutional challenges that influence how climate education is integrated into schools and higher education. Many countries across the continent continue to grapple with curriculum overload, limited teacher preparation in environmental and climate science, inadequate instructional resources, and insufficient institutional support for interdisciplinary sustainability education (Rosalyn McKeown, 2014; Selby & Kagawa, 2013). These systemic constraints often hinder teachers' and institutions' ability to translate climate change policies into meaningful classroom practices. Consequently, situating Climate Change Education within the specific realities of African education systems is essential for developing contextually relevant professional development models that support teachers in both schools and higher education institutions.

Teachers play a central role in translating curriculum intentions into classroom practice and shaping students' climate knowledge, environmental attitudes, and sustainability behaviours. Their effectiveness in implementing CCE depends not only on subject-matter expertise but also on pedagogical competence, professional confidence, access to instructional resources, and sustained professional learning opportunities (Evans et al., 2021). However, research across developing and developed contexts consistently indicates that many teachers feel insufficiently prepared to teach climate change concepts due to limited training, inadequate curriculum support, and uncertainty about appropriate pedagogical strategies for addressing complex and sometimes controversial environmental topics (Boateng, 2021; Ochieng & Adebayo, 2023). Strengthening teacher professional development, therefore, represents a key pathway for improving the effectiveness of Climate Change Education initiatives.

Traditional professional development models that rely primarily on short-term workshops or one-off training sessions often produce limited instructional change because they do not provide teachers with sustained opportunities to experiment with new practices, reflect on classroom experiences, or receive ongoing professional support (Darling-Hammond et al., 2017). Contemporary scholarship increasingly emphasises reflective professional learning as a critical mechanism for promoting sustained pedagogical transformation. Reflective practice enables teachers to examine their instructional decisions, evaluate classroom outcomes, identify knowledge or practice gaps, and adapt their teaching approaches to meet student needs better (Farrell, 2018; Hatton & Smith, 1995). Through reflective inquiry, teachers move beyond routine instructional habits toward intentional, evidence-informed professional learning.

The theoretical positioning of this chapter is grounded in reflective practice theory and experiential learning theory, which conceptualise professional learning as a cyclical process linking experience, reflection, conceptual understanding, and instructional experimentation (Kolb, 1984; Schön, 1983). From this perspective, teacher learning occurs most effectively when professional development experiences are integrated with classroom practice and supported by structured opportunities for reflection, peer collaboration, and mentoring. Reflective professional development is particularly relevant for Climate Change Education because the rapidly evolving nature of climate science, the interdisciplinary character of sustainability issues, and the contextual variability of environmental challenges require teachers to continuously adapt their instructional approaches and contextualise learning within local realities (Mogren et al., 2019).

In African educational contexts, the need for reflective professional development is especially pronounced. Teachers frequently operate within resource-constrained environments characterised by large class sizes, limited instructional materials, and competing curriculum demands. These contextual factors often limit opportunities for sustained professional learning and experimentation with innovative instructional strategies. Reflective professional learning models that incorporate collaborative professional learning communities, peer mentoring, and locally relevant instructional resources offer promising approaches for supporting teacher capacity development in such contexts (Ochieng & Adebayo, 2023). By enabling teachers to analyse classroom experiences, adapt pedagogical strategies, and integrate locally meaningful climate examples into instruction, reflective practice strengthens both teacher competence and the contextual relevance of Climate Change Education. Evidence from Ghana further shows that access to quality instructional resources and supportive learning environments significantly shapes teachers' capacity to implement effective science instruction (Asano et al., 2019).

This chapter, therefore, positions reflective professional development as a central mechanism for strengthening teacher competence and instructional effectiveness in Climate Change Education. It synthesises theoretical perspectives and empirical research on teacher professional learning, reflective practice frameworks, pedagogical integration strategies, and the contextual challenges affecting CCE implementation, with particular emphasis on the development of educational systems. Through this synthesis, the chapter proposes a conceptual model illustrating how reflective professional development processes can enhance pedagogical transformation, improve climate literacy instruction, and promote student engagement in sustainability action. By linking reflective practice theory with climate education scholarship, the chapter contributes to the growing body of literature examining how teacher professional learning can support educational responses to global sustainability challenges.

Reflective professional development also aligns with the principles of Participatory Action Research (PAR), which emphasises collaborative inquiry, critical reflection, and collective problem-solving among practitioners. PAR positions teachers not merely as implementers of curriculum but as co-researchers who systematically examine their practice to improve educational outcomes (Orlando Fals-Borda, 1987; Kemmis & McTaggart, 2005). Within the context of Climate Change Education, participatory approaches enable teachers to explore pedagogical strategies collaboratively, reflect on classroom experiences, and adapt instructional practices to address emerging environmental challenges. Integrating PAR principles into reflective professional development, therefore, supports continuous teacher learning and contextually responsive climate education practices.

CONCEPTUAL FOUNDATIONS OF CLIMATE CHANGE EDUCATION (CCE)

Climate Change Education (CCE) has emerged as a central pillar of education for sustainable development, reflecting growing global recognition that addressing climate change requires not only scientific and technological innovation but also informed and engaged citizens capable of making responsible environmental decisions. CCE seeks to equip learners with the knowledge, competencies, attitudes, and values necessary to understand climate systems, evaluate environmental risks, and participate actively in mitigation and adaptation efforts (Mochizuki & Bryan, 2015). As climate change increasingly affects ecological systems, economic development, and social stability, education systems are being called upon to integrate climate literacy across disciplines and levels of education.

At its core, Climate Change Education extends beyond the transmission of scientific information to include the development of higher-order thinking skills, ethical reasoning, and action-oriented learning competencies. Contemporary educational frameworks conceptualise CCE as a multidimensional construct that integrates knowledge of climate science, environmental literacy, sustainability values, and participatory competencies required for collective climate action (Stevenson et al., 2017). These dimensions align closely with broader sustainability education agendas, which emphasise the development of learners capable of analysing complex socio-environmental systems, evaluating trade-offs, and contributing to sustainable development initiatives at local and global levels (UNESCO, 2021).

One foundational component of CCE is climate science literacy, which involves understanding the scientific principles underlying climate systems, including greenhouse gas emissions, atmospheric processes, global warming trends, and the environmental consequences of climate variability. Scientific literacy provides the cognitive foundation necessary for interpreting climate data, evaluating environmental policies, and understanding the causes and consequences of climate change (IPCC, 2022). Without a solid grounding in scientific understanding, learners may struggle to evaluate climate-related information critically or to engage effectively in climate-related decision-making.

A second essential dimension of CCE is environmental literacy, which encompasses awareness of ecological systems, understanding of human–environment interactions, and the capacity to make environmentally responsible decisions. Environmental literacy goes beyond factual knowledge to include affective and behavioural components, encouraging learners to develop attitudes and values supportive of sustainability practices (Fien, 2018). Educational programmes that integrate environmental literacy encourage students to examine local environmental challenges, explore community-based sustainability solutions, and recognise the interconnectedness of ecological, economic, and social systems.

A third dimension is sustainability competence, which refers to the skills and capabilities required to address complex environmental problems through collaborative, interdisciplinary approaches. Sustainability competence includes systems thinking, problem-solving, anticipatory thinking, and the ability to participate in collective decision-making processes aimed at promoting sustainable development (Rieckmann, 2018). These competencies enable learners to move beyond theoretical understanding toward practical engagement with sustainability challenges, supporting the transition from knowledge acquisition to behavioural action.

An additional defining characteristic of contemporary CCE frameworks is their emphasis on action-oriented learning. Educational approaches that incorporate experiential projects, community engagement initiatives, and participatory environmental activities enable learners to apply knowledge in real-world contexts, strengthening both understanding and motivation for environmental stewardship (Rousell & Cutter-Mackenzie-Knowles, 2020). Action-oriented learning also reinforces the civic dimension of climate education by encouraging learners to participate in community-level sustainability initiatives such as conservation projects, climate advocacy campaigns, or local environmental monitoring activities.

Pedagogically, Climate Change Education is commonly associated with learner-centred instructional approaches, including inquiry-based and project-based learning, as well as interdisciplinary teaching methods. Inquiry-based learning encourages learners to investigate environmental questions, collect and analyse data, and develop evidence-based explanations, thereby strengthening scientific reasoning and critical thinking skills (Stevenson et al., 2017). Project-based learning engages students in sustained problem-solving activities related to real-world climate issues, fostering collaboration, creativity, and applied learning. Interdisciplinary approaches enable the integration of climate topics across subject areas, demonstrating the interconnected nature of environmental challenges and promoting a holistic understanding of sustainability issues. Integrating digital tools can also strengthen climate change education by enabling teachers to access updated scientific data, simulations, and multimedia learning resources. Studies in Ghanaian secondary schools show that ICT-supported instruction improves student engagement and conceptual understanding when teachers are adequately trained to integrate technology into their pedagogy (Awoniyi et al., 2025).

The effectiveness of Climate Change Education is influenced by contextual relevance. Educational programmes that incorporate local environmental knowledge, indigenous ecological practices, and community-based sustainability initiatives are more likely to engage learners and support meaningful climate action (Mogren et al., 2019). In African contexts, for example, integrating local agricultural practices, water resource management strategies, and indigenous environmental knowledge systems into classroom instruction enhances the relevance of climate education. It strengthens learners' capacity to connect global climate issues with local environmental realities. Contextually responsive CCE approaches also promote community engagement and support the co-production of knowledge between schools and local communities.

Despite the growing recognition of CCE as a critical educational priority, the implementation of climate education across many education systems remains uneven. Challenges such as limited curriculum integration, insufficient teacher preparation, and inadequate instructional resources often constrain effective classroom implementation (Evans et al., 2021). Empirical studies from several African countries highlight the challenges associated with implementing Climate Change Education. For example, studies in South Africa indicate that although climate change is incorporated into aspects of the national curriculum, many teachers report insufficient training and lack confidence in teaching climate-related topics (Lotz-Sisitka et al., 2015). Similarly, research conducted in Ghana suggests that while climate change concepts are embedded within science and social studies curricula, teachers often lack access to appropriate instructional resources and professional development opportunities needed to effectively implement these topics in classrooms (Yaw Osei, 2017). Across East Africa, studies in Kenya and Uganda have also highlighted the limited integration of climate education into teacher training programmes, which constrains educators' ability to facilitate experiential, action-oriented climate learning (Daniel Ochieng, 2019). Addressing these challenges requires coordinated efforts involving curriculum reform, teacher professional development, institutional support, and policy frameworks that prioritise sustainability education. Teachers play a particularly important role in this process, as their knowledge, pedagogical competence, and instructional practices determine how climate education is translated into classroom learning experiences.

The conceptual foundations of Climate Change Education, therefore, highlight the integration of scientific understanding, environmental literacy, sustainability competencies, and action-oriented learning as mutually reinforcing dimensions of effective climate instruction. Advancing these dimensions requires education systems to strengthen teacher competence and instructional capacity, since the translation of conceptual frameworks into classroom practice depends largely on teachers' preparedness, pedagogical choices, and access to sustained professional learning opportunities.

METHODS

Research Design

This chapter employed a systematic literature review methodology to synthesise existing research on Climate Change Education (CCE), teacher professional development, and reflective practice in teacher learning. Systematic literature reviews provide a structured and transparent approach for identifying, evaluating, and synthesising relevant research evidence within a particular field (Hannah Snyder, 2019). Unlike traditional narrative reviews, systematic reviews follow clearly defined procedures for literature identification, screening, and analysis in order to minimise bias and enhance the reliability of findings.

The review also followed general principles of systematic evidence synthesis recommended by David Moher et al. (2009) in the PRISMA (Preferred Reporting Items for Systematic Reviews and Meta-Analyses) guidelines. These guidelines provide methodological standards for documenting the search process, screening criteria, and selection of studies included in systematic reviews.

The systematic review approach was considered appropriate because the objective of this chapter is to synthesise existing scholarship and develop a conceptual framework explaining how reflective professional development can strengthen teacher competence in Climate Change Education. In synthesising the literature, particular attention was given to studies that emphasised reflective practice, collaborative inquiry, and participatory professional learning approaches. These perspectives align with participatory action research traditions that position teachers as reflective practitioners engaged in continuous cycles of inquiry and improvement.

Literature Search Strategy

A comprehensive literature search was conducted across major academic databases to identify peer-reviewed studies related to Climate Change Education, teacher professional development, and reflective practice in teaching. The primary databases used for the search included:

- Scopus
- Web of Science
- ERIC (Education Resources Information Centre)
- Google Scholar

These databases were selected because they provide extensive coverage of research in education, sustainability education, and teacher professional learning.

The search process involved combinations of the following keywords and search strings:

- "climate change education" and "teacher professional development"
- "reflective practice" and "teacher learning"
- "sustainability education" and "teacher competence"
- "environmental education" and "teacher training"
- "reflective professional development" and "teaching"

Boolean operators such as AND and OR were used to broaden or refine search results. The search focused primarily on publications from 2000 to 2024 to capture contemporary research on climate education and teacher professional learning, while also incorporating foundational theoretical works on reflective practice.

Inclusion and Exclusion Criteria

To ensure the relevance and quality of the literature included in the review, specific inclusion and exclusion criteria were applied.

Inclusion Criteria

Studies were included if they:

1. Focused on Climate Change Education, environmental education, or sustainability education.
2. Examined teacher professional development, teacher competence, or teacher learning.
3. Addressed reflective practice or experiential learning in teacher education.
4. Were published in peer-reviewed journals, academic books, or reputable institutional reports.
5. Were written in English.

Exclusion Criteria

Studies were excluded if they:

- Focused exclusively on climate science without educational implications.
- Did not address teachers or teacher education.
- Were opinion articles without scholarly evidence.
- Were non-academic publications lacking peer review?

These criteria ensured that the review focused specifically on research relevant to teacher development and the implementation of Climate Change Education.

Study Screening and Selection

The study selection process followed the PRISMA screening procedure. Initially, the database search yielded a broad pool of publications on climate education and teacher professional learning. Duplicate records were removed, after which titles and abstracts were screened to determine relevance to the research focus.

Articles that met the initial screening criteria were subjected to full-text review to determine their relevance to the themes of Climate Change Education, reflective practice, and teacher professional development. Studies that did not sufficiently address these themes were excluded from the final review.

This systematic screening process ensured that the literature included in the analysis represented high-quality and relevant research on the intersection of climate education and teacher professional learning.

Data Analysis and Synthesis

The selected studies were analysed using thematic synthesis, which involves identifying recurring concepts and patterns across studies in order to generate broader insights and conceptual understanding (Braun & Clarke, 2006). Through iterative reading and coding of the literature, several key themes were identified:

1. Conceptual Foundations of Climate Change Education
2. Teacher competence and preparedness for climate education
3. Professional development approaches for sustainability education
4. Reflective practice frameworks in teacher learning
5. Institutional and contextual factors influencing professional development

These themes guided the synthesis of the literature and informed the organisation of the chapter.

Development of the Conceptual Framework

Following the thematic synthesis, insights from the reviewed studies were integrated to develop a Reflective Professional Development Model for Climate Change Education. Conceptual model development is a common outcome of systematic literature synthesis when the objective is to integrate existing research findings into a coherent explanatory framework (Snyder, 2019).

The proposed model integrates theoretical perspectives from:

- reflective practice theory (Schön, 1983),
- experiential learning theory (Kolb, 1984),
- teacher professional development research (Darling-Hammond et al., 2017),
- and climate change education scholarship (Stevenson et al., 2017).

By synthesising these perspectives, the model conceptualises teacher learning as a cyclical process linking professional development inputs, reflective learning processes, pedagogical transformation, and student climate literacy outcomes.

Limitations of the Review

Although systematic literature reviews provide comprehensive insights into existing scholarship, certain limitations should be acknowledged. First, the review relies on previously published studies rather than primary empirical data. Second, the selection of studies may be influenced by the availability of research within the chosen databases and language restrictions. Future empirical studies are required to test and validate the conceptual framework proposed in this chapter across different educational contexts.

TEACHER COMPETENCE IN CLIMATE CHANGE EDUCATION

Teachers play a pivotal role in determining the effectiveness of Climate Change Education (CCE) because they serve as the primary mediators between curriculum policy and classroom practice. Their competence influences how climate concepts are presented, how learners engage with sustainability issues, and how educational programmes translate into meaningful behavioural outcomes. Teacher competence in CCE extends beyond mastery of climate science content to include pedagogical knowledge, environmental attitudes, and the professional capacity to design contextually relevant learning experiences that promote climate literacy and action-oriented learning (Evans et al., 2021).

One of the central dimensions of teacher competence in Climate Change Education is content knowledge. Teachers require a strong conceptual understanding of climate systems, including the causes and consequences of climate change, mitigation and adaptation strategies, and the socio-economic dimensions of environmental sustainability (IPCC, 2022). Adequate content knowledge enables teachers to explain complex climate processes accurately, address misconceptions, and support learners in interpreting scientific evidence. However, research across many educational contexts indicates that teachers often report gaps in climate-related knowledge, particularly where climate education has only recently been introduced into national curricula (Boateng, 2021). These knowledge gaps can limit teachers' confidence and willingness to integrate climate topics into classroom instruction.

In addition to subject-matter expertise, pedagogical competence is essential for effective CCE implementation. Climate change is an interdisciplinary topic that requires teaching approaches that integrate scientific, social, ethical, and economic perspectives. Teachers must therefore possess pedagogical strategies that promote inquiry-based learning, collaborative problem solving, project-based activities, and experiential learning opportunities that engage students in analysing real-world environmental challenges (Stevenson et al., 2017). Pedagogical competence also involves the ability to scaffold complex concepts, facilitate classroom discussions on potentially controversial environmental issues, and support students in connecting global climate challenges with local community contexts. Studies involving preservice teachers also indicate that instructional self-efficacy and learning dispositions are closely associated with the development of instructional competence (Amponsah et al., 2021).

A third critical dimension of teacher competence is attitudinal readiness, which includes teachers' beliefs, perceptions, and motivation regarding climate education. Teachers who perceive climate education as relevant to learners' lives and future societal needs are more likely to integrate sustainability topics into their teaching and adopt innovative instructional practices (Ochieng & Adebayo, 2023). Conversely, teachers who feel uncertain about their knowledge or perceive climate change as politically sensitive may avoid engaging deeply with the topic. Positive teacher attitudes toward environmental sustainability are, therefore, an important determinant of instructional commitment and classroom innovation.

Teacher competence in Climate Change Education is also shaped by professional learning capacity, which refers to the ability to engage in continuous professional development, reflective practice, and collaborative learning. Reflective teachers who regularly evaluate their instructional strategies, analyse student responses, and experiment with new pedagogical approaches are more likely to adapt successfully to emerging curriculum priorities such as sustainability education (Farrell, 2018). Professional learning communities, peer mentoring systems, and collaborative lesson-planning initiatives provide important platforms for teachers to share experiences, exchange instructional resources, and strengthen collective expertise in climate education.

In many developing educational contexts, teacher competence in CCE is influenced by structural and institutional factors, including access to professional development opportunities, availability of teaching and learning materials, curriculum support, and leadership commitment to sustainability education (Evans et al., 2021). Where education systems provide limited training or insufficient instructional resources, teachers may struggle to translate climate education policies into classroom practice. Conversely, supportive institutional environments that prioritise teacher development, provide access to instructional materials, and encourage collaborative professional learning can significantly enhance teacher capacity to implement climate education effectively.

Another important aspect of teacher competence involves the ability to contextualise climate education within local environmental realities. Teachers who integrate locally relevant examples—such as community-level environmental challenges, local agricultural practices, water resource issues, or indigenous ecological knowledge—help learners connect theoretical climate concepts to everyday experiences (Mogren et al., 2019). Contextualised instruction enhances student engagement, strengthens environmental awareness, and promotes the development of practical sustainability skills that are directly applicable to local contexts.

Furthermore, teacher competence in Climate Change Education includes the ability to assess student learning that captures not only cognitive understanding but also attitudes, values, and action-oriented competencies. Traditional assessment approaches that focus solely on factual knowledge may not adequately capture the broader educational objectives of sustainability education. Teachers, therefore, require assessment strategies that evaluate critical thinking, collaborative problem-solving, and participation in environmental projects, enabling a more comprehensive understanding of student learning outcomes.

Strengthening teacher competence in Climate Change Education requires coordinated efforts involving teacher education institutions, professional development systems, and policy frameworks that prioritise sustainability education. Preservice teacher preparation programmes should incorporate climate education content, sustainability pedagogy, and experiential learning opportunities to ensure that new teachers enter the profession with foundational climate literacy competencies. In-service professional development programmes should provide ongoing training, mentoring, and reflective learning opportunities that enable teachers to update their knowledge, refine pedagogical approaches, and adapt instruction to evolving climate education priorities (Darling-Hammond et al., 2017).

Teacher competence in Climate Change Education should thus be understood as a multidimensional construct encompassing content knowledge, pedagogical expertise, attitudinal readiness, contextual responsiveness, and reflective professional learning capacity. Strengthening these interconnected competencies through sustained professional development and supportive institutional environments remains essential for improving instructional effectiveness and advancing climate literacy outcomes across educational systems.

PROFESSIONAL DEVELOPMENT FOR CLIMATE CHANGE EDUCATION

Professional development (PD) plays a central role in strengthening teachers' capacity to implement Climate Change Education (CCE) effectively. As climate change continues to emerge as a priority within global and national education agendas, teachers require sustained opportunities to update their knowledge, develop pedagogical skills, and adapt instructional practices to evolving sustainability learning goals. Professional development initiatives provide the structured learning environments through which teachers acquire new competencies, experiment with innovative instructional strategies, and engage in collaborative professional learning processes that support long-term instructional improvement (Darling-Hammond et al., 2017).

A key objective of professional development in Climate Change Education is to strengthen teachers' knowledge of climate content. Because climate science evolves rapidly as new research findings emerge, teachers require continuous access to updated scientific information, climate data, and interdisciplinary knowledge linking environmental science with social, economic, and policy perspectives (IPCC, 2022). Professional development programmes that provide access to climate experts, research-informed instructional materials, and evidence-based climate education resources help teachers build the conceptual understanding necessary to teach climate-related topics confidently and accurately. Strengthening content knowledge is particularly important in contexts where teachers' initial training did not include specialised climate education components.

Beyond content knowledge, professional development must also focus on pedagogical innovation. Climate Change Education requires instructional approaches that promote inquiry, experiential learning, project-based activities, and collaborative problem solving. These learner-centred pedagogical strategies encourage students to investigate real-world environmental challenges, develop evidence-based solutions, and engage in sustainability actions within their communities (Stevenson et al., 2017). Professional development programmes that model these instructional strategies through workshops, demonstration lessons, and classroom-based experimentation provide teachers with practical opportunities to translate pedagogical theory into classroom practice. Empirical evidence from Ghanaian higher education contexts further demonstrates that blended learning environments can enhance instructional innovation and professional learning engagement among educators (Kyei-Akuoko et al., 2020).

Another essential dimension of professional development for CCE is the promotion of reflective professional learning. Research consistently indicates that professional development programmes produce more sustainable instructional change when teachers are provided with opportunities to reflect on their classroom experiences, analyse instructional outcomes, and refine teaching practices through ongoing feedback processes (Farrell, 2018). Reflective professional learning encourages teachers to move beyond passive participation in training sessions toward active engagement with instructional improvement processes. Activities such as reflective journaling, peer observation, collaborative lesson planning, and professional learning communities enable teachers to transform professional development experiences into sustained pedagogical transformation.

Their duration and continuity also influence the effectiveness of professional development initiatives. Short-term workshops or isolated training sessions often have limited long-term impact because they do not provide teachers with sufficient opportunities to apply new knowledge, receive feedback, or engage in continuous learning cycles (Darling-Hammond et al., 2017). Sustained professional development models that include mentoring, follow-up sessions, collaborative inquiry, and classroom-embedded learning experiences are more likely to produce meaningful instructional change. Long-term professional learning programmes allow teachers to experiment with instructional innovations, monitor student learning outcomes, and refine teaching strategies over time.

In many educational systems, particularly in developing regions, professional development for Climate Change Education faces structural challenges, including limited funding, insufficient training opportunities, heavy teacher workloads, and inadequate institutional support. These constraints often restrict teachers' access to sustained learning opportunities and limit their ability to implement innovative instructional approaches (Evans et al., 2021). Addressing these challenges requires coordinated policy interventions that prioritise teacher professional development within national sustainability education strategies, allocate resources for training programmes, and establish institutional mechanisms for ongoing teacher support.

Collaborative professional development models have gained increasing recognition as effective approaches for strengthening teacher capacity in climate education. Professional learning communities (PLCs), peer mentoring systems, and collaborative lesson study initiatives enable teachers to share instructional experiences, exchange teaching resources, and collectively address classroom challenges related to climate education implementation (Šarić & Šteh, 2017). Collaborative learning environments also foster professional motivation and reduce the isolation that teachers may experience when introducing new curricular priorities such as sustainability education. Through collaborative professional learning, teachers develop shared expertise and collective responsibility for improving instructional quality. School-based professional development provides opportunities for continuous learning within teachers' immediate work contexts and enables collaborative reflection on instructional practices (Adjei-Boateng & Cobbinah, 2021).

Another important component of effective professional development for CCE is the integration of contextually relevant instructional resources. Training programmes that incorporate local environmental data, community-based sustainability initiatives, indigenous ecological knowledge, and locally relevant case studies enhance the practical relevance of climate education and support teachers in connecting global climate concepts with local environmental realities (Mogren et al., 2019). Contextually responsive professional development strengthens teachers' ability to design learning experiences that resonate with students' lived experiences and encourage meaningful community engagement.

Digital technologies are also increasingly shaping professional development opportunities in Climate Change Education. Online training platforms, virtual professional learning communities, and digital instructional resources provide teachers with access to global knowledge networks and innovative teaching tools that support the implementation of climate education (UNESCO, 2021). Technology-enhanced professional development expands access to training opportunities, particularly for teachers in remote or underserved regions, and facilitates continuous professional learning beyond traditional workshop-based models.

Professional development initiatives that combine sustained learning opportunities, reflective engagement, collaborative inquiry, and contextually relevant instructional support create the conditions necessary for meaningful instructional transformation. Such integrated professional learning systems enable teachers to translate climate education policies into classroom practice while continuously refining pedagogical approaches in response to emerging sustainability learning needs. Effective professional development initiatives also depend on supportive school leadership and supervisory structures. Research in Ghanaian basic schools shows that school improvement support officers and headteachers play a critical role in guiding teachers' instructional practices and facilitating ongoing professional learning within schools (Adjei-Boateng, Cobbinah, & Adamba, 2025).

REFLECTIVE PRACTICE FRAMEWORKS IN TEACHER PROFESSIONAL LEARNING

Reflective practice has become a central concept in teacher professional learning, particularly within contemporary educational reforms that emphasise continuous improvement, learner-centred pedagogy, and adaptive instructional strategies. In the context of Climate Change Education (CCE), reflective practice is especially important because teachers must frequently interpret evolving scientific information, address complex socio-environmental issues, and adapt instructional approaches to diverse classroom and community contexts. Reflective practice has been widely recognised as a powerful mechanism for professional learning because it allows teachers to examine their instructional experiences and refine pedagogical strategies critically. Evidence from Ghana indicates that structured reflective activities such as journal writing enable teachers to analyse classroom practices, identify instructional challenges, and develop more responsive teaching approaches (Adjei-Boateng, 2025). Reflective professional learning provides teachers with structured processes for analysing their instructional experiences, identifying areas for improvement, and developing more effective pedagogical strategies over time (Farrell, 2018).

The theoretical foundations of reflective practice can be traced to the work of Schön (1983), who conceptualised professional learning as a dynamic process involving reflection-in-action and reflection-on-action. Reflection-in-action refers to teachers' ability to make immediate instructional adjustments during teaching based on student responses, classroom dynamics, and emerging learning needs. Reflection-on-action involves retrospective analysis of teaching experiences, enabling educators to evaluate instructional effectiveness, identify strengths and limitations, and plan improvements for future practice. These reflective processes support the development of professional judgement and adaptive teaching skills essential for addressing complex instructional challenges such as those encountered in Climate Change Education.

Another influential framework for understanding reflective professional learning is Kolb's experiential learning cycle, which conceptualises learning as a continuous process involving four interconnected stages: concrete experience, reflective observation, abstract conceptualisation, and active experimentation (Kolb, 1984). Within this framework, teachers first engage in instructional experiences, reflect on their outcomes, develop theoretical insights from those reflections, and subsequently test revised strategies in classroom practice. The cyclical nature of experiential learning highlights the importance of integrating professional development experiences with classroom practice, ensuring that reflection leads to practical pedagogical innovation rather than remaining a purely theoretical exercise.

Gibbs' reflective cycle provides another structured model widely used in teacher professional development. This model guides reflection through stages that include describing the experience, exploring feelings, evaluating outcomes, analysing underlying factors, drawing conclusions about lessons learned, and planning actions for future improvement. Structured reflection models such as Gibbs' cycle help teachers move beyond descriptive reflection toward critical analysis and intentional instructional change. In Climate Change Education contexts, such frameworks enable teachers to analyse how students respond to inquiry-based climate lessons, evaluate the effectiveness of project-based activities, and refine strategies for promoting student engagement in sustainability initiatives.

The ALACT reflective model (Action, Looking Back, Awareness, Creating Alternatives, Trial) offers another systematic approach to reflective professional learning, emphasising iterative cycles of professional action and reflection. Teachers implement instructional strategies, review classroom outcomes, identify key insights, generate alternative approaches, and test revised instructional practices in subsequent teaching episodes. This structured reflective process encourages continuous professional growth and supports the integration of innovative pedagogical practices within everyday teaching routines (Korthagen, 2001). In the context of Climate Change Education, the ALACT model is particularly useful for helping teachers refine inquiry-based activities, community engagement projects, and experiential learning strategies.

Reflective practice frameworks are most effective when embedded within collaborative professional learning environments. Professional learning communities (PLCs), peer mentoring systems, and lesson study groups provide opportunities for teachers to share reflective insights, discuss instructional challenges, and collaboratively develop improved teaching strategies (Šarić & Šteh, 2017). Collaborative reflection deepens professional learning by exposing teachers to diverse perspectives, fostering critical dialogue, and supporting collective problem-solving. Such collaborative structures are particularly valuable in implementing interdisciplinary subjects such as Climate Change Education, where teachers may need to integrate content from multiple subject areas and adapt instruction to local environmental contexts. Research conducted in Ghana also indicates that teachers' conceptions of inquiry significantly influence the extent to which inquiry-based instructional approaches are implemented in science classrooms (Mohammed & Amponsah, 2021).

In addition to improving instructional practices, reflective practice frameworks foster teachers' professional identity and self-efficacy. Teachers who engage in regular reflection become more aware of their instructional beliefs, pedagogical assumptions, and decision-making processes, enabling them to align teaching practices more closely with educational goals related to sustainability and environmental responsibility. Reflective professional learning also supports teacher motivation and commitment to instructional innovation by fostering a sense of professional autonomy and continuous growth.

Despite the recognised benefits of reflective practice, implementing structured reflective professional learning within educational systems can present challenges. Time constraints, heavy teaching workloads, limited institutional support, and insufficient training in reflective methodologies may reduce teachers' opportunities to engage in meaningful reflection. Addressing these challenges requires education systems to institutionalise reflective professional learning by allocating dedicated time for collaborative reflection, incorporating reflective activities into professional development programmes, and providing mentoring support that guides teachers in applying reflective frameworks effectively.

When embedded within collaborative professional learning environments and supported by institutional policies that prioritise teacher learning, reflective practice frameworks provide powerful mechanisms for continuous instructional improvement. Their structured application enables teachers to adapt pedagogical approaches, integrate interdisciplinary climate content effectively, and sustain long-term professional growth.

INTEGRATING REFLECTION INTO PROFESSIONAL DEVELOPMENT MODELS FOR CLIMATE CHANGE EDUCATION

Integrating reflective practice into professional development (PD) models is a critical strategy to ensure that Climate Change Education (CCE) initiatives lead to sustained instructional transformation rather than temporary adjustments. While traditional professional development programmes often focus on delivering new knowledge or instructional techniques, research increasingly demonstrates that meaningful changes in teaching practices occur when teachers are provided with opportunities to apply new learning in classroom contexts, reflect on their experiences, and refine instructional strategies through iterative learning cycles (Darling-Hammond et al., 2017). Embedding reflective processes within PD structures, therefore, strengthens the long-term effectiveness of teacher learning initiatives in climate education.

One effective approach to integrating reflection into PD models is classroom-embedded professional learning, in which teachers implement new instructional strategies in their classrooms and then engage in guided reflection sessions to evaluate outcomes and identify areas for improvement. This approach ensures that professional development is closely aligned with classroom realities, enabling teachers to contextualise climate education strategies within their specific teaching environments. Reflection sessions may include structured discussions, peer feedback, and facilitated analysis of student learning outcomes, allowing teachers to connect theoretical training with practical classroom experiences (Farrell, 2018).

Collaborative professional learning communities (PLCs) provide another powerful mechanism for embedding reflection into professional development programmes. PLCs create structured environments in which teachers meet regularly to share experiences, analyse classroom challenges, and collectively design instructional improvements (Šarić & Šteh, 2017). In the context of Climate Change Education,

PLCs enable teachers from different subject areas to collaborate on developing interdisciplinary climate lessons, exchange locally relevant teaching resources, and discuss strategies to address student misconceptions about environmental issues. Collaborative reflection also promotes collective professional responsibility and strengthens institutional commitment to sustainability education.

Mentoring and coaching systems further enhance the integration of reflection within PD models. Experienced educators or instructional coaches can guide teachers through reflective processes by facilitating lesson observations, providing feedback on instructional practices, and supporting teachers in developing action plans for instructional improvement. Coaching models that incorporate reflective dialogue encourage teachers to critically examine their teaching decisions, identify instructional strengths and challenges, and experiment with alternative pedagogical approaches (Evans et al., 2021). Such mentoring systems are particularly beneficial for teachers who are newly implementing Climate Change Education topics and may require ongoing professional guidance.

Another strategy for integrating reflection into PD models involves using reflective documentation tools, such as teaching portfolios, reflective journals, and digital learning logs. These tools encourage teachers to systematically record instructional experiences, analyse student responses, and document professional learning progress over time. Reflective documentation supports deeper professional learning by enabling teachers to track instructional changes, evaluate the effectiveness of new teaching strategies, and identify emerging professional development needs. Digital platforms can further support reflective documentation by allowing teachers to share reflections within online professional learning communities, thereby expanding opportunities for collaborative feedback and peer learning (UNESCO, 2021). Collaborative instructional approaches have also been shown to facilitate conceptual change and deepen learners' scientific understanding when supported by reflective teaching practices.

Action research approaches also provide an effective framework for integrating reflection into professional development. Through classroom-based action research, teachers identify instructional challenges, design interventions, collect evidence on student learning outcomes, and reflect on the effectiveness of implemented strategies. Action research not only strengthens teachers' research literacy but also fosters a culture of inquiry-driven professional learning that aligns closely with the goals of Climate Change Education. Teachers who engage in action research develop stronger analytical skills, increased instructional confidence, and a deeper understanding of how pedagogical innovations influence student learning (Mogren et al., 2019).

Institutional leadership plays a significant role in supporting the integration of reflective practice within professional development systems. School leaders who allocate time for collaborative reflection, provide resources for professional learning communities, and encourage teacher-led inquiry initiatives create organisational environments that sustain reflective professional learning. Institutional policies that formally recognise reflective activities as part of teachers' professional responsibilities further strengthen the integration of reflection into professional development programmes and promote long-term instructional innovation (Ochieng & Adebayo, 2023).

The integration of reflective practice into PD models is particularly important for Climate Change Education because climate-related topics often require teachers to address interdisciplinary content, evolving scientific knowledge, and locally specific environmental challenges. Reflective professional learning enables teachers to continuously adapt instructional approaches, incorporate new climate information into lessons, and respond effectively to students' emerging questions and concerns about environmental issues. By supporting continuous pedagogical refinement, reflective PD models enhance the quality and relevance of climate education instruction.

Designing professional development programmes around reflective learning cycles, therefore, ensures that teacher learning extends beyond participation in training to sustained classroom innovation. By linking professional learning experiences directly to instructional experimentation and reflective evaluation, education systems can strengthen the long-term effectiveness of Climate Change Education initiatives.

CHALLENGES AND ENABLING CONDITIONS FOR REFLECTIVE PROFESSIONAL DEVELOPMENT

The successful implementation of reflective professional development (PD) for Climate Change Education (CCE) depends on a complex interaction of institutional, structural, pedagogical, and contextual factors. While reflective professional learning offers significant potential for improving teacher competence and instructional quality, several challenges continue to limit its widespread adoption in many educational systems. Understanding these challenges, alongside the enabling conditions that support effective implementation, is essential for designing sustainable teacher development programmes that strengthen climate education practices.

One of the most common challenges affecting reflective professional development is limited institutional support. In many education systems, teacher professional development programmes are often structured around short-term workshops that prioritise information delivery rather than sustained reflective learning. Without institutional mechanisms that allocate time for collaborative reflection, mentoring, and classroom-based experimentation, teachers may struggle to engage in meaningful reflective practice (Darling-Hammond et al., 2017). Heavy teaching workloads, administrative responsibilities, and limited scheduling flexibility further constrain teachers' opportunities to engage in reflective professional learning.

Another major challenge relates to resource constraints, particularly in developing educational contexts. Limited access to teaching materials, climate-related instructional resources, and digital learning tools can reduce teachers' ability to implement innovative climate education strategies introduced during professional development programmes (Evans et al., 2021). In schools where basic instructional resources are insufficient, teachers may prioritise examination-oriented teaching practices over inquiry-based or experiential learning approaches that require additional materials and preparation time. These resource limitations can also affect the sustainability of professional development initiatives, particularly when funding for long-term training programmes is inconsistent.

Teacher preparedness and confidence also influence the effectiveness of reflective professional development. Teachers who lack foundational knowledge of climate science or feel uncertain about interdisciplinary teaching approaches may find it difficult to engage fully in reflective professional learning. In some cases, teachers may perceive reflective activities such as journaling, peer observation, or collaborative inquiry as additional workload requirements rather than opportunities for professional growth, particularly when institutional support is weak. Addressing these challenges requires professional development programmes that combine content knowledge strengthening with structured reflective learning opportunities, ensuring that teachers develop both conceptual understanding and reflective professional learning skills.

Despite these challenges, several enabling conditions can significantly enhance the effectiveness of reflective professional development initiatives. One critical enabling factor is supportive leadership at both school and policy levels. Educational leaders who prioritise teacher learning, allocate time for collaborative professional development, and provide resources for reflective activities create organisational environments that sustain reflective practice. Leadership commitment also signals the importance of climate education within institutional priorities, encouraging teachers to invest time and effort in developing new instructional competencies (Ochieng & Adebayo, 2023).

Collaborative professional learning structures represent another important enabling condition. Professional learning communities (PLCs), peer mentoring systems, and lesson study groups provide structured platforms for teachers to engage in shared reflection, discuss instructional challenges, and exchange innovative teaching strategies (Šarić & Šteh, 2017). Collaborative learning environments enhance professional motivation, reduce teacher isolation, and support the development of collective expertise in Climate Change Education. These collaborative structures are particularly effective when participation is sustained over time and linked directly to classroom practice.

The availability of contextually relevant instructional resources also plays a significant role in enabling reflective professional development. Training programmes that provide teachers with locally relevant climate data, community-based case studies, and adaptable teaching materials enable educators to design meaningful learning experiences that connect global climate issues with local environmental realities (Mogren et al., 2019). Access to such resources strengthens teachers' confidence in implementing climate education lessons and supports the practical application of professional learning outcomes.

Policy support and curriculum alignment further enhance the sustainability of reflective professional development initiatives. When national education policies explicitly incorporate Climate Change Education into curriculum frameworks and teacher professional standards, professional development programmes gain institutional legitimacy and long-term strategic direction. Curriculum alignment ensures that teachers' professional learning activities are directly linked to instructional expectations, assessment frameworks, and national sustainability education priorities (UNESCO, 2021).

Digital technologies are increasingly emerging as an additional enabling condition for reflective professional development. Online professional learning platforms, virtual communities of practice, and digital collaboration tools enable teachers to participate in reflective learning activities beyond physical training sessions, expanding access to professional learning opportunities and facilitating continuous peer interaction. Technology-enhanced reflective learning environments are particularly valuable for teachers in geographically remote areas who may have limited access to face-to-face professional development programmes.

Addressing structural constraints while strengthening enabling conditions, such as leadership support, collaborative learning structures, policy alignment, and resource provision, is fundamental to sustaining reflective professional development systems. Education systems that successfully coordinate these factors are more likely to achieve sustained improvements in teacher competence and in the implementation of climate education.

Integrative Evidence Synthesis

Emerging evidence from Ghanaian science education research further reinforces the importance of reflective professional development, instructional resources, and inquiry-oriented pedagogical practices in strengthening teaching effectiveness and learner outcomes. Studies have shown that the availability and effective utilisation of quality teaching and learning resources significantly enhance instructional delivery and student engagement in science classrooms (Asano et al., 2019). Research involving preservice science teachers also indicates that instructional self-efficacy, emotional intelligence, and learning dispositions play an important role in shaping and developing teacher competence and instructional readiness (Amponsah et al., 2021). Similarly, investigations into blended learning environments demonstrate that technology-supported instructional systems can promote instructional innovation and support continuous professional learning among educators (Kyei-Akuoko et al., 2020). Evidence further suggests that teachers' conceptions of inquiry influence the extent to which inquiry-based science teaching approaches are implemented in classroom practice (Mohammed & Amponsah, 2021), while collaborative instructional strategies grounded in reflective practice have been shown to facilitate conceptual change and deeper learner understanding in science education contexts. Collectively, these findings underscore the critical role of reflective professional learning, collaborative pedagogy, and supportive instructional environments in strengthening teacher capacity for effective Climate Change Education implementation.

Building on this growing body of scholarship, the present chapter synthesises conceptual and empirical insights to advance a reflective professional development framework for strengthening Climate Change Education implementation across diverse educational contexts. By integrating evidence from prior studies on instructional resources, teacher competence, inquiry-oriented pedagogy, and collaborative teaching strategies, the chapter provides a consolidated perspective on how reflective professional learning can support sustained instructional transformation and improved sustainability learning outcomes.

CONCEPTUAL MODEL OF REFLECTIVE PROFESSIONAL DEVELOPMENT FOR CLIMATE CHANGE EDUCATION

The effectiveness of Climate Change Education (CCE) depends largely on the quality of teacher professional learning systems that support continuous instructional improvement. To guide the integration of reflective practice into teacher development initiatives, this chapter proposes a Reflective Professional Development Model for Climate Change Education, which conceptualises teacher learning as a cyclical process linking professional development inputs, reflective inquiry, pedagogical transformation, and student climate learning outcomes (Darling-Hammond et al., 2017; Farrell, 2018). The model provides a conceptual framework for understanding how structured, reflective professional learning can strengthen instructional practices and enhance the implementation of climate education across educational contexts.

To illustrate the interaction between professional learning, reflective inquiry, instructional transformation, and climate learning outcomes, this chapter presents a Reflective Professional Development Model for Climate Change Education. The model conceptualises teacher professional learning as a continuous cyclical process in which professional development inputs support reflective practice, leading to pedagogical innovation and improved student climate literacy outcomes (Kolb, 1984; Schön, 1983). It further emphasises the role of institutional, policy, and resource-related enabling conditions in sustaining long-term professional learning and instructional improvement. Figure 1 illustrates the proposed conceptual framework guiding reflective professional development for Climate Change Education.

Figure 1. Reflective Professional Development Model for Climate Change Education

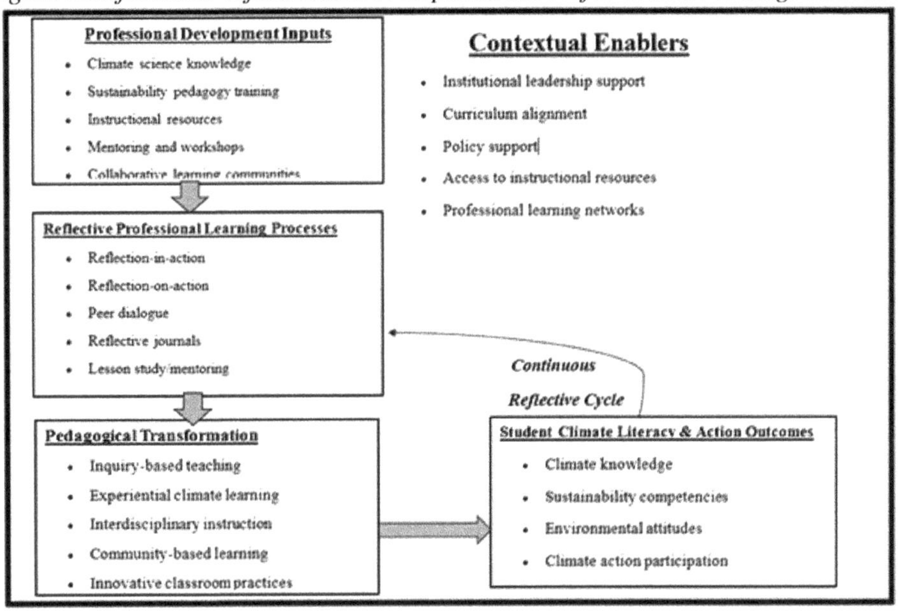

Figure Note. *The model illustrates the cyclical relationship between professional development inputs, reflective professional learning processes, pedagogical transformation, and student climate learning outcomes, supported by contextual enablers such as institutional leadership, curriculum alignment, policy support, and access to instructional resources (Ochieng & Adebayo, 2023). The proposed Reflective Professional Development Model also reflects the cyclical processes associated with participatory action research. Action research frameworks emphasise iterative cycles of planning, action, observation, and reflection that enable practitioners to continuously refine their practice (Kemmis & McTaggart, 2005). Similarly, the model conceptualises teacher learning in Climate Change Education as an ongoing reflective cycle in which professional development experiences inform classroom experimentation, reflective evaluation, and subsequent pedagogical improvement.*

At the first stage of the model are Professional Development Inputs, which include training workshops, mentoring programmes, collaborative learning communities, instructional resource provision, and access to climate science knowledge. These inputs provide teachers with the foundational competencies necessary to implement Climate Change Education, including knowledge of climate content, sustainability pedagogy, interdisciplinary instructional strategies, and assessment approaches that support environmental learning outcomes (Evans et al., 2021).

The second stage focuses on Reflective Professional Learning Processes, in which teachers engage in structured reflective practices to examine instructional experiences, analyse classroom outcomes, and identify areas for pedagogical improvement. Reflective processes may include reflective journaling, peer observation, collaborative lesson study, mentoring dialogues, and professional learning community discussions (Farrell, 2018). Through these reflective activities, teachers transform professional development experiences into professional knowledge, enabling them to refine instructional strategies and adapt teaching approaches to diverse classroom contexts.

The third stage involves Pedagogical Transformation and Classroom Implementation, during which teachers apply insights generated through reflective learning to design and implement improved instructional practices, such as inquiry-based climate investigations, project-based sustainability learning, experiential environmental activities, and interdisciplinary climate lessons (Stevenson et al., 2017). Pedagogical transformation also involves integrating locally relevant environmental examples that connect global climate issues with local community experiences (Mogren et al., 2019).

The fourth stage highlights Student Climate Literacy and Sustainability Outcomes, including improved knowledge of climate science, environmental awareness, critical thinking skills, and participation in sustainability initiatives (UNESCO, 2021). These outcomes align with the broader educational goals of Climate Change Education, which aim to prepare learners to understand environmental challenges and engage in climate action within their communities.

Surrounding the entire reflective professional development cycle are Contextual Enablers, including institutional leadership support, curriculum alignment, access to instructional resources, collaborative professional learning structures, and supportive policy frameworks (Ochieng & Adebayo, 2023). These enabling conditions create organisational environments that sustain reflective professional learning and ensure the long-term effectiveness of Climate Change Education initiatives.

A defining feature of the proposed conceptual model is its cyclical structure, which emphasises continuous professional learning rather than linear training processes. Classroom implementation generates new teaching experiences that feed back into reflective learning processes, leading to ongoing instructional refinement and continuous teacher growth. By reinforcing continuous professional learning cycles, the model highlights the importance of sustained reflective engagement in achieving long-term improvements in climate education practices (Kolb, 1984).

IMPLICATIONS FOR POLICY AND TEACHER EDUCATION

The integration of reflective professional development into Climate Change Education (CCE) systems carries important implications for education policy, teacher education institutions, and school-level professional learning structures. As climate change becomes an increasingly urgent societal concern, education systems must adopt coordinated strategies that ensure teachers are adequately prepared to deliver effective climate education and foster sustainability-oriented learning among students. Aligning policy frameworks, teacher preparation programmes, and professional development systems is therefore essential for strengthening the long-term impact of climate education initiatives.

At the policy level, strengthening Climate Change Education requires coordinated efforts among ministries of education, teacher-training institutions, and schools. Policies that promote continuous professional development in climate and sustainability education enable teachers to develop the pedagogical competencies needed for effective climate instruction (UNESCO, 2017). In the African context, institutional policies should prioritise teacher training, curriculum integration, and the provision of locally relevant teaching resources to support the successful implementation of Climate Change Education across both basic and higher education systems. At the policy level, governments and education authorities should formally incorporate Climate Change Education within national curriculum frameworks, teacher professional standards, and sustainability education policies. Explicit curriculum integration ensures that climate education is not treated as an optional or supplementary topic but as a core component of interdisciplinary learning across subject areas. Policy-level recognition also facilitates the allocation of funding for instructional resources, professional development programmes, and research initiatives that support the implementation of climate education within schools. Clear policy directives provide strategic guidance for teacher education institutions and school systems, enabling coordinated implementation of climate literacy initiatives.

The Reflective Professional Development Model proposed in this chapter has important implications for institutional policy within both schools and higher education systems in Africa. Universities and teacher education institutions play a critical role in preparing teachers to address climate change through interdisciplinary and action-oriented pedagogies. Research suggests that integrating sustainability and climate education into teacher education curricula can strengthen teachers' capacity to address environmental challenges within local contexts (Arjen Wals, 2015). By embedding reflective and participatory learning approaches into teacher training programmes, educational institutions can support educators in developing locally relevant climate teaching practices. Teacher education institutions play a crucial role in preparing educators to implement Climate Change Education effectively. Preservice teacher preparation programmes should integrate climate science literacy, sustainability pedagogy, and interdisciplinary teaching approaches into teacher education curricula. Embedding climate education within subject-methods courses, practicum experiences, and curriculum design modules ensures that new teachers enter the profession with the foundational competencies necessary to deliver climate-related instruction. Preservice programmes should also incorporate reflective professional learning activities, such as reflective teaching portfolios, supervised teaching reflections, and collaborative lesson planning exercises, to strengthen future teachers' reflective practice skills. The findings suggest that schools and teacher education institutions could adopt participatory action research approaches to support reflective professional development in Climate Change Education. Through collaborative inquiry, lesson experimentation, and reflective dialogue, teachers can collectively refine pedagogical strategies that foster climate literacy and sustainability awareness among students.

In-service teacher professional development systems must also be strengthened to support continuous teacher learning in climate education. Education authorities should prioritise long-term professional development programmes that combine workshops, mentoring systems, classroom-based experimentation, and collaborative professional learning communities. Sustained professional learning models enable teachers to update their knowledge, experiment with innovative instructional approaches, and refine teaching practices through ongoing reflective processes. Policy frameworks that allocate dedicated time for professional learning within teachers' workloads can significantly increase participation in reflective professional development initiatives.

Institutional leadership at the school level represents another important factor influencing the success of climate education implementation. School leaders who prioritise teacher learning, encourage collaborative professional development, and support interdisciplinary instructional innovation create organisational environments that foster continuous professional growth. Leadership support may include providing resources for professional learning communities, facilitating partnerships with environmental organisations, and encouraging teacher-led sustainability initiatives within schools. Such leadership actions strengthen institutional commitment to climate education and promote sustained instructional innovation.

Another important policy implication relates to the provision of instructional resources and infrastructure. Effective Climate Change Education requires access to teaching materials, environmental data resources, digital learning tools, and locally relevant instructional content that supports inquiry-based and experiential learning approaches. Policymakers should invest in the development and distribution of contextually relevant climate education materials, ensuring that schools across different regions have equitable access to resources that support effective instructional practices. Partnerships between ministries of education, environmental agencies, universities, and non-governmental organisations can support the development of high-quality instructional resources and teacher training initiatives.

Collaboration across sectors also represents a critical policy priority. Climate change is a cross-sectoral challenge that intersects with environmental management, agriculture, health, and economic development. Education policies that promote collaboration between schools, community organisations, environmental agencies, and research institutions can enhance the relevance of climate education programmes and provide teachers with access to expertise, field-based learning opportunities, and community engagement initiatives. Such partnerships strengthen the practical dimension of Climate Change Education and support learners' participation in local sustainability initiatives.

Finally, education systems should establish monitoring and evaluation mechanisms that assess the effectiveness of professional development programmes in improving teacher competence and instructional practices in climate education. Systematic evaluation processes enable policymakers and teacher education institutions to identify strengths and weaknesses in professional learning initiatives, refine programme design, and ensure that teacher development strategies align with evolving sustainability education goals. Evidence-based policy development enhances the long-term effectiveness of climate education initiatives and supports continuous improvement in teacher professional learning systems.

Coherent alignment between policy frameworks, teacher education programmes, professional learning systems, and institutional leadership structures is therefore central to strengthening Climate Change Education implementation. Such systemic coordination enables sustained teacher development and supports the broader goal of embedding sustainability learning across education systems.

CONCLUSION

Climate change represents one of the most significant global challenges confronting education systems in the twenty-first century, requiring transformative approaches that prepare learners to understand complex environmental processes and participate actively in sustainability initiatives. Climate Change Education (CCE) has therefore emerged as an essential component of contemporary education aimed at developing climate literacy, environmental responsibility, and action-oriented competencies among learners. However, the effectiveness of CCE largely depends on teachers' competence, preparedness, and professional learning experiences in translating curriculum goals into classroom practice. This chapter, therefore, contributes not only to the conceptual understanding of teacher reflections on professional development but also offers a framework to inform curriculum implementation for climate change education. By guiding teacher learning, curriculum design, and school-level practices, the framework can support education systems in preparing learners with the knowledge, skills, and dispositions needed to respond to climate change challenges.

This chapter has emphasised the importance of reflective professional development as a central mechanism for strengthening teacher capacity in Climate Change Education. Teachers require not only knowledge of climate science but also pedagogical competence, interdisciplinary instructional skills, and the professional ability to continuously adapt teaching strategies to evolving climate education needs. Reflective professional learning processes—such as collaborative inquiry, mentoring, lesson study, and professional learning communities—enable teachers to critically examine their instructional experiences, identify areas for improvement, and refine their teaching practices through continuous learning cycles. These reflective processes support sustained pedagogical transformation and enhance teachers' confidence in implementing innovative instructional approaches.

The chapter also highlighted that effective professional development for climate education must move beyond isolated training workshops toward sustained, collaborative, and contextually responsive learning models. Long-term professional learning initiatives that combine training, classroom-based experimentation, reflective inquiry, and ongoing mentoring are more likely to produce meaningful instructional change and improved student learning outcomes. Such approaches enable teachers to connect professional development experiences directly to classroom practice, thereby strengthening the translation of climate education policies into effective instructional implementation.

Furthermore, the success of reflective professional development initiatives is influenced by broader institutional and policy environments. Supportive leadership, curriculum integration, availability of instructional resources, collaborative professional learning structures, and national policy recognition of Climate Change Education all contribute to creating enabling conditions for sustained teacher learning. Education systems that invest in these enabling conditions enhance the long-term sustainability of climate education initiatives and strengthen teachers' capacity to deliver meaningful sustainability instruction.

The conceptual model proposed in this chapter underscores the cyclical nature of reflective professional learning, demonstrating how professional development inputs, reflective inquiry, pedagogical transformation, and contextual support structures interact to influence student climate learning outcomes. By conceptualising teacher professional learning as an iterative and collaborative process, the model provides a practical framework for designing professional development programmes that promote continuous instructional improvement in Climate Change Education.

Ultimately, preparing learners to respond effectively to climate challenges requires education systems to prioritise teacher development as a strategic component of sustainability education. Teachers who are equipped with strong climate literacy, reflective professional learning skills, and supportive institutional environments are better positioned to cultivate environmentally literate learners capable of contributing to climate mitigation, adaptation, and sustainable development initiatives. Strengthening reflective professional development systems, therefore, represents a critical pathway for enhancing the effectiveness of Climate Change Education and supporting global efforts to address the climate crisis through education.

RECOMMENDATIONS

Based on the conceptual analysis and empirical synthesis presented in this chapter, several recommendations are proposed to strengthen reflective professional development for Climate Change Education (CCE).

First, teacher education institutions should systematically integrate Climate Change Education into both preservice and in-service teacher preparation programmes. Training should include climate science literacy, sustainability competencies, inquiry-based pedagogy, and interdisciplinary instructional strategies that enable teachers to translate climate education concepts into meaningful classroom learning experiences. Embedding reflective practice within teacher education curricula will ensure that teachers develop the capacity for continuous professional learning and instructional improvement.

Second, education authorities should prioritise sustained and continuous professional development models rather than short-term training workshops. Professional development initiatives should incorporate mentoring systems, collaborative professional learning communities, classroom-based experimentation, and structured reflective activities such as reflective journals, peer observations, and lesson study cycles. Long-term professional learning structures enhance teachers' ability to apply newly acquired knowledge, evaluate instructional outcomes, and refine pedagogical practices over time.

Third, policymakers should strengthen institutional support systems that enable teachers to engage in reflective professional learning. Schools should allocate dedicated time for collaborative reflection, provide access to instructional resources, and encourage interdisciplinary collaboration among teachers. Supportive leadership practices that recognise professional learning as an essential component of instructional improvement can significantly enhance the sustainability of professional development initiatives.

Fourth, curriculum planners and educational policymakers should ensure that Climate Change Education is explicitly integrated into national curriculum frameworks across subject areas. Formal curriculum inclusion will legitimise the allocation of instructional time for climate-related topics and guide teacher professional development priorities. Curriculum alignment with sustainability education goals also ensures coherence between policy expectations, teacher training programmes, and classroom instructional practices.

Fifth, education systems should invest in developing and distributing contextually relevant instructional materials that support the implementation of climate education. Locally relevant climate case studies, environmental data resources, and experiential learning tools can enhance the practical relevance of climate education and enable teachers to connect global climate concepts with local environmental challenges. Access to such resources strengthens teachers' confidence in implementing climate-related instruction and improves student engagement in sustainability learning activities.

Finally, education systems should establish monitoring and evaluation mechanisms that assess the effectiveness of professional development programmes in improving teacher competence and classroom implementation of Climate Change Education. Continuous evaluation will enable policymakers and teacher education institutions to refine professional development strategies, identify emerging training needs, and ensure that teacher professional learning systems remain responsive to evolving climate education priorities.

Collectively, these recommendations highlight the importance of aligning teacher education, professional development systems, curriculum frameworks, and institutional leadership support to strengthen the implementation of reflective professional development for Climate Change Education. When these elements are coordinated effectively, education systems can enhance teacher capacity, improve instructional quality, and foster climate-literate learners who can contribute to sustainable development efforts.

Implementing these recommendations requires coordinated efforts across policy, institutional, and classroom levels to ensure that teacher professional learning systems remain reflective, sustained, and contextually responsive. When effectively implemented, such reforms can significantly enhance instructional quality and support the development of climate-literate learners prepared to contribute to sustainable development initiatives.

REFERENCES

Adjei-Boateng, E. (2025). *Reflective practice in teacher education: Insights from Ghana.* Journal of Teacher Education and Practice.

Adjei-Boateng, E., & Cobbinah, J. E. (2021). School-based professional development and teacher learning in Ghanaian basic schools. *International Journal of Educational Development*, *82*, 102384.

Adjei-Boateng, E., Cobbinah, J. E., & Adamba, C. (2025). The role of school leadership in supporting teacher professional learning: Evidence from Ghana. *Educational Management Administration & Leadership.*

Amponsah, K. D., Boateng, E. A., & Mohammed, S. M. (2021). Preservice teachers' instructional self-efficacy and learning dispositions toward science teaching. *Journal of Science Teacher Education, 32*(4), 451–472.

Asano, K., Amponsah, K. D., & Boateng, E. A. (2019). Instructional resources and science teaching in Ghanaian schools: A mixed-methods study. *International Journal of Science Education, 41*(12), 1645–1663.

Awoniyi, F. C., Yeboah, R., Amponsah, K. D., & Adjei-Boateng, E.. (2025). Perceived Effects of ICT Integration on Teaching and Learning in Ghanaian Senior High Schools: A Mixed-Methods Study. *Education Quarterly Eeview.*

Boateng, E. A. (2021). Teachers' preparedness for climate change education: A study of primary and secondary educators in Ghana. *Environmental Education Research, 27*(8), 1145–1162.

Braun, V., & Clarke, V. (2006). Using thematic analysis in psychology. *Qualitative Research in Psychology, 3*(2), 77–101. DOI: 10.1191/1478088706qp063oa

Darling-Hammond, L., Hyler, M. E., & Gardner, M. (2017). *Effective teacher professional development.* Learning Policy Institute., DOI: 10.54300/122.311

Evans, L. S., Amponsah, K. D., & Boateng, E. A. (2021). Teacher competence and professional development for sustainability education in developing contexts. *Sustainability, 13*(15), 8412. DOI: 10.3390/su13158412

Fals-Borda, O. (1987). The application of participatory action research in Latin America. *International Sociology, 2*(4), 329–347. DOI: 10.1177/026858098700200401

Farrell, T. S. C. (2018). *Reflective practice in ELT: A guide for teachers and teacher educators.* Equinox Publishing.

Fien, J. (2018). Environmental literacy and sustainability education: Theoretical perspectives and practical applications. In *Issues and Trends in Education for Sustainable Development* (pp. 61–78). UNESCO.

Hatton, N., & Smith, D. (1995). Reflection in teacher education: Towards definition and implementation. *Teaching and Teacher Education, 11*(1), 33–49. DOI: 10.1016/0742-051X(94)00012-U

IPCC. (2022). *Climate Change 2022: Mitigation of climate change. Contribution of Working Group III to the Sixth Assessment Report of the Intergovernmental Panel on Climate Change.* Cambridge University Press., DOI: 10.1017/9781009157926

Kemmis, S., & McTaggart, R. (2005). Participatory action research: Communicative action and the public sphere. In Denzin, N. K., & Lincoln, Y. S. (Eds.), *The Sage Handbook of Qualitative Research* (3rd ed., pp. 559–603). Sage Publications.

Kolb, D. A. (1984). *Experiential learning: Experience as the source of learning and development.* Prentice-Hall.

Korthagen, F. A. (2001). *Linking practice and theory: The pedagogy of realistic teacher education.* Lawrence Erlbaum Associates. DOI: 10.4324/9781410600523

Kyei-Akuoko, E., Boateng, E. A., & Amponsah, K. D. (2020). Blended learning and professional development in Ghanaian higher education. *Journal of Online Learning Research and Practice, 8*(1), 45–62.

Lotz-Sisitka, H., Wals, A. E., Kronlid, D., & Blackmore, C. (2015). Transformative, transgressive social learning: Rethinking higher education pedagogy in light of global sustainability challenges. *Current Opinion in Environmental Sustainability, 16*, 73–80. DOI: 10.1016/j.cosust.2015.07.018

McKeown, R. (2014). *The ESD schoolman's guide to climate change education.* UNESCO.

Mochizuki, Y., & Bryan, A. (2015). Climate change education in the context of education for sustainable development. *Journal of Education for Sustainable Development, 9*(1), 4–26. DOI: 10.1177/0973408215569109

Mogren, A., Gericke, N., & Scherp, H. (2019). Whole school approaches to climate change education: A systematic review. *Environmental Education Research, 25*(8), 1121–1140. DOI: 10.1080/13504622.2018.1446104

Mohammed, S. M., & Amponsah, K. D. (2021). Teachers' and educational administrators' conceptions of inquiry: Do they promote or constrain inquiry-based science teaching? *Journal of Curriculum and Teaching*, *10*(3), 1–15. DOI: 10.5430/jct.v10n3p58

Moher, D., Liberati, A., Tetzlaff, J., & Altman, D. G. (2009). Preferred reporting items for systematic reviews and meta-analyses: The PRISMA statement. *PLoS Medicine*, *6*(7), e1000097. DOI: 10.1371/journal.pmed.1000097 PMID: 19621072

Ochieng, D. (2019). Climate education in teacher training: Perspectives from East Africa. *International Journal of Sustainability in Higher Education*, *20*(4), 612–630.

Ochieng, J., & Adebayo, A. (2023). Teacher professional development and climate education implementation in African schools. *International Journal of Educational Development*, *95*, 102654. DOI: 10.1016/j.ijedudev.2022.102654

Osei, Y. (2017). Integrating climate change into the Ghanaian school curriculum: Challenges and opportunities. *Journal of Education and Practice*, *8*(12), 110–122.

Rieckmann, M. (2018). Learning to transform the world: Key competencies in education for sustainable development. In *Issues and Trends in Education for Sustainable Development* (pp. 39–59). UNESCO.

Rousell, D., & Cutter-Mackenzie-Knowles, A. (2020). A systematic review of climate change education: Giving children and young people a voice and a hand in the future. *Educational Research Review*, *31*, 100343. DOI: 10.1016/j.edurev.2020.100343

Šarić, M., & Šteh, B. (2017). Professional learning communities and teacher collaboration: A mechanism for pedagogical transformation. *Educational Studies*, *43*(2), 161–175. DOI: 10.1080/03055698.2016.1248905

Schön, D. A. (1983). *The reflective practitioner: How professionals think in action*. Basic Books.

Selby, D., & Kagawa, F. (2013). *Sustainability frontiers: Critical and transformative voices from the borderlands of sustainability education*. Barbara Budrich Publishers.

Snyder, H. (2019). Literature review as a research methodology: An overview and guidelines. *Journal of Business Research*, *104*, 333–339. DOI: 10.1016/j.jbusres.2019.07.039

Stevenson, R., Nicholls, J., & Whitehouse, H. (2017). *What is climate change education? Curriculum, pedagogy and climate change education*. In *The SAGE Handbook of Curriculum, Pedagogy and Assessment* (pp. 539–556). SAGE Publications. DOI: 10.4135/9781473921405.n34

UNEP. (2021). *Africa's vulnerability to climate change: Adaptation and mitigation strategies*. United Nations Environment Programme.

UNESCO. (2021). *Education for sustainable development: A roadmap*. UNESCO Publishing.

Chapter 8
Fostering Sustainability Through Integrated STEM Education:
Policy, Pedagogical, and Systemic Perspectives From Ghana and Africa

Kwaku Darko Amponsah
University of Ghana, Ghana

Emmanuel Adjei-Boateng
http://orcid.org/0000-0003-1786-8116
University of Ghana, Ghana

Phyllis Bernice Kwarteng Donkor
http://orcid.org/0000-0003-3249-3025
University of Ghana, Ghana

ABSTRACT

This literature review synthesises the philosophical and conceptual foundations, international practices, and contextual relevance of promoting sustainability through STEM education in Ghana and Africa. It argues that addressing complex socio-environmental challenges such as climate change, resource scarcity, and social inequality requires integrating sustainability principles into STEM education. The review highlights constructivism, inquiry-based learning, design thinking, and systems thinking as key pedagogical approaches that support active, contextualised, and interdisciplinary problem-solving. Evidence from international case studies demonstrates the effectiveness of Phenomenon-Based Learning, Project-Based Learning,

DOI: 10.4018/979-8-3373-6975-4.ch008

and Maker Education in developing sustainability competencies. In Africa, sustainability demands are intensified by limited resources, inadequate teacher capacity, and gender imbalances in STEM, requiring coordinated policy action, curriculum reform, sustained investment, and continuous teacher professional development

INTRODUCTION

Education systems across the world are increasingly confronted with the responsibility of preparing learners to understand and respond to complex sustainability challenges, including climate change, biodiversity loss, environmental degradation, food insecurity, public-health vulnerabilities, energy transition pressures, and widening socio-economic inequalities (Nusche, et al., 2024). These challenges are multidimensional and interconnected, requiring innovative and interdisciplinary solutions that combine scientific knowledge, engineering design, technological innovation, and data-informed decision-making. As a result, the role of science, technology, engineering, and mathematics (STEM) education has expanded beyond disciplinary knowledge transmission to include the development of competencies that enable learners to analyse real-world sustainability problems and generate contextually relevant solutions through collaborative, inquiry-driven, and problem-solving learning processes (Dare, et al., 2021; Nusche, et al., 2024). Education systems are therefore increasingly expected to cultivate learners who are not only scientifically literate but also capable of applying knowledge creatively and responsibly to address pressing socio-environmental challenges.

Globally, governments and international organisations increasingly recognise STEM education as a strategic driver of sustainable development and economic transformation. The United Nations' 2030 Agenda for Sustainable Development emphasises the importance of strengthening education systems capable of producing innovative and sustainability-conscious citizens who can contribute to achieving the Sustainable Development Goals (SDGs), particularly those related to climate action, sustainable industry, clean energy, sustainable cities, and responsible consumption (United Nations, 2015). Education for Sustainable Development (ESD) frameworks further emphasise the integration of sustainability competencies—such as systems thinking, anticipatory thinking, interdisciplinary collaboration, ethical reasoning, and responsible decision-making—into educational curricula and teaching practices across all levels of education (Advance HE, 2025). These competencies enable learners to understand the interconnections between environmental, social, and economic systems and to evaluate the long-term consequences of human activities. STEM education provides a particularly suitable platform for achieving this integration because it equips learners with analytical tools, technological capabilities, and engineering design approaches required to develop innovative solutions to complex sustainability problems.

In recent years, the concept of integrated STEM education has gained prominence as an instructional approach that moves beyond teaching science, technology, engineering, and mathematics as isolated disciplines toward interdisciplinary learning environments in which learners engage in authentic, real-world problem solving. Integrated STEM education emphasises project-based learning, inquiry-based investigations, engineering design challenges, and community-connected innovation projects that encourage learners to apply knowledge across disciplinary boundaries while strengthening critical thinking, creativity, and collaborative problem-solving skills (Kelley & Knowles, 2016; Lee & Lee, 2025). When sustainability concepts are embedded within such interdisciplinary STEM learning environments, students are more likely to develop competencies required to understand complex socio-ecological systems and design contextually relevant solutions. Empirical evidence across diverse educational contexts indicates that sustainability-focused STEM programmes improve learners' environmental awareness, systems reasoning, scientific literacy, and applied problem-solving capabilities (Homer et al., 2026; Tafese & Kopp, 2025). Despite the growing global emphasis on sustainability-integrated STEM education, the distribution of research evidence and implementation experiences remains uneven. Much of the existing empirical and conceptual scholarship originates from developed educational systems, where infrastructure, teacher preparation systems, funding mechanisms, and institutional support structures often differ significantly from those found in developing regions. In many African countries, education systems are simultaneously addressing multiple priorities, including expanding access to quality STEM education, strengthening teacher preparation and professional development, reducing gender disparities in STEM participation, improving digital and laboratory infrastructure, and enhancing curriculum relevance to labour-market demands. These challenges occur within broader national contexts characterised by rapid urbanisation, environmental pressures, youth population growth, and increasing demand for technological innovation to support sustainable economic development. Consequently, integrating sustainability into STEM education in African contexts requires pedagogically sound, institutionally feasible, and responsive approaches.

Ghana represents a particularly important case for examining the intersection of sustainability education and STEM reform. Over the past decade, national policy initiatives have increasingly prioritised the expansion of STEM education, including the establishment of specialised STEM secondary schools, the development of regional STEM centres, the promotion of digital learning technologies, and curriculum reforms emphasising competency-based learning. These initiatives reflect growing recognition of STEM education's role in supporting national industrialisation, technological innovation, and workforce development. Empirical research in Ghanaian secondary schools also shows that instructional effectiveness and student academic performance are strongly influenced by teacher preparedness, school leadership, and availability of teaching resources, highlighting the importance of institutional support for STEM reform implementation (Mensah et al., 2024). At the same time, Ghana faces pressing sustainability challenges, including climate variability, water resource management pressures, deforestation, coastal erosion, waste management concerns, renewable energy transition needs, and agricultural productivity challenges associated with changing environmental conditions. Addressing these issues requires a workforce equipped with strong scientific, technological, and engineering competencies to develop locally relevant sustainability innovations. Education systems, therefore, play a critical role in ensuring that learners acquire not only disciplinary knowledge but also interdisciplinary sustainability competencies that enable them to contribute meaningfully to national development priorities and community-level problem solving.

Although research on STEM education in Ghana and across Africa has expanded in recent years, existing scholarship often focuses on individual components of education reform—such as teacher professional development, curriculum implementation, infrastructure provision, or technology integration—without sufficiently examining how these elements interact within broader education systems to support sustainability outcomes. Similarly, while global frameworks provide valuable conceptual guidance for integrating sustainability into STEM education, their direct application within African contexts may be constrained by differences in institutional capacity, policy coordination, funding availability, and community engagement structures. A comprehensive synthesis that integrates global evidence with regional contextual analysis is therefore essential for guiding policymakers, curriculum developers, teacher educators, and researchers seeking to strengthen sustainability-oriented STEM education systems across Africa.

This chapter addresses this need by synthesising global and African scholarship on sustainability-integrated STEM education and examining how interdisciplinary STEM approaches can support the development of sustainability competencies among learners in Ghana and comparable African contexts. The chapter reviews theoretical foundations underpinning sustainability-oriented STEM learning, analyses international models and pedagogical innovations, and examines the institutional, policy, and implementation contexts shaping STEM education development across Africa. Building on this synthesis, the chapter proposes a context-responsive systemic framework that conceptualises sustainability-integrated STEM education as an interconnected ecosystem linking national policy priorities, curriculum integration, teacher education and professional development, inquiry-driven pedagogy, and learner competency development to community-level sustainability innovation outcomes. By situating this framework within the institutional, socio-economic, and environmental realities of Ghana and the wider African region, the chapter provides theoretically grounded and policy-relevant guidance for strengthening the role of STEM education as a strategic driver of sustainable development, workforce transformation, and locally grounded innovation.

In addition to providing a conceptual synthesis, the chapter offers practical insights for education leaders, curriculum planners, teacher educators, and development partners seeking to strengthen the alignment between STEM education reforms and sustainability-oriented national development strategies. By highlighting the interdependence of policy frameworks, curriculum systems, teacher capacity development, and pedagogical innovation, the chapter emphasises that sustainable transformation of STEM education requires coordinated systemic action rather than isolated programme-level interventions. Such coordinated approaches are particularly important in developing contexts where resource constraints, institutional fragmentation, and implementation challenges often limit the effectiveness of educational reforms. Through its integrated analytical approach, the chapter advances scholarly understanding of how sustainability-oriented STEM education can serve as a transformative driver of educational, technological, and socio-economic development in Ghana and across Africa.

Globally, education systems have undergone significant disruptions and transformations because of the COVID-19 pandemic, which revealed substantial inequalities in access to digital technologies, infrastructure, and quality learning opportunities. These disruptions have intensified the global focus on strengthening education systems to address complex societal challenges such as climate change, environmental degradation, food security, and sustainable resource management. In this context, integrated STEM education is increasingly recognised as an effective approach for developing learners' interdisciplinary problem-solving competencies required to address sustainability challenges (Tan et al., 2025).

In many African countries, the implementation of integrated STEM education is hindered by structural constraints, including limited technological infrastructure, shortages of qualified STEM educators, and rigid, examination-focused curricula. Nevertheless, the continent also presents considerable opportunities for contextualised STEM education that links scientific learning to real-world community challenges, including water management, renewable energy adoption, sustainable agriculture, and environmental conservation (Agai, 2024; Bardoe et al., 2023; Chisom et al., 2024). This chapter examines sustainability-oriented STEM education from both a continental African perspective and within the specific context of Ghanaian policy and education, highlighting how integrated STEM approaches can foster sustainable development competencies among learners.

CONCEPTUAL AND THEORETICAL FOUNDATIONS

Integrated STEM and Sustainability Education

STEM education has evolved significantly over the past two decades from an approach centred on discrete disciplinary instruction to an integrated framework that emphasises interdisciplinary problem-solving, contextual learning, and the application of knowledge to real-world challenges. Integrated STEM education promotes the purposeful blending of science, technology, engineering, and mathematics concepts within authentic problem contexts, enabling learners to engage in inquiry, experimentation, modelling, and engineering design processes that mirror really scientific and technological practice (Kelley & Knowles, 2016). This interdisciplinary orientation strengthens learners' capacity to transfer knowledge across domains, fostering deeper conceptual understanding and enabling them to address complex societal problems.

Parallel to the development of integrated STEM education, sustainability education has gained global prominence through international initiatives such as Education for Sustainable Development (ESD), which emphasise the cultivation of competencies that enable individuals and communities to contribute to environmentally responsible, socially equitable, and economically viable development (Tarlochan et al., 2025). Sustainability education focuses on developing learners' abilities to understand the interconnections among ecological, social, and economic systems; anticipate future challenges; evaluate alternative solutions; and participate in collective action to improve societal well-being. Integrating sustainability within STEM education, therefore, represents a strategic convergence of two complementary educational movements: STEM education, which equips learners with analytical and technological problem-solving capabilities, and sustainability education, which emphasises ethical reasoning, systems awareness, and responsible decision-making.

The integration of sustainability themes into STEM education enables transforming classroom learning from abstract knowledge acquisition into meaningful engagement with authentic socio-environmental challenges. For example, sustainability-oriented STEM learning may involve projects focused on renewable energy systems, water resource management, waste-reduction technologies, climate adaptation strategies, or sustainable agricultural practices. Such learning experiences provide learners with opportunities to apply interdisciplinary knowledge to design, test, and refine solutions relevant to their communities. Research indicates that when STEM instruction is linked to sustainability challenges, learners demonstrate increased motivation, improved systems reasoning, stronger environmental awareness, and enhanced collaborative problem-solving abilities (Cordaro, et al, 2025; Tarlochan et al., 2025). These outcomes highlight the potential of sustainability-integrated STEM education to prepare learners not only for participation in technology-driven economies but also for responsible citizenship within rapidly changing socio-ecological systems.

Constructivist Foundations of Sustainability-Focused STEM Learning

Constructivist learning theory provides a central theoretical foundation for sustainability-oriented STEM education. Constructivist perspectives emphasise that learners actively construct knowledge through meaningful experiences, inquiry processes, and collaborative interactions, rather than passively receiving information transmitted by instructors. Within sustainability-integrated STEM contexts, constructivist learning environments encourage students to investigate real-world problems, gather evidence, develop explanatory models, and refine their understanding through reflection and dialogue (Crawford, 2014; Zajda, 2021; Bou Saad et al., 2025). These processes align closely with the competencies required for sustainability problem solving, including critical analysis, adaptive reasoning, and the capacity to evaluate multiple perspectives when addressing complex societal issues.

Inquiry-based learning represents a key pedagogical expression of constructivist theory within STEM education. Inquiry-oriented instructional approaches involve learners in posing questions, conducting investigations, analysing data, and constructing evidence-based explanations, thereby strengthening scientific reasoning and conceptual understanding (Pedaste et al., 2015). In sustainability-focused STEM education, inquiry-based learning can be applied to investigating local environmental issues such as water quality, soil degradation, biodiversity conservation, or energy consumption patterns. By engaging learners in authentic investigations linked to their communities, inquiry-based approaches enhance both cognitive learning outcomes and the development of environmental responsibility.

Design Thinking and Engineering Design as Innovation Frameworks

Design thinking and engineering design frameworks provide additional conceptual foundations for sustainability-integrated STEM education by emphasising iterative problem solving, user-centred innovation, and solution prototyping. Design thinking approaches typically involve stages such as problem identification, empathy with stakeholders, ideation, prototyping, testing, and refinement, enabling learners to develop creative and contextually responsive solutions to sustainability challenges. Engineering design processes similarly encourage learners to conceptualise technological solutions, evaluate constraints, test prototypes, and optimise designs based on performance data. These approaches foster innovative competencies essential for addressing sustainability challenges that require technological and engineering interventions, including renewable energy systems, water purification technologies, and sustainable infrastructure development.

Research indicates that integrating engineering design challenges into STEM curricula enhances learners' engagement, creativity, and persistence in problem solving while strengthening their understanding of how scientific principles inform technological applications (Bybee, 2020). When applied in sustainability contexts, design thinking and engineering design activities offer learners opportunities to develop practical innovation skills while gaining awareness of the environmental and social implications of technological solutions. Such learning experiences contribute to the development of responsible innovation competencies that combine technical expertise with ethical and environmental considerations.

Systems Thinking as a Core Sustainability Competency

System thinking represents a critical competency underpinning sustainability education and integrated STEM learning. Sustainability challenges often involve complex interactions among environmental, social, economic, and technological systems, requiring individuals to understand feedback loops, interdependencies, and long-term consequences of interventions. System thinking enables learners to move beyond linear cause-and-effect reasoning to recognise the dynamic relationships shaping sustainability challenges and the potential unintended consequences of policy or technological decisions (Wiek et al., 2011). Within STEM education, system-thinking approaches may involve modelling ecological systems, analysing energy networks, or examining interactions among urban infrastructure, environmental resources, and population dynamics.

Embedding systems thinking within STEM education supports learners' ability to analyse complex sustainability challenges holistically and to design solutions that consider environmental, economic, and social dimensions simultaneously. Educational interventions that incorporate systems modelling, scenario analysis, and interdisciplinary case studies have been shown to enhance learners' capacity to evaluate sustainability problems and develop integrative problem-solving strategies (Rieckmann, 2012). Consequently, systems thinking is widely regarded as a foundational competency for sustainability-integrated STEM education.

Sociocultural and Contextual Perspectives in Sustainability STEM

Sociocultural learning theories further emphasise that knowledge construction occurs within cultural, institutional, and community contexts, highlighting the importance of context-responsive STEM education that reflects learners lived experiences and local realities. In sustainability-oriented STEM education, incorporating indigenous knowledge systems, community practices, and locally relevant environmental challenges enhances the relevance and authenticity of learning experiences. For example, integrating traditional agricultural knowledge, local conservation practices, or community water-management systems into STEM learning activities can enrich learners' understanding of sustainability while fostering respect for diverse knowledge systems (Aikenhead & Ogawa, 2007). Such approaches support inclusive education by recognising the value of multiple epistemologies and enabling learners to connect formal scientific knowledge with community-based practices.

Sociocultural perspectives also highlight the role of collaborative learning environments in promoting sustainability competencies. Group-based inquiry projects, community partnerships, and participatory research activities encourage learners to engage collectively in problem solving, reflecting the collaborative nature of sustainability innovation in real-world contexts. These approaches foster communication skills, teamwork, and social responsibility, which are essential competencies for addressing sustainability challenges that require coordinated community and institutional action.

Synthesis: Toward a Sustainability-Integrated STEM Learning Paradigm

Collectively, the conceptual and theoretical perspectives discussed above—integrated STEM education, constructivist learning theory, inquiry-based learning, design thinking, engineering design, systems thinking, and sociocultural learning frameworks—provide a comprehensive foundation for sustainability-integrated STEM education (Toma, et al., 2024; Portillo-Blanco et al., 2025). These perspectives converge in emphasising interdisciplinary problem solving, learner-centred inquiry, contextualised innovation, and collaborative engagement with real-world sustainability challenges. Integrating these theoretical principles into education systems enables the development of learners who possess not only disciplinary STEM knowledge but also the sustainability competencies required to design innovative solutions to environmental, technological, and socio-economic challenges.

Within African and Ghanaian educational contexts, applying these theoretical frameworks requires careful consideration of institutional capacities, cultural knowledge systems, and resource realities. Context-responsive sustainability-integrated STEM education must therefore combine global theoretical insights with locally grounded pedagogical approaches that reflect community priorities and national development needs (Tarlochan, et al., 2025). The conceptual foundation outlined in this section provides the basis for examining global implementation models and contextual policy considerations discussed in the subsequent sections of this chapter.

GLOBAL MODELS

Across the world, education systems increasingly recognise sustainability-integrated STEM education as a strategic mechanism for preparing learners to address complex environmental, technological, and socio-economic challenges (Jangde & Ahmad, 2025. Governments and educational institutions have implemented curriculum reforms, pedagogical innovations, and school–community partnerships to embed sustainability competencies into STEM learning (Gouëdard et al., 2020). These global initiatives provide important insights into how interdisciplinary STEM education can foster systems thinking, innovation, and real-world problem-solving competencies that support sustainable development.

Curriculum Integration Models

Several countries have introduced curriculum reforms that explicitly integrate sustainability themes into STEM education (Oyenuga, et al., 2025). Finland's phenomenon-based learning model is widely cited for promoting interdisciplinary exploration of real-world themes such as climate change, renewable energy, and sustainable urban systems. By organising learning around societal phenomena rather than traditional subject boundaries, the approach strengthens systems thinking and collaborative problem solving.

Similarly, the United States' Next Generation Science Standards (NGSS) incorporate sustainability concepts through cross-cutting themes, such as environmental systems and human impacts on Earth systems, and emphasise performance-based learning and engineering design practices. Students engage in sustainability-related investigations, including renewable-energy modelling, ecosystem monitoring, and water-quality analysis, thereby strengthening scientific literacy and applied problem-solving skills (Bybee, 2020).

Singapore provides another example of strong curriculum–policy alignment, where interdisciplinary, design-based STEM programmes are supported by national innovation strategies that link education to economic development. Investments in innovative laboratories and maker-education environments provide students with hands-on opportunities to design sustainability-focused technological solutions.

Pedagogical Innovations

Project-based learning (PBL) remains one of the most widely adopted instructional approaches supporting sustainability-focused STEM education. Sustainability-oriented PBL initiatives typically involve community-connected projects such as designing renewable energy systems, water filtration technologies, or waste management solutions. Such experiences enhance student engagement, interdisciplinary knowledge application, and collaborative problem-solving competencies (Rizki & Suprapto, 2024).

Maker education and engineering design studios further strengthen experiential STEM learning. Access to digital fabrication tools, robotics, and prototyping materials enables students to design and test technological solutions to address environmental challenges. These learning environments enhance technological literacy, creativity, and entrepreneurial thinking while connecting classroom instruction to real-world innovation processes (Li et al., 2020).

Inquiry-based environmental monitoring programmes also support sustainability learning by engaging students in citizen science initiatives such as biodiversity monitoring, air quality measurement, and climate data collection. Participation in such activities enables learners to apply scientific methods to real environmental investigations while contributing to the generation of community knowledge and environmental stewardship.

Community-Connected and Place-Based Models

Community-connected STEM education models emphasise linking classroom learning to local sustainability challenges. Place-based education initiatives encourage students to investigate environmental and socio-economic issues affecting their communities, thereby strengthening the relevance of STEM learning and promoting civic engagement (Mercier et al., 2025). Programmes implemented in countries such as Canada, Australia, India, Brazil, and South Africa demonstrate that sustainability-integrated STEM education can support both student learning and community problem solving by engaging learners in designing locally relevant technological solutions.

Teacher Professional Development Models

Teacher capacity development is central to the success of sustainability-integrated STEM education. Countries such as the United Kingdom and Australia have developed national professional development programmes supporting interdisciplinary curriculum design and sustainability-focused project-based learning. Research shows that sustained collaborative professional learning models—such as professional learning communities and coaching-based training—are more effective than short-term workshops in supporting long-term instructional change (Antinluoma et al., 2021).

Lessons for African and Ghanaian Systems

Global experiences suggest several lessons for strengthening sustainability-oriented STEM education. First, curriculum frameworks should explicitly integrate sustainability competencies across STEM disciplines. Second, pedagogical approaches such as project-based learning, inquiry-based instruction, and engineering design challenges are essential for enabling learners to apply interdisciplinary knowledge to real-world sustainability problems. Third, sustained teacher professional development systems are necessary for equipping educators with interdisciplinary teaching skills. Finally, strong policy coordination linking curriculum reform, infrastructure investment, and innovation ecosystems enhances the long-term sustainability of STEM education initiatives (Castaño et al., 2025).

For Ghana and other African countries, adapting these models requires context-responsive strategies that reflect local institutional capacities, resource availability, and community sustainability priorities. While high-income countries often rely on advanced technological infrastructure, developing education systems may prioritise low-cost innovation approaches, community-based projects, and partnerships with local industries and research institutions (Oyenuga et al., 2025). Understanding how global sustainability-integrated STEM models can be adapted to African contexts provides a foundation for the regional analysis presented in the next section.

Recent scholarship underscores the growing prioritisation of sustainability-oriented STEM education within global policy frameworks (Nusche et al., 2024; Marginson et al., 2024). International organisations, including UNESCO and the OECD, have highlighted the importance of STEM education in equipping learners to address complex environmental and socio-economic challenges related to sustainable development (Fazio, 2024; OECD, n.d.). Nevertheless, scholars warn that models originating in highly industrialised contexts are not always directly transferable to developing regions without careful adaptation to local educational, economic, and technological conditions (Marginson et al., 2024). Therefore, effective implementation of integrated STEM education in African contexts necessitates pedagogical approaches that are locally grounded and that link STEM learning to community-based sustainability challenges and indigenous knowledge systems.

AFRICAN AND GHANAIAN CONTEXT

The Evolving Landscape of STEM Education in Africa

Across sub-Saharan Africa, governments increasingly recognise the strategic importance of STEM education in driving technological advancement, economic diversification, and sustainable development (African Union, 2022; Dare, et al., 2021). Rapid population growth, urbanisation, climate vulnerability, and resource-management challenges have intensified the need for a workforce capable of applying scientific and technological knowledge to address development priorities (United Nations, 2015). Consequently, several African countries have introduced national STEM policies, curriculum reforms, and specialised STEM schools to strengthen science and technology capacity among young learners (Nusche, et al., 2024).

Despite growing policy commitment, implementation remains uneven due to structural constraints, including shortages of laboratory facilities, limited access to instructional technologies, and insufficient teacher preparation for inquiry-based instruction (UNESCO, 2020). Gender disparities in STEM participation also persist, particularly at advanced levels of education, limiting the potential talent pool available for scientific and technological innovation (African Union, 2022). These systemic challenges highlight the importance of strengthening teacher education systems, infrastructure provision, and policy coordination mechanisms to support effective STEM education delivery.

Sustainability Challenges Requiring STEM-Based Solutions

The urgency of strengthening sustainability-oriented STEM education in Africa is closely linked to the region's exposure to environmental and socio-economic challenges that require science- and technology-driven solutions. Climate variability has significantly affected agricultural productivity, water availability, and ecosystem stability across many African countries, increasing vulnerability among rural and urban populations (United Nations, 2015). Rapid urbanisation has intensified challenges related to waste management, energy demand, air quality, and infrastructure sustainability, further reinforcing the need for STEM-driven innovation (Nusche, et al., 2024).

Addressing these challenges requires the development of locally relevant technologies, climate-resilient agricultural practices, renewable-energy innovations, and sustainable urban planning solutions that depend heavily on strong STEM education systems capable of producing skilled professionals and innovative thinkers (Dare, et al., 2021). Integrating sustainability perspectives into STEM education ensures that technological development aligns with environmental stewardship and inclusive socio-economic growth.

The Ghanaian Context: Policy Initiatives and Educational Reforms

Ghana provides an important case for examining the intersection of sustainability education and STEM reform. Over the past decade, national policy initiatives have prioritised the expansion of STEM education through the establishment of specialised STEM secondary schools, regional STEM centres, and competency-based curriculum reforms emphasising critical thinking and problem-solving (Ministry of Education, Ghana, 2019; UNESCO, 2020). These initiatives reflect growing recognition of STEM education as a foundation for industrialisation, technological innovation, and workforce development.

At the same time, Ghana faces pressing sustainability challenges, including climate variability, deforestation, coastal erosion, waste management, and the need for a renewable energy transition (United Nations, 2015). Addressing these issues requires a workforce equipped with strong scientific, technological, and engineering competencies to develop locally relevant sustainability innovations. Education systems, therefore, play a critical role in ensuring that learners acquire interdisciplinary sustainability competencies alongside disciplinary knowledge (Nusche, et al., 2024).

Teacher Capacity and Institutional Challenges

Teacher capacity remains a significant factor influencing the effectiveness of sustainability-oriented STEM education in Ghana (Mohammed & Amponsah, 2021). Many teachers were trained within traditional discipline-based systems that emphasised theoretical instruction rather than interdisciplinary inquiry and applied problem solving, creating challenges for implementing integrated STEM pedagogies (Gardner, 1995; Falasi, 2024). Professional development programmes aimed at strengthening pedagogical content knowledge, interdisciplinary teaching skills, and inquiry-based instructional approaches are, therefore, essential for successful reform implementation (Dare, et al., 2021). Shifting teachers from traditional instruction to inquiry-based sustainability education requires structured reflective practices, such as the use of reflective journals, which have been shown to help pre-service teachers in Ghana navigate the challenges of classroom instruction during their field experiences (Adjei-Boateng, 2025).

Infrastructure limitations also present challenges for effective STEM education delivery. Limited access to laboratories, digital technologies, and instructional materials in some schools constrains opportunities for hands-on experimentation and engineering design activities, which are essential for sustainability-focused learning (African Union, 2022). Addressing these challenges requires coordinated investment in educational infrastructure, teacher professional development, and curriculum support systems.

A significant factor affecting the implementation of STEM education in Ghana is the influence of national assessment systems. High-stakes examinations have historically prioritised content mastery and recall, which may discourage teachers from adopting inquiry-based, project-based, and interdisciplinary approaches associated with integrated STEM education (Akon-Yamga et al., 2024). When teachers are pressured to focus on examination outcomes, classroom instruction often emphasises syllabus coverage at the expense of exploratory learning activities that foster experimentation, problem-solving, and collaborative investigation (Akyeampong et al., 2018; West African Examinations Council (WAEC), 2022). The success of systemic STEM reform is not only a matter of classroom pedagogy but also depends on the quality of external supervision and the perspectives of school leaders on improvement support mechanisms (Adjei-Boateng et al., 2025).

Assessment Systems and Inquiry-Based STEM Learning in Ghana

A significant factor affecting the implementation of STEM education in Ghana is the influence of national assessment systems. High-stakes examinations have historically prioritised content mastery, procedural knowledge, and factual recall, which often discourages teachers from employing inquiry-based, project-based, and interdisciplinary pedagogical methods associated with integrated STEM education (Falasi, 2024). When teachers are compelled to focus on preparing students for national examinations such as the West African Senior School Certificate Examination (WASSCE), classroom instruction tends to emphasise syllabus coverage and examination preparation at the expense of exploratory learning activities that foster experimentation, problem-solving, and collaborative investigation (Akyeampong et al., 2018; West African Examinations Council (WAEC], 2022).

Examination-driven instructional cultures limit learners' opportunities to engage in authentic scientific inquiry and interdisciplinary problem-solving, both of which are essential to sustainability-focused STEM education (Bybee, 2020; Rizki et & Suprapto, 2024). However, assessment systems can facilitate STEM reform when evaluation frameworks are aligned with competencies such as critical thinking, creativity, and the practical application of knowledge. In Ghana, the Standards-Based Curriculum developed by the National Council for Curriculum and Assessment emphasises competency development, inquiry-based learning, and the application of knowledge to real-life contexts (NaCCA, 2019). Aligning national assessment practices with these competencies is likely to encourage teachers to implement more student-centred and interdisciplinary pedagogies that advance sustainability-oriented STEM education.

Emerging Opportunities for Sustainability-Focused STEM Education

Despite existing constraints, several emerging opportunities create favourable conditions for strengthening sustainability-oriented STEM education in Ghana and across Africa. Increasing access to digital technologies and mobile learning platforms provides new avenues for delivering STEM learning resources, virtual laboratories, and collaborative innovation projects (Asoma et al., 2024; Li et al., 2020). While digital platforms offer new avenues for STEM delivery, recent mixed-methods evidence from Ghanaian senior high schools suggests that the perceived effectiveness of these tools is closely tied to how well they are integrated into daily pedagogical routines (Asoma et al., 2024). Partnerships among universities, industry organisations, and community groups are expanding opportunities for experiential STEM learning linked to real-world sustainability challenges (Rizki & Suprapto, 2024).

Growing policy attention to green economies and sustainable development further underscores the importance of integrating sustainability competencies within STEM education systems (United Nations, 2015). Aligning STEM curricula with national sustainability priorities can prepare learners for emerging workforce opportunities in renewable energy, environmental management, and technological innovation sectors (Nusche, et al., 2024).

Gender and Inclusion in Sustainability-Focused STEM Education

Gender disparities continue to pose significant challenges in STEM education globally and across numerous African countries (Amponsah et al., 2014). Female students remain underrepresented in advanced STEM subjects because of socio-cultural expectations, limited access to role models, and structural barriers within educational systems (UNESCO, 2020). In various contexts, girls frequently report lower confidence in science and mathematics despite having academic abilities comparable to those of their male counterparts. These disparities influence participation in STEM pathways and ultimately affect the representation of women in science-related careers.

The integration of sustainability themes into STEM education presents promising opportunities to address gender participation gaps. Research indicates that connecting STEM learning to real-world social and environmental challenges can increase engagement and motivation among female learners (Chisango & Mavhunga, 2024). Community-centred sustainability projects provide meaningful contexts that enable female learners to recognise the relevance of STEM knowledge to daily life and the community. Within Ghana, initiatives to promote girls' participation in STEM have increasingly focused on inclusive pedagogical approaches, mentorship programmes, and contextualised problem-based learning. Incorporating sustainability-oriented STEM projects into school curricula may help reduce gender disparities while also enhancing learners' environmental awareness and civic responsibility. Fostering an inclusive STEM environment for female learners also requires addressing gender equity in school leadership, as the instructional practices of both male and female headteachers play a critical role in shaping the educational climate (Abonyi et al., 2024).

Technological Infrastructure as a Barrier to Integrated STEM Education

Although integrated STEM education is receiving increased attention across Africa, inadequate technological infrastructure continues to impede effective implementation (Chisom et al., 2024). Schools in rural and underserved communities frequently face unreliable electricity, limited internet connectivity, insufficient laboratory equipment, and restricted access to digital learning resources. These infrastructural limitations hinder teachers' capacity to implement inquiry-based learning, digital simulations, maker education, and project-based STEM activities, all of which are fundamental to integrated STEM pedagogy (Fitzgerald et al., 2019).

Maker-centred STEM learning environments, which often utilise robotics, fabrication tools, coding platforms, and digital prototyping technologies, depend on stable electricity and reliable internet access. In many African education systems, these resources are distributed unevenly, creating disparities between urban and rural schools. In the absence of sufficient technological support, teachers are often unable to transition from traditional lecture-based instruction to more hands-on and inquiry-driven STEM learning experiences.

Addressing technological infrastructure challenges requires coordinated policy interventions, including investments in school connectivity, the provision of digital learning tools, teacher training in technology integration, and partnerships with private and community stakeholders (Inter-agency Task Force on Financing for Development, n.d.). Enhancing technological infrastructure is critical to supporting sustainability-focused STEM education that equips learners to participate in knowledge-driven economies and to engage in environmentally responsible innovation.

Synthesis

The African and Ghanaian contexts illustrate that strengthening sustainability-oriented STEM education requires systemic approaches addressing policy alignment, curriculum integration, teacher preparation, infrastructure development, and community engagement simultaneously (Dare, et al., 2021). While global models provide valuable guidance, effective implementation in African education systems must consider local institutional capacities, socio-economic conditions, and development priorities. Context-responsive sustainability-integrated STEM education, therefore, involves adapting interdisciplinary pedagogical approaches, fostering partnerships with community stakeholders, and aligning education reforms with national sustainability strategies.

TEACHER PREPARATION

The Role of Teachers in Sustainability STEM Implementation

Teachers play a central role in determining the success of sustainability-integrated STEM education because they translate curriculum intentions into classroom practice and shape the learning experiences through which students develop sustainability competencies (Scheie et al., 2025; Anning, 2024). Effective sustainability-oriented STEM instruction requires educators to move beyond lecture-based teaching toward learner-centred pedagogies that emphasise inquiry, collaboration, engineering design, and community-based problem solving. However, many teacher education systems have historically emphasised discipline-based instruction with a limited focus on interdisciplinary teaching and sustainability competencies, creating capacity gaps that must be addressed through targeted reforms (Nusche, et al., 2024).

Reorienting Pre-Service Teacher Education

Pre-service teacher education institutions play a foundational role in preparing future educators to implement sustainability-oriented STEM instruction. Integrating interdisciplinary STEM frameworks into teacher preparation curricula enables prospective teachers to experience inquiry-based learning, design thinking, and collaborative problem-solving before applying these approaches in their professional practice (Kelley & Knowles, 2016). Research on Ghanaian pre-service science teachers also indicates that diverse cognitive abilities and learning styles influence how prospective teachers understand and apply scientific concepts. This suggests that teacher education programmes must incorporate pedagogical approaches that accommodate multiple intelligences and diverse learning strategies (Budu, Kemetse, & Amponsah, 2022). In this context, integrated STEM education is increasingly recognised as an effective approach for developing learners' interdisciplinary problem-solving competencies required to address sustainability challenges (Marginson et al., 2024). Partnerships between teacher education institutions, schools, and community organisations can further strengthen practicum experiences by exposing student teachers to community-based sustainability projects and innovation initiatives. Such experiences enhance teachers' capacity to design contextually relevant STEM learning activities that connect disciplinary knowledge with societal challenges.

Continuous Professional Development for Practising Teachers

Ongoing professional development is essential for supporting practising teachers in transitioning from traditional instructional approaches to interdisciplinary sustainability-oriented STEM pedagogy. Sustained professional learning models—such as professional learning communities, coaching-based training, and collaborative lesson-design workshops—have demonstrated greater effectiveness than short-term workshops in supporting long-term instructional change. Professional development programmes should focus on interdisciplinary curriculum design, inquiry-based instructional strategies, engineering design processes, and assessment approaches that capture problem-solving competencies (Bybee, 2020).

Training that supports the use of locally available materials and low-cost technologies is particularly valuable in resource-constrained educational contexts, enabling teachers to conduct sustainability-focused STEM investigations even where advanced laboratory facilities are limited (UNESCO, 2020).

Strengthening Pedagogical Content Knowledge and Institutional Support

Pedagogical content knowledge (PCK) is critical for enabling teachers to present STEM concepts in ways that connect with sustainability challenges and support student inquiry (Park & Oliver, 2008). Collaborative planning sessions, interdisciplinary teaching teams, and cross-departmental curriculum design workshops can support teachers in developing integrated instructional strategies while strengthening coordination across subject areas (Nusche, et al., 2024). Institutional support from school leadership, access to instructional resources, and supportive policy environments further influence teachers' ability to implement sustainability-focused STEM instruction effectively.

Digital learning platforms also provide opportunities for expanding access to teacher professional development through online training modules, virtual professional learning communities, and shared instructional resource repositories (Li et al., 2020). Blended professional development models combining face-to-face and online learning can extend the reach of teacher training initiatives while reducing logistical constraints.

Synthesis: Building Teacher Capacity for Transformative STEM Education

Strengthening teacher preparation and professional development systems is essential for achieving sustainable transformation in STEM education. Teachers equipped with interdisciplinary knowledge, inquiry-based pedagogical skills, and sustainability competencies are better positioned to design learning experiences that connect STEM concepts to real-world challenges and foster student innovation (Kelley & Knowles, 2016). Coordinated reforms across pre-service education, in-service professional development, institutional support systems, and policy frameworks are therefore necessary for building education systems capable of producing learners equipped to address emerging sustainability challenges.

STUDENT LEARNING

Developing Sustainability Competencies through STEM Learning

One of the primary objectives of sustainability-integrated STEM education is to develop learners who possess the knowledge, skills, and dispositions necessary to address complex socio-environmental challenges (Wiek et al., 2011; Dare, et al., 2021). Sustainability competencies extend beyond disciplinary understanding to include systems thinking, problem-solving abilities, collaborative skills, ethical reasoning, and the capacity to design innovative solutions to community and global challenges (Rieckmann, 2012). Integrated STEM education provides a powerful platform for cultivating these competencies by emphasising interdisciplinary learning, inquiry-driven investigations, and engineering design processes that mirror real-world problem-solving practices (Kelley & Knowles, 2016).

When students engage in sustainability-focused STEM activities, they apply scientific and technological knowledge to analyse real-world issues such as renewable energy, waste management, climate adaptation, water resource conservation, and sustainable agriculture (Bybee, 2020). These applied learning experiences enable students to understand the relevance of STEM knowledge while strengthening their ability to evaluate multiple solution pathways and assess the environmental and social implications of technological innovations (Quintana-Cifuentes et al., 2019). Research indicates that students who participate in interdisciplinary STEM learning programmes demonstrate stronger problem-solving abilities, improved conceptual understanding, and increased motivation to pursue STEM-related careers compared with those exposed to traditional discipline-based instruction (Rizki & Suprapto, 2024).

Inquiry-Based and Project-Based Learning Effects on Student Outcomes

Inquiry-based learning and project-based learning (PBL) represent two instructional approaches that significantly influence student learning outcomes within sustainability-oriented STEM education (Pedaste et al., 2015; Krajcik & Merritt, 2019). Inquiry-based learning engages students in posing research questions, conducting investigations, analysing data, and constructing evidence-based explanations, thereby strengthening scientific reasoning and analytical thinking skills (Crawford, 2014). Sustainability-related inquiry projects may involve environmental monitoring, community-based research, or laboratory experiments examining ecological processes. Through these experiences, students develop investigative skills and gain a deeper understanding of how scientific knowledge informs sustainability decision-making (McNeill et al., 2022).

Project-based learning complements inquiry approaches by engaging students in extended design and problem-solving activities that require interdisciplinary collaboration (Saavedra & Rapaport, 2024). Sustainability-focused PBL projects often involve designing technological or engineering solutions addressing real-world challenges, such as developing solar-powered devices, creating water-filtration systems, or designing waste-reduction strategies for schools and communities. Participation in such projects enhances students' collaboration skills, creativity, and persistence in solving complex problems while fostering a sense of ownership over their learning experiences (Hariharasakthisudhan et al., 2025). These outcomes contribute to the development of innovative competencies essential for sustainable development.

Building Systems Thinking and Innovation Skills

Systems thinking is a central competency in sustainability education, enabling learners to recognise the interconnections among environmental, technological, economic, and social systems (Wiek et al., 2011). Integrated STEM education supports systems thinking by encouraging students to examine relationships among variables, analyse feedback loops, and consider long-term consequences of interventions (Rieckmann, 2012). Activities such as modelling ecological systems, analysing energy consumption patterns, or evaluating the environmental impact of technological solutions help learners develop holistic perspectives on sustainability challenges (Salonen & Aarnio-Linnanvuori, 2020).

Innovation skills also emerge as important outcomes of sustainability-integrated STEM education. Engineering design challenges and maker-based learning activities encourage students to conceptualise, prototype, test, and refine technological solutions to sustainability problems (Bybee, 2020). These experiences foster creativity, adaptability, and entrepreneurial thinking, which are essential for addressing emerging technological and environmental challenges (Li et al., 2020). Students exposed to such innovation-oriented learning environments are more likely to develop confidence in applying STEM knowledge to real-world contexts and to pursue careers in science, engineering, and sustainability-related fields (Rizki & Suprapto, 2024).

Equity, Inclusion, and Participation in Sustainability STEM Learning

Sustainability-integrated STEM education also plays an important role in promoting equity and inclusion within education systems (Zeidler, 2016). Interdisciplinary, project-based learning approaches often provide diverse entry points for student participation, enabling learners with different learning styles and interests to engage meaningfully in STEM activities (Yeboah et al., 2025). Community-connected sustainability projects can increase participation among underrepresented groups by highlighting the social relevance of STEM learning and demonstrating its potential impact on local communities (Van Wart et al., 2020). Programmes that emphasise collaborative learning, mentorship, and inclusive instructional strategies help reduce gender disparities and improve access to STEM opportunities for diverse student populations (Nusche, et al., 2024).

Digital technologies further support inclusive STEM learning by expanding access to virtual laboratories, simulation tools, and collaborative online platforms that enable students to participate in inquiry-based learning experiences regardless of geographic location (Li et al., 2020). Integrating digital learning tools within sustainability-focused STEM instruction can therefore enhance both learning outcomes and access to quality educational resources, particularly in resource-constrained educational environments (UNESCO, 2020).

Preparing Learners for Green Economies and Future Workforce Needs

As countries transition toward green economies and sustainable development pathways, the demand for workers with sustainability-oriented STEM competencies continues to increase (United Nations, 2015). Education systems, therefore, play a critical role in ensuring that learners acquire interdisciplinary sustainability competencies alongside disciplinary knowledge (Nusche, et al., 2024; Hariharasakthisudhan, et al., 2025; UNESCO, 2026). Sustainability-integrated STEM education equips students with interdisciplinary problem-solving skills, technological literacy, and innovation competencies that align with these evolving workforce demands (Bybee, 2020).

Experiential learning opportunities, including internships, community-based innovation projects, and partnerships with industry and research institutions, further enhance students' readiness for sustainability-oriented careers (Scheie et al., 2025). Such initiatives provide learners with exposure to real-world applications of STEM knowledge and strengthen their understanding of the connections between education, technological innovation, and sustainable economic development.

Synthesis: Student Competency Development as the Core Outcome of Sustainability STEM

The development of sustainability competencies among learners represents a central outcome of integrated STEM education reforms (Wiek et al., 2011). Through inquiry-based investigations, project-based learning activities, engineering design challenges, and community-connected innovation experiences, students acquire the knowledge, skills, and dispositions required to address complex sustainability challenges (Kelley & Knowles, 2016). Education systems that prioritise interdisciplinary STEM learning environments are better positioned to cultivate learners capable of contributing to technological innovation, environmental stewardship, and sustainable economic growth (Dare, et al., 2021). The next section examines the policy, institutional, and governance ecosystems that support the implementation of sustainability-integrated STEM education initiatives.

POLICY AND INSTITUTIONAL ECOSYSTEMS

Policy Frameworks for Sustainability STEM Advancement

The effective implementation of sustainability-integrated STEM education depends on coherent national policy frameworks that align education reforms with broader development priorities (Dare, et al., 2021). Governments play a central role in establishing curriculum standards, funding mechanisms, and teacher professional development systems that support interdisciplinary STEM learning and the development of sustainability competencies (Nusche, et al., 2024). Policy frameworks that integrate sustainability objectives into national education planning ensure that STEM education directly supports national commitments to climate action, renewable energy development, environmental conservation, and green economic transitions (United Nations, 2015).

Education for Sustainable Development (ESD) initiatives further encourage cross-sectoral collaboration among education ministries, environmental agencies, research institutions, and industry stakeholders, strengthening the integration of sustainability competencies across education systems (UNESCO, 2020). Such coordination enhances the relevance and long-term sustainability of STEM education reforms.

Institutional Governance and Multi-Stakeholder Partnerships

Institutional governance structures play a critical role in translating policy intentions into effective implementation. Ministries of education, curriculum development agencies, teacher education institutions, universities, and research organisations must collaborate to design coherent strategies supporting interdisciplinary STEM instruction and sustainability-focused learning (Global Partnership for Education, n.d.). Universities contribute through research, teacher training, and innovation initiatives, while industry partnerships provide opportunities for internships, experiential learning, and technology-driven sustainability projects (Li et al., 2020).

Multi-stakeholder partnerships involving government agencies, private-sector organisations, non-governmental organisations, and community groups enhance programme scalability and sustainability by connecting classroom learning with real-world innovation ecosystems (Rizki & Suprapto, 2024). Such collaborative governance models strengthen the integration of education, technological innovation, and economic development.

Funding and Infrastructure Support

Sustained funding mechanisms are essential for supporting infrastructure development, professional teacher development, curriculum innovation, and research initiatives (Nusche, et al., 2024). Investments in laboratories, digital technologies, and innovation-oriented learning environments enable schools and universities to provide experiential STEM learning opportunities essential for sustainability-focused instruction. Financing models that combine government funding, international development assistance, and private-sector partnerships can support the long-term sustainability of STEM education reforms (United Nations, 2015).

Monitoring, Evaluation, and Accountability

Monitoring and evaluation systems are necessary for assessing programme effectiveness and guiding continuous improvement of sustainability-integrated STEM education initiatives (Dare, et al., 2021). Evaluation frameworks that track student learning outcomes, teacher participation in professional development, STEM enrolment trends, and curriculum integration provide evidence for refining policies and allocating resources. Participatory evaluation approaches involving schools, universities, and community stakeholders enhance accountability and promote the dissemination of effective practices across education systems.

Governance Implications for Ghana and African Contexts

For Ghana and other African countries, strengthening governance ecosystems requires coordinated action across curriculum authorities, teacher education institutions, research organisations, and development partners to align STEM education reforms with national sustainability priorities (African Union, 2022). Addressing regional disparities in infrastructure access and teacher capacity is essential for ensuring equitable participation in STEM learning opportunities. Partnerships between universities, innovation hubs, and local industries can further support community-based sustainability innovation initiatives while strengthening workforce readiness.

Synthesis: Governance Ecosystems as Enablers of Sustainability STEM Transformation

Coherent policy frameworks, collaborative institutional partnerships, sustained funding mechanisms, and robust monitoring systems collectively function as foundational enablers of sustainability-integrated STEM education (Wiegant et al., 2024). Education systems that align governance structures with sustainability priorities are better positioned to cultivate learners equipped with the competencies required to support technological innovation, environmental stewardship, and sustainable economic development.

SUSTAINABILITY-INTEGRATED STEM EDUCATION ECOSYSTEM FRAMEWORK

The successful implementation of sustainability-focused STEM education requires coordinated interactions among policy structures, teacher professional capacity, pedagogical practices, and learner competency development. Contemporary research increasingly emphasises that sustainability education initiatives are most effective when implemented through systemic frameworks that align institutional support, teacher development, interdisciplinary pedagogy, and learner-centred instructional approaches (Rieckmann, 2012; Dare, et al., 2021). Such system-level integration enables education systems to move beyond isolated curriculum interventions toward sustainable instructional transformation that supports long-term sustainable learning outcomes.

To illustrate the interaction between policy alignment, teacher capacity development, pedagogical integration, and learner sustainability outcomes, this chapter proposes a Sustainability-Integrated STEM Education Ecosystem Framework. The framework conceptualises sustainability-focused STEM education as a systemic process in which institutional foundations support teacher professional learning, leading to pedagogical transformation and the development of learner sustainability competencies (Tarlochan, et al., 2025). It further emphasises the role of contextual ecosystem enablers in sustaining long-term implementation of sustainability-integrated STEM education. Figure 1 illustrates the proposed conceptual framework guiding the implementation of sustainability-focused STEM education.

Figure 1. Sustainability-Integrated STEM Education Ecosystem Framework

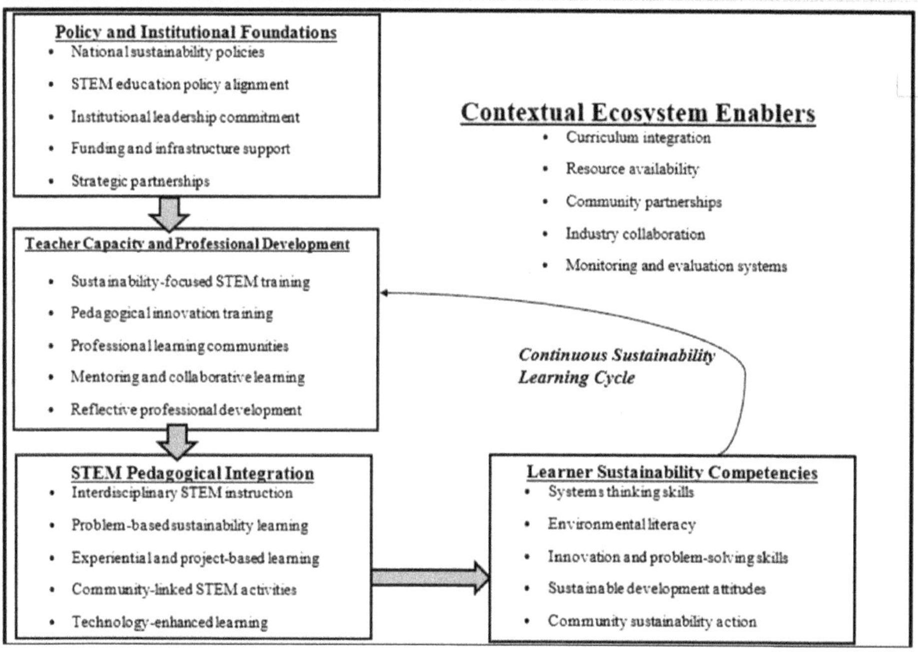

Figure Note. *The framework illustrates the interaction between policy foundations, teacher capacity development, STEM pedagogical integration, and learner sustainability competencies, supported by contextual ecosystem enablers that sustain long-term sustainability-focused STEM education implementation (Dare, et al., 2021).*

Figure 1 presents a vertically integrated ecosystem showing how policy foundations support teacher professional development, which in turn enables pedagogical integration and the development of learner sustainability competencies. The framework highlights sustainability-focused STEM education as a systemic process requiring coordinated institutional, instructional, and learner-level interventions to achieve long-term sustainability education outcomes (Rieckmann, 2012).

At the first level of the framework are Policy and Institutional Foundations, which include national sustainability policies, STEM education reforms, institutional leadership commitment, funding structures, and strategic partnerships. Policy alignment ensures that sustainability education priorities are embedded within curriculum standards, teacher professional development systems, and institutional planning processes, thereby creating enabling environments for long-term implementation (Dare, et al., 2021). Institutional leadership plays a particularly important role in supporting teacher development initiatives, allocating resources for interdisciplinary STEM programmes, and facilitating partnerships with environmental agencies and community organisations.

The second level focuses on Teacher Capacity and Professional Development, which represents a central mechanism for implementing sustainability-focused STEM education. Teachers require sustained opportunities to develop knowledge of sustainability, interdisciplinary pedagogical competence, and the capacity to design problem-based and experiential STEM learning activities. Professional learning communities, mentoring systems, collaborative lesson study, and reflective professional development programmes strengthen teachers' instructional competence and support continuous professional learning (Hendrickx et al., 2025). Building teacher capacity also enhances instructional confidence and encourages the adoption of innovative teaching strategies that integrate sustainability issues across STEM disciplines.

The third level of the framework involves STEM Pedagogical Integration, where teachers translate professional learning into classroom practice through interdisciplinary instruction, project-based sustainability learning, inquiry-driven STEM investigations, and community-linked experiential activities. Such pedagogical approaches enable learners to engage actively with real-world sustainability challenges, apply scientific and technological knowledge to problem-solving, and develop collaborative innovation skills (Tarlochan, et al., 2025). Technology-enhanced instructional tools further support the integration of sustainability concepts within STEM classrooms by providing access to data-driven learning resources, simulation tools, and collaborative digital learning environments.

The fourth level highlights Learner Sustainability Competencies, including systems-thinking skills, environmental literacy, problem-solving abilities, sustainability-oriented attitudes, and participation in community sustainability initiatives. The development of these competencies represents a key objective of sustainability-integrated STEM education, preparing learners to address complex environmental challenges and contribute to sustainable development efforts (Rieckmann, 2012). Learner participation in community-based sustainability projects further strengthens experiential learning and promotes the application of classroom knowledge to real-world contexts.

Surrounding the vertical structure of the framework are Contextual Ecosystem Enablers, including curriculum integration, resource availability, community partnerships, industry collaboration, and monitoring and evaluation systems. These enabling conditions ensure that sustainability-focused STEM initiatives are supported by institutional infrastructure, stakeholder engagement, and continuous programme evaluation processes (Dare, et al., 2021). Strong ecosystem support enhances programme sustainability and strengthens the long-term impact of STEM-based sustainability education initiatives.

A defining feature of the framework is its continuous sustainability learning cycle, represented by feedback processes that link learner outcomes to teacher professional development and instructional innovation. As learners engage in sustainability-focused STEM activities, new insights generated from classroom practice inform subsequent professional learning cycles, supporting ongoing instructional refinement and continuous improvement in sustainability education implementation (Cordaro et al., 2025). This cyclical perspective reinforces the importance of sustained professional learning systems in achieving long-term educational transformation.

CHALLENGES AND OPPORTUNITIES

Structural and Institutional Challenges

Despite increasing global commitment to sustainability-oriented STEM education, several structural and institutional challenges continue to constrain effective implementation, particularly in developing contexts. Limited infrastructure and instructional resources—including inadequate laboratory facilities, restricted access to digital technologies, and insufficient instructional materials—reduce opportunities for experiential STEM learning that is essential for sustainability-focused instruction (Nusche, et al., 2024). Resource disparities between urban and rural schools further contribute to unequal access to quality STEM education.

Teacher preparation also remains a significant challenge. Many educators were trained within discipline-based instructional systems and may lack the interdisciplinary pedagogical knowledge required to implement integrated STEM instruction or sustainability-focused project-based learning (Ofsted, 2024). Institutional barriers, such as rigid curricula, examination-driven assessment systems, and limited time for interdisciplinary learning, can further hinder the adoption of innovative instructional approaches (Dare, et al., 2021).

Equity and Access Concerns

Equity-related challenges continue to affect participation in sustainability-integrated STEM education. Disparities in resource allocation across regions can limit access to STEM learning opportunities and qualified teachers (African Union, 2022). Gender disparities in STEM participation also persist across several contexts, particularly at advanced levels of education, reducing the diversity of talent available for sustainability innovation. Digital access inequalities may further restrict participation in technology-enabled STEM learning initiatives, particularly in areas with limited internet connectivity or technological infrastructure (UNESCO, 2020).

Emerging Opportunities for System Transformation

Despite these constraints, several emerging opportunities create favourable conditions for strengthening sustainability-integrated STEM education. Advances in digital technologies, including virtual laboratories and simulation tools, offer new opportunities to expand experiential learning even in resource-constrained environments (Li et al., 2020). Partnerships among educational institutions, research organisations, industry, and community groups also support sustainability-focused innovation ecosystems that connect classroom learning with real-world problem solving (Rizki & Suprapto, 2024).

Growing global policy attention to climate action, renewable energy development, and green economic transitions has increased demand for sustainability-oriented STEM competencies, creating incentives for governments to invest in STEM education reforms aligned with national sustainability priorities (United Nations, 2015). International development programmes supporting teacher professional development, infrastructure investment, and STEM capacity building further enhance opportunities for expanding sustainability-focused education initiatives.

Implications for Ghana and African Contexts

For Ghana and other African countries, leveraging these opportunities requires coordinated policy action, targeted infrastructure investment, and strengthened institutional partnerships. Expanding collaborations among universities, innovation hubs, private-sector organisations, and community groups can support experiential learning opportunities and locally grounded sustainability innovation initiatives. Addressing infrastructure disparities, strengthening teacher professional development systems, and promoting inclusive participation in STEM education represent critical steps toward maximising the long-term impact of sustainability-integrated STEM education reforms.

RECOMMENDATIONS

Strengthening Policy Alignment and Strategic Planning

Governments and education authorities should explicitly integrate sustainability competencies into national STEM education strategies to ensure alignment with broader environmental and socio-economic development priorities (Dare, et al., 2021). Curriculum frameworks should embed interdisciplinary sustainability themes, systems-thinking competencies, and engineering design practices across STEM subjects rather than treating sustainability education as a supplementary topic (Bybee, 2020). Coordinated policy collaboration among ministries responsible for education, environment, science and technology, and economic development can further strengthen implementation effectiveness (United Nations, 2015).

Long-term strategic planning should also include sustainable funding mechanisms that support STEM infrastructure expansion, digital learning technologies, and innovation-oriented learning environments (Nusche, et al., 2024). Public–private partnerships and innovation grants can provide the resources needed to scale sustainability-focused STEM initiatives.

Reforming Teacher Education and Professional Development

Teacher preparation and professional development systems should be reoriented to support interdisciplinary STEM pedagogy and sustainability-focused instructional practices. Pre-service teacher education programmes should incorporate sustainability education modules, interdisciplinary coursework, and practicum experiences requiring prospective teachers to design inquiry-based STEM lessons addressing real-world sustainability challenges (Ofsted, 2024). For practicing teachers, sustained collaborative professional learning models—such as professional learning communities and coaching-based training—can support the gradual adoption of innovative instructional practices (Nusche, et al., 2024).

Blended professional development models combining face-to-face and online learning platforms can expand access to teacher training opportunities, particularly in geographically dispersed education systems (Li et al., 2020).

Enhancing Infrastructure and Learning Resources

Investment in laboratory facilities, digital learning technologies, engineering design equipment, and maker-education spaces is essential for enabling experiential sustainability-focused STEM learning (Li et al., 2020). Education authorities should prioritise equitable distribution of infrastructure resources, particularly in underserved regions where access to STEM facilities may be limited (African Union, 2022). Open-access digital learning resources, virtual laboratories, and sustainability-focused instructional materials can further expand access to quality STEM education (UNESCO, 2020).

Promoting Community Partnerships and Innovation Ecosystems

Partnerships among schools, universities, industry organisations, research institutions, and community groups can strengthen sustainability-integrated STEM education by connecting classroom learning with real-world innovation challenges (Rizki & Suprapto, 2024). Community-based sustainability projects offer learners opportunities to apply STEM knowledge to local environmental and technological challenges while supporting community development. Innovation hubs and university–industry collaboration platforms can further enhance student engagement in sustainability-focused research and entrepreneurial activities.

Strengthening Monitoring, Evaluation, and Research Systems

Education authorities should establish monitoring and evaluation frameworks to track student sustainability competencies, teacher participation in professional development, infrastructure utilisation, and curriculum integration outcomes (Dare, et al., 2021). Evidence generated through evaluation studies can guide policy refinement and resource allocation decisions. Supporting collaborative research initiatives examining the impact of interdisciplinary STEM learning on student outcomes and workforce development can further strengthen evidence-based policymaking.

Synthesis: Toward Sustainable Transformation of STEM Education

Achieving sustainable transformation in STEM education requires coordinated action across policy alignment, teacher education reform, infrastructure investment, community partnerships, and evaluation systems. Education systems that adopt integrated approaches linking these components are better positioned to cultivate learners equipped with the competencies necessary to address environmental, technological, and socio-economic challenges (Wiek et al., 2011). Strategic investments in teacher capacity development, infrastructure expansion, and policy coordination will therefore be essential for positioning sustainability-integrated STEM education as a driver of innovation and sustainable development.

CONCLUSION

Education systems worldwide are undergoing significant transformation as societies confront increasingly complex environmental, technological, and socio-economic challenges. Sustainability-integrated STEM education has emerged as a strategic approach to equipping learners with the interdisciplinary competencies needed to address these challenges while supporting national development priorities and global sustainability goals. By combining scientific knowledge, technological innovation, engineering design, and systems-thinking approaches, integrated STEM education provides a platform for learners to develop the analytical, creative, and collaborative skills necessary for sustainable problem-solving.

The analysis presented in this chapter demonstrates that sustainability-oriented STEM education is not merely a curriculum innovation but a systemic transformation requiring coordinated action across policy frameworks, curriculum design, teacher preparation systems, infrastructure development, and community partnerships. Policy alignment establishes national priorities, funding mechanisms, and institutional support structures that enable education systems to effectively integrate sustainability competencies into STEM learning. Curriculum reforms that embed interdisciplinary problem-solving, inquiry-based learning, and engineering design processes provide the instructional foundation for developing sustainability competencies. In contrast, sustained teacher professional development initiatives ensure educators have the pedagogical expertise to implement innovative instructional approaches.

The chapter further highlights the importance of contextual adaptation in implementing sustainability-integrated STEM education, particularly within African and Ghanaian education systems. While global models offer valuable insights into effective curriculum integration and pedagogical practices, successful implementation in developing contexts requires careful consideration of institutional capacities, resource availability, and sociocultural environments. Investments in teacher education reform, infrastructure development, and digital learning technologies are essential for enabling schools to provide experiential STEM learning opportunities that connect classroom instruction with real-world sustainability challenges. Strengthening partnerships among education institutions, industry organisations, research centres, and community stakeholders further enhances the relevance and impact of STEM education initiatives by fostering locally grounded innovation ecosystems.

Student learning outcomes serve as the central measure of the effectiveness of sustainability-integrated STEM education. Interdisciplinary inquiry-based and project-based learning environments enable students to develop sustainability competencies, such as systems thinking, innovation, ethical reasoning, and collaborative problem-solving. These competencies are increasingly important for preparing learners to participate in emerging green economies, where demand for sustainability-oriented STEM expertise continues to grow. Education systems that prioritise experiential learning, community-based innovation projects, and the integration of digital learning are better positioned to cultivate learners capable of contributing to technological advancement and environmental stewardship.

The integrated conceptual framework proposed in this chapter emphasises that sustainable transformation in STEM education occurs when policy alignment, curriculum integration, teacher capacity development, pedagogical innovation, and learner competency development function as mutually reinforcing components of a coordinated education ecosystem. Monitoring, evaluation, and feedback mechanisms are also essential for sustaining long-term reform effectiveness by supporting continuous improvement and evidence-based decision-making. By adopting systemic approaches that align institutional structures, instructional practices, and sustainability priorities, education systems can strengthen STEM education's role as a catalyst for national development and long-term socio-economic resilience.

Looking ahead, continued research and policy commitment will be necessary to strengthen the long-term impact of sustainability-integrated STEM education initiatives. Comparative and context-responsive studies can provide insights into effective implementation strategies across diverse educational settings, while locally grounded research can inform reforms tailored to specific institutional environments. For Ghana and other African countries, sustained investment in teacher capacity development, the expansion of innovation-focused learning environments, and the strengthening of community-connected STEM initiatives will be critical to ensuring that education systems produce graduates capable of addressing pressing sustainability challenges.

In conclusion, sustainability-integrated STEM education represents a transformative pathway for preparing learners to contribute to environmentally responsible, technologically advanced, and socially inclusive development. By aligning education systems with sustainability priorities, nations can position STEM education as a central driver of innovation, workforce readiness, and long-term development in an increasingly complex global landscape.

REFERENCES

Abonyi, U. K., Boateng, F. K., Adjei-Boateng, E., & Ansaah, E. (2024). Promoting gender equity in school leadership appointments: A comparison of male and female headteachers' instructional leadership practices in Ghana. *Educational Management Administration & Leadership*, *52*(3), 610–628. DOI: 10.1177/17411432221090746

Adjei-Boateng, E. (2025). Examining pre-service teachers' journal reflection experiences during supported teaching in schools (STS) in Ghana: Benefits and challenges. *African Journal of Empirical Research*, *6*(3), 749–760. https://ajernet .net. DOI: 10.51867/ajernet.6.3.57

Adjei-Boateng, E., Cobbinah, J. E., & Adamba, C. (2025). Headteachers' and teachers' perspectives on the work of school improvement support officers in selected basic schools in Ghana. *Teachers and Teaching*, •••, 1–19. DOI: 10.1080/13540602.2025.2526637

Advance, H. E. (2025). *Framework for education for sustainable development*. https://www.advance-he.ac.uk

African Union. (2022). *Continental education strategy for Africa (CESA 16–25): Emphasising STEM for economic and sustainable development*. African Union Commission.

Agai, M. J. (2024). The role of STEM education in teaching and learning in South Africa's underprivileged societies. *Innovare Journal of Education*, *12*(2), 50350. Advance online publication. DOI: 10.22159/ijoe.2024v12i2.50350

Aikenhead, G. S., & Ogawa, M. (2007). Indigenous knowledge and science revisited. *Cultural Studies of Science Education*, *2*(3), 539–620. DOI: 10.1007/s11422-007-9067-8

Akon-Yamga, G., Funkor, G., Tsey, K., Kingsford-Adaboh, R., Quaye, W., & Amoni Ntewusu, D. (2024). Perspectives from students and teachers about the challenges of teaching and learning STEM subjects in Ghana. *Frontiers in Education*, *9*, 1288413. DOI: 10.3389/feduc.2024.1288413

Akyeampong, K., Lussier, K., Pryor, J., & Westbrook, J. (2018). Improving teaching and learning of basic maths and reading in Africa: Does teacher preparation count? *International Journal of Educational Development*, *58*, 12–23. DOI: 10.1016/j. ijedudev.2017.08.003

Amponsah, K. D., Mensah, F., & Mensah, A. (2014). Constraints experienced by female students pursuing science and technology-based university programs in their learning of science in Ghana. *International Journal of Research Studies in Education*, *3*(4), 1–12. DOI: 10.5861/ijrse.2014.684

Anning, A. S. (2024). Investigating STEM teachers' personal and sociocultural particularities in the Ghanaian context. *International Journal of Educational Research*, *125*, 102341. DOI: 10.1016/j.ijer.2024.102341

Antinluoma, M., Ilomäki, L., & Toom, A. (2021). Practices of professional learning communities. *Frontiers in Education*, *6*, 617613. DOI: 10.3389/feduc.2021.617613

Asoma, R. K., Agyei, D. D., Assamah, G., & Amponsah, K. D. (2024). Examining teachers' perceptions of the impact of the Government of Ghana's Wi-Fi technology program on teaching practices: An empirical study from senior high schools. *Cogent Education*, *11*(1), 2302764. Advance online publication. DOI: 10.1080/2331186X.2024.2302764

Bardoe, D., Hayford, D., Bio, R. B., & Gyabeng, J. (2023). Challenges to the implementation of STEM education in the Bono East Region of Ghana. *Heliyon*, *9*(10), e20416. DOI: 10.1016/j.heliyon.2023.e20416 PMID: 37822609

Bou Saad, R., Llorens Garcia, A., & Cabre, J. M. (2025). Mapping constructivist active learning for STEM: Toward sustainable education. *Sustainability (Basel)*, *17*(13), 6225. DOI: 10.3390/su17136225

Budu, G. B., Kemetse, J. K., & Amponsah, K. D. (2022). A Ghanaian study on multiple intelligences of pre-service science teachers in selected colleges of education. *Education Quarterly Eeview*, *5*(1), 15–27. DOI: 10.31014/aior.1993.05.01.414

Bybee, R. W. (2020). The role of STEM education in sustainable development. *Journal of Research in Science Teaching*, *57*(5), 752–765. DOI: 10.1002/tea.21640

Castaño, C., Caballero, R., Noguera, J. C., Chen Austin, M., Bernal, B., Jaén-Ortega, A. A., & Ortega-Del-Rosario, M. D. L. Á. (2025). Developing sustainability competencies through active learning strategies across school and university settings. *Sustainability (Basel)*, *17*(19), 8886. DOI: 10.3390/su17198886

Chisango, G., & Mavhunga, E. (2024). Gender participation and engagement in STEM learning in African schools. *International Journal of STEM Education*, *11*(1), 1–15. DOI: 10.1186/s40594-024-00472-9

Chisom, O. N., Unachukwu, C. C., & Osawaru, B. (2024). STEM education advancements in African contexts: A comprehensive review. *World Journal of Advanced Research and Reviews*, *21*(1), 145–160. DOI: 10.30574/wjarr.2024.21.1.2719

Cordaro, J. A., Murphy, C., & Redman, E. (2025). Bridging STEM education and sustainability: Insights from Pennsylvania educators. *Education Sciences*, *15*(3), 282. DOI: 10.3390/educsci15030282

Crawford, B. A. (2014). From inquiry to scientific practices in the science classroom. In Lederman, N. G., & Abell, S. K. (Eds.), *Handbook of research on science education* (1st ed., Vol. 2, pp. 515–541). Routledge.

Dare, E. A., Keratithamkul, K., Hiwatig, B. M., & Li, F. (2021). Beyond content: The role of STEM disciplines, real-world problems, 21st century skills, and STEM careers within science teachers' conceptions of integrated STEM education. *Education Sciences*, *11*(11), 737. DOI: 10.3390/educsci11110737

Falasi, M. (2024). Innovative pedagogies: A comparative analysis of traditional and modern teaching methods. *Academy of Educational Leadership Journal*, *28*(S1), 1–2.

Fazio, X. (2024). STEM education: International perspectives in STEM education. In *Global perspectives on STEM education* (pp. xx–xx). Springer., DOI: 10.1007/978-3-031-60676-2_6

Fitzgerald, M., Danaia, L., & McKinnon, D. H. (2019). Barriers inhibiting inquiry-based science teaching and potential solutions: Perceptions of positively inclined early adopters. *Research in Science Education*, *49*(2), 543–566. DOI: 10.1007/s11165-017-9623-5

Global Partnership for Education. (n.d.). *Coordinated action to transform education: What's in it for ministries of education?* https://www.globalpartnership.org

Gouëdard, P., Pont, B., Hyttinen, S., & Huang, P. (2020). *Curriculum reform: A literature review to support effective implementation* (OECD Education Working Paper No. 239). Organisation for Economic Co-operation and Development. DOI: 10.1787/5cc2d673-en

Hariharasakthisudhan, P., Logesh, K., Sathickbasha, K., & Kannan, S. (2025). Enhancing project-based learning in engineering education: A hybrid DT-CDIO-RA framework for sustainable product design. *Discover Education*, *4*(1), 138. Advance online publication. DOI: 10.1007/s44217-025-00579-3

Hendrickx, M. M. H. G., Thurlings, M. C. G., & Den Brok, P. (2025). Teachers' collaborative knowledge building in professional learning communities: Connecting interaction patterns to learning gains. *European Journal of Psychology of Education*, *40*(1), 39. DOI: 10.1007/s10212-024-00938-y

Homer, S. T., Chuen, P. W., & Chuah, C. (2026). Sustainable development education through student engagement: Findings from a comprehensive study. *Quality & Quantity, 60*(1), 1–31. DOI: 10.1007/s11135-025-02237-0

Inter-agency Task Force on Financing for Development. (n.d.). *Digital infrastructure investment: Collaborative approaches and mechanisms to finance universal connectivity*. United Nations.

Jangde, S., & Ahmad, M. A. (2025). Integrating sustainability into equal education opportunities. In Parray, J. A., Shameem, N., & Haghi, A. K. (Eds.), *Environmental landscape and sustainable biodiversity for healthy green growth* (pp. xx–xx). Springer., DOI: 10.1007/978-3-031-75405-0_10

Kelley, T. R., & Knowles, J. G. (2016). A conceptual framework for integrated STEM education. *International Journal of STEM Education, 3*(1), 11. DOI: 10.1186/s40594-016-0046-z

Krajcik, J., & Merritt, J. (2019). Designing project-based science learning experiences to promote sustainable development. In Leal Filho, W. (Ed.), *Encyclopedia of sustainability in higher education*. Springer., DOI: 10.1007/978-3-319-63951-2_130-1

Lee, M. Y., & Lee, J. S. (2025). Project-based learning as a catalyst for integrated STEM education. *Education Sciences, 15*(7), 871. DOI: 10.3390/educsci15070871

Li, Y., Wang, K., Xiao, Y., & Froyd, J. E. (2020). Research and trends in STEM education: A systematic review of journal publications. *International Journal of STEM Education, 7*(1), 11. DOI: 10.1186/s40594-020-00207-6

Marginson, S., Tytler, R., Freeman, B., & Roberts, K. (2024). *STEM education and global innovation systems*. Routledge. https://www.routledge.com/STEM-education-and-global-innovation-systems/Marginson-Tytler-Freeman-Roberts/p/book/9780367746353

McNeill, K. L., Pimentel, D. S., & Pimentel, E. S. (2022). Inquiry-based instruction: A systematic review of its effects on student science learning. *Review of Educational Research, 92*(5), 705–740. DOI: 10.3102/00346543221086147

Mensah, R. O., Amponsah, K. D., Adiza Babah, P., & Jibril, H. S. (2024). Factors affecting students' academic performance and teachers' efficiency in Ghana; a case study of Wa senior high school. *Cogent Arts & Humanities, 11*(1), 2412944. Advance online publication. DOI: 10.1080/23311983.2024.2412944

Mohammed, S. M., & Amponsah, K. D. (2021). Junior high school teachers' attitudes toward inquiry-based science teaching: Enabling or disabling dispositions? *Journal of Education and Training Studies, 9*(7), 41–54. DOI: 10.11114/jets.v9i7.5266

National Council for Curriculum and Assessment. (2019). *Standards-based curriculum for basic schools*. https://nacca.gov.gh

Nusche, D., Fuster Rabella, M., & Lauterbach, S. (2024). *Rethinking education in the context of climate change: Leverage points for transformative change* (OECD Education Working Paper No. 307). Organisation for Economic Co-operation and Development. DOI: 10.1787/abc5a69e-en

Organisation for Economic Co-operation and Development (OECD). (n.d.). *Future of education and skills 2030/2040*. https://www.oecd.org/en/about/projects/future-of-education-and-skills-2030.html

Park, S., & Oliver, J. S. (2008). Revisiting the conceptualisation of pedagogical content knowledge (PCK): PCK as a conceptual tool to understand teachers as professionals. *Research in Science Education*, *38*(3), 261–284. DOI: 10.1007/s11165-007-9049-6

Pedaste, M., Mäeots, M., Siiman, L. A., de Jong, T., van Riesen, S. A., Kamp, E. T., Manoli, C. C., Zacharia, Z. C., & Tsourlidakis, H. (2015). Phases of inquiry-based learning: Definitions and the inquiry cycle. *Educational Research Review*, *14*, 47–61. DOI: 10.1016/j.edurev.2015.02.003

Portillo-Blanco, A., Guisasola, J., & Zuza, K. (2025). Integrated STEM education: Addressing theoretical ambiguities and practical applications. *Frontiers in Education*, *10*, 1568885. DOI: 10.3389/feduc.2025.1568885

Quintana-Cifuentes, P., González, F., & Sjøberg, S. (2019). Systems thinking and sustainability competencies in STEM education: A systematic review. *Environmental Education Research*, *25*(10), 1435–1453. DOI: 10.1080/13504622.2019.1671839

Rieckmann, M. (2012). Future-oriented higher education: Which key competencies should be fostered through university teaching and learning? *Futures*, *44*(2), 127–135. DOI: 10.1016/j.futures.2011.09.005

Rizki, I., & Suprapto, N. (2024). Project-oriented problem-based learning through SR-STEM to foster students' critical thinking skills in renewable energy material. *Journal of Science Education and Technology*, *33*(4), 526–541. DOI: 10.1007/s10956-024-10102-2

Saavedra, A. R., & Rapaport, A. (2024). Key lessons from research about project-based teaching and learning. *Phi Delta Kappan*, *105*(5), 19–25. DOI: 10.1177/00317217241230780

Salonen, A. O., & Aarnio-Linnanvuori, K. (2020). Phenomenon-based learning and the circular economy: The Finnish approach to systems thinking. *Sustainability*, *12*(4), 1362. DOI: 10.3390/su12041362

Scheie, E., Arntzen, M., & Haug, B. S. (2025). Teachers' implementation of self-designed interdisciplinary curriculum units in education for sustainable development. *Research in Science Education*, *55*(4), 793–815. DOI: 10.1007/s11165-025-10279-5

Tafese, M. B., & Kopp, E. (2025). Embedding social sustainability in education: A thematic review of practices and trends across educational pathways from a global perspective. *Sustainability (Basel)*, *17*(10), 4342. DOI: 10.3390/su17104342

Tan, C. Y., Jang, S. T., Lam, S. M., Qi An, A., & Lo, U. K. V. (2025). Teaching and learning challenges due to the COVID-19 pandemic: A systematic review. *Educational Research Review*, *47*, 100667. DOI: 10.1016/j.edurev.2025.100667

Tarlochan, F., Alduais, A., Chaaban, Y., & Du, X. (2025). Integrating sustainability into STEM education and career development: A scientometric and narrative review. *International Journal of STEM Education*, *12*(1), 62. DOI: 10.1186/s40594-025-00582-y

Toma, R., Yánez-Pérez, I., & Meneses-Villagrá, J. (2024). Towards a Socio-Constructivist Didactic Model for Integrated STEM Education. *Interchange*, *55*(1), 75–91. DOI: 10.1007/s10780-024-09513-2

UNESCO. (2020). *Education for sustainable development: A roadmap*. UNESCO. https://unesdoc.unesco.org/ark:/48223/pf0000374802

UNESCO. (2026, April 8). *What you need to know about education for sustainable development*. https://www.unesco.org/en/education/sustainable-development/what-you-need-to-know

United Nations. (2015). *Transforming our world: The 2030 agenda for sustainable development*. https://sustainabledevelopment.un.org/post2015/transformingourworld

Van Wart, A., O'Brien, T. C., Varvayanis, S., Alder, J., Greenier, J., Layton, R. L., Stayart, C. A., Wefes, I., & Brady, A. E. (2020). Applying experiential learning to career development training for biomedical graduate students and postdocs: Perspectives on program development and design. *CBE Life Sciences Education*, *19*(3), es7. DOI: 10.1187/cbe.19-12-0270 PMID: 32822277

West African Examinations Council (WAEC). (2022). *Regulations and syllabuses for the West African Senior School Certificate Examination*. RULES AND REGULATIONS – The West African Examinations Council.

Wiegant, D., Dewulf, A., & Van Zeben, J. (2024). Alignment mechanisms to effectively govern the sustainable development goals. *World Development*, *182*, 106721. DOI: 10.1016/j.worlddev.2024.106721

Wiek, A., Withycombe, L., & Redman, C. L. (2011). Key competencies in sustainability: A reference framework for academic program development. *Sustainability Science*, *6*(2), 203–218. DOI: 10.1007/s11625-011-0132-6

Yeboah, R., Amponsah K. D., Commey-Mintah, P., Sedofia, J., & Donkor (Opare), P. B. K. (2025). Game-based learning in Ghanaian primary schools: listening to the views of teachers. *Education 3-13, 51*(2). DOI: 10.1080/03004279.2023.2171269

Zajda, J. (2021). Constructivist Learning Theory and Creating Effective Learning Environments. In *Globalisation and Education Reforms. Globalisation, Comparative Education and Policy Research* (Vol. 25). Springer., DOI: 10.1007/978-3-030-71575-5_3

Zeidler, D. L. (2016). STEM education: A deficit framework for the twenty-first century? A sociocultural socioscientific response. *Cultural Studies of Science Education*, *11*(1), 11–26. DOI: 10.1007/s11422-014-9578-z

Chapter 9
Assessing Barriers and Opportunities for Integrating Climate Change Education in Marine Science Content Among Secondary School Teacher Training Programs

Alex Kanyimba
The International University of Management, Namibia

Timoteus Kadhila
http://orcid.org/0000-0001-5566-2288
The International University of Management, Namibia

ABSTRACT

Despite Namibia's dependence on the resources harvested off the Benguela Current ecosystems, climate change education in Namibia primarily focuses on land with minimal integration of marine content. This chapter examines barriers and opportunities for integrating climate change education in marine science content among secondary school teacher training programmes in Namibia. The research employs a qualitative, interpretivist approach involving document analysis, interviews, and case studies of existing climate education initiatives. The findings reveal significant

DOI: 10.4018/979-8-3373-6975-4.ch009

gaps in teachers' ocean climate literacy, a lack of resources and limited understanding of the ocean's role in climate regulation. There is an opportunity to introduce context-specific marine learning activities into the secondary school curriculum due to policy frameworks and institutional structures supporting climate change education in Namibia. There is a strong policy and institutional support and growing interest in marine and environmental studies. The Namibia's Education for Sustainable Development (ESD) strategy for 2030 is actively working on a template to integrate climate awareness into formal and non-formal education, including teacher education. The chapter suggests teacher development strategies that can enhance climate teaching and promote marine ecological connections following transformative learning pedagogies. The recommendations focus on curriculum reform, improving teacher competence, and utilising Namibia's unique maritime and coastal settings as starting points for teaching about climate change.

INTRODUCTION AND BACKGROUND

Namibia is highly vulnerable to climate change, with projections showing faster-than-average temperature increases, leading to frequent droughts, floods, and extreme heat events (Spear, 2018). The strategies to address these changes rest on mitigation and adaptation. Mitigation focuses on reducing the root causes by lowering greenhouse gas emissions, while adaptation involves adjusting to the current and future impacts that are already occurring or unavoidable. Effective climate action requires both approaches, as they are complementary and address different aspects of the climate change scenarios.

In Southern Africa, and Namibia in particular, the urgency to integrate climate change education (CCE) into the curriculum has never been greater (see the work of Kanyimba, Nantinda and Namutenya (2014). Education is a critical tool for both climate change mitigation and adaptation, as it can change behaviours, build green skills, and increase resilience. Education is increasingly recognized as a critical tool for empowering communities to mitigate and adapt to these impacts. Climate change education helps people understand and address the impacts of the climate crisis, empowering them with the knowledge, skills, values and attitudes needed to act as agents of change.

Marine science education in Namibia is becoming an important focus for climate change education programmes (Blue Marine Foundation, 2024). The report by the Foundation Tara Ocean Org (n.d) explains why it is crucial for climate change education initiatives to concentrate on marine environments. The first reason is that the ocean plays a vital role in regulating the climate producing oxygen and absorbing approximately 90% of the excess heat from human greenhouse gas emissions. The second reason is human dependency. It emphasises the crucial role the ocean plays in human life, providing food (with nearly half the world's population relying on fish for protein), livelihoods for millions, and essential resources. Additionally, teaching about the ocean helps students to grasp the Interconnectedness of all ecosystems and how climate change impacts them, from local communities to global systems. Simultaneously, the climate crisis is negatively affecting access to education and learning. Extreme weather events, exacerbated by climate change, pose a threat to schooling, learning and wellbeing. These impacts disproportionately affect poorer students.

The data and information presented in this chapter adopt a qualitative and interpretivist approach drawing on conceptual data sources. The chapter also includes the following data collection methods: Document analysis of curriculum frameworks and national education policies in Namibia.

The remainder of this chapter follows this sequence: Firstly, the chapter presents the theoretical framework followed climate change in marine spaces and its effects. Next is the policy frameworks and institutional structures supporting climate change education in Namibia. The next subsections touches on barriers as well as the opportunities or integrating climate change education in the marine science curricula. The final subsection presents the conclusions and the way forward, including the recommendations for infusing climate change education in the Marine science curriculum at teacher education levels

THEORECTICAL FRAMEWORK

The study to assess barriers and opportunities for integrating climate change education in marine science content among secondary school teacher training programmes in Namibia is guided by the Transformative Learning Theory (Mezirow, 1997). Transformative learning theory, primarily developed by Jack Mezirow, is an educational framework focused on how learners critically examine and change their deeply held assumptions, beliefs, and worldviews. The focus of the transformative learning theory is on transformation "the process of effecting change in a frame of reference". The analysis of the foregoing assertion is that about the move for educators to not only question their assumptions but also act based on a new, more inclusive perspective. Figure 1 shows the Transformative learning process and stages that may be adopted in climate change education.

Figure 1. Transformative learning process in climate change education

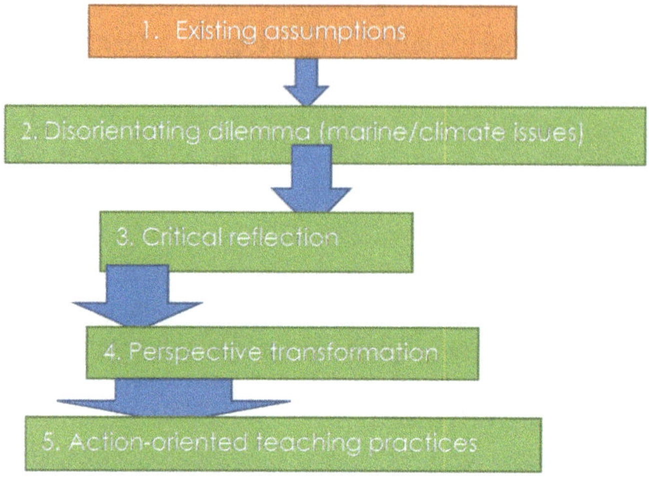

Source: Authors' own design (2026)

The relevance of this theory in the study of climate change education goes beyond content transmission to transformative experiences that reshape learners' worldviews and promote critical reflection and action. Within teacher training programmes, transformative learning enables pre-service and in-service teachers to:

- Critically reflect on their understanding of climate change and human-environment interactions;
- Question dominant narratives related to resource use, development, and environmental responsibility;
- Develop new perspectives that recognize climate change as a socio-ecological challenge rather than a purely scientific issue; and
- Translate new understandings into pedagogical practices that encourage learner agency, ethical reasoning, and climate action.

By integrating marine science content such as ocean circulation, coastal ecosystems, fisheries, and climate-ocean interactions, teacher education programmes can create disorienting dilemmas (e.g., declining fish stocks, sea-level rise, marine pollution) that prompt reflective learning. This process fosters transformative shifts in teachers' professional identities, enabling them to act as change agents who promote climate awareness and sustainability in secondary schools across Namibia.

CLIMATE CHANGE AND ITS EFFECTS ON THE NAMIBIAN OCEAN SPACE

Climate Change in a marine context can be defined as the long-term changes in ocean temperature, sea level, ocean chemistry, and marine weather patterns that affect marine organisms, fisheries, and coastal ecosystems (IPCC, 2019). Namibia's marine environment is part of the Benguela Current Large Marine Ecosystem (BCLME), which is known as one of the most productive yet climate-sensitive ocean systems globally (Barnard, 2014; Hutchings et al. 2006). This productivity supports fisheries, biodiversity, and socio-economic development. However, climate change manifested through rising sea temperatures, weakened upwelling, ocean acidification, and shifting current systems poses serious risks to this ecosystem (IPCC, 2019). Table 1 presents the physical and impact on species resulting from climate change on the Namibian Ocean.

Table 1. Summary of Climate change impacts on the Namibian Ocean

Climate change impact	Elaboration
Benguela Current System Disruptions	The cold, nutrient-rich Benguela Current is undergoing shifts, with reduced upwelling potentially causing lower nutrient levels and reduced fishery yields.
Marine Ecosystem Shifts	Rising ocean temperatures and deoxygenated water are linked to increased jellyfish populations, which impact fish stocks. The spawning success of species like pilchard is compromised by these changing conditions.
Ocean Acidification	The marine ecosystem is facing increased acidity, which alters water quality and threatens marine biodiversity.
Sea Level Rise and Coastal Erosion	The coastline is experiencing increased erosion, with areas like the Pelican Peninsula in Walvis Bay showing significant narrowing (390m between 1984 and 2014). Sea levels are projected to continue rising, threatening coastal infrastructure in Walvis Bay, Swakopmund, and Lüderitz.
Migration of Cold-Water Species	Species dependent on oxygen-rich, cold water such as hake and sardine experience stress as warming and low-oxygen zones expand (Hutchings et al. 2006; BCC, 2018). This results in reduced growth rates, lower reproductive success, increased mortality and distributional shifts (IPCC, 2019).
Decline in Marine Fish Stocks	The reduced nutrient supply decreases phytoplankton production, leading to declines in key fish species such as sardine, anchovy, and horse mackerel (Boyer et al. 2001).
Changes in Migration Patterns and Spawning Cycles	Climate change alters fish distribution, abundance, and changes in migratory behaviour in the BCLME (Jarre et al. 2013; Timmermann & Friedrich, 2016) complicate fisheries management and affect resource availability.

Source: Based on (FAO, 2020; World Bank, 2022), Moltke & Blythe, 2020)

In summary, physical changes associated with climate change such as rising temperatures, altered rainfall patterns, ocean warming, sea-level rise, and increased frequency of extreme weather events have direct and indirect impacts on food security, economic stability, and social well-being. Climate change education is an essential factor in the increasingly urgent global fight against climate change. Knowledge regarding this phenomenon helps teachers to understand and integrate global warming, encourages them to change their behavior and assists them in adapting to what is already a global emergency. Overall, these interconnected impacts demonstrate that physical climate and marine changes are not isolated environmental phenomena but systemic challenges that affect food systems, economic resilience, social stability, and sustainable development. Understanding these linkages is essential for effective climate change education and for preparing teachers and learners to respond to climate-related challenges in Namibia.

THE NAMIBIAN CONTEXT: POLICY FRAMEWORKS AND INSTITUTIONAL STRUCTURES SUPPORTING CLIMATE CHANGE EDUCATION

The discussion of policy frameworks and institutional structures supporting climate change education in Namibia will focus on international agreements and frameworks as well as the national policies that support climate change education in Namibia

International Agreements and Frameworks

Climate change education in Namibia is anchored in several influential global policy instruments. This section will focus on the United Nations Framework Convention on Climate Change (UNFCCC) and the Paris Agreement.

UNFCCC

The United Nations Framework Convention on Climate Change (UNFCCC) obliges parties to promote climate education, training and public awareness (UNFCCC, 1992). Article 6 of the UNFCC focuses on education, training and public awareness and presents the following items.

(a) Promote and facilitate at the national and as appropriate, sub-regional and regional levels, and in accordance with national laws and regulations, and within their respective capacities:
   (i) The development and implementation of educational and public awareness programmes on climate change and its effects;
   (ii) Public access to information on climate change and its effects;

(iii) Public participation in addressing climate change and its effects and developing adequate responses; and

(iv) Training of scientific, technical and managerial personnel.

(b) Cooperate in and promote, at the international level, and, where appropriate, using existing bodies:

  (i) The development and exchange of educational and public awareness material on climate change and its effects; and

  (ii) The development and implementation of education and training programmes, including the strengthening of national institutions and the exchange or secondment of personnel to train experts in this field, in particular for developing countries.

As shown above, Article 6 of the UNFCC emphasises the importance of educational and public awareness programmes on climate change and its impacts. It also specifically states the necessity for public participation, access to information, and educational initiatives to enhance societal resilience. Article 6 also emphasise the importance of cooperation and promoting educational opportunities using existing resources. Additionally, it calls for education and training programmes, including the enhancement of national institutions. This implies the need to strengthen teacher education institutions as well.

The Paris Agreement

The Paris Agreement reinforces these mandates by emphasizing the role of climate education in enhancing adaptive capacity and promoting sustainable development (UNFCCC, 2016). The review of the Paris agreement document shows that the document focuses on a number of areas including mitigation and nationally determined contributions. The document further outlines that in the NDCs, countries must communicate actions they will take to reduce their greenhouse gas emissions in order to reach the goals of the Paris Agreement and their NDCs actions they will take to build resilience to adapt to the impacts of climate change. Thus, the publication by Streck, Keenlyside and von Unger (2016) asserts that the Paris Agreement places great emphasis on climate-related capacity-building for developing countries and requests all developed countries to enhance support for capacity-building actions in developing countries.

Namibia ratified the Paris Agreement in 2016. This entails that Namibia is committed to ambitious climate action as a highly vulnerable, arid nation and aims to reduce greenhouse gas emissions by 91% by 2030 by focusing on renewable energy, sustainable agriculture, and conservation.

National Policies That Support Climate Change Education in Namibia

Namibia has developed a robust policy environment to address climate change and its challenges. The key policies that are to be discussed in the section include Namibia Vision 2030, the sixth National Development Plan for Namibia (NDP 6), the National Policy on Climate Change (NPCC), and the Blue Economy Policy for Namibia.

The Namibia Vision 2030

The Namibia Vision 2030 outlines the broad thrusts and strategies for the nation's long-term development agenda. The vision for Namibia is to become a prosperous and industrialised nation, developed by her human resources, enjoying peace, harmony and political stability by the year 2030 (Government of the Republic of Namibia, 2004). The Namibia Vision 2030 prides itself on the principle of sustainable development. The following are the driving forces of the development trajectory:

- Education, Science and Technology
- Health and Development
- Sustainable Agriculture,
- Peace and Social Justice
- Gender Equality

The review of Namibia's Vision 2030 shows that it aims for a prosperous, industrialized nation, with the blue economy, fisheries, marine mining, and tourism serving as a critical pillar for sustainable development and economic growth. It is suggested in reports that by 2030, Namibia aims to protect 30% of its marine areas, focusing on sustainable management, and aligning with SDG 14. This goal is part of a broader commitment to sustainable ocean management, supported by international partnerships such as the 'High Level Panel for A Sustainable Ocean Economy' (Government of the Republic of Namibia, 2004a).

The National Development Plan for Namibia (NDP 6)

Namibia has committed to achieve national goals by 2030, however, to get to these national goals, National Development Plans (NDPs) have been developed. The NDPs are implemented after every five years, with the first National Development Plan (NDP 1) launched in 1995 to the year 2000, the second National Development Plan (NDP 2) was in force from 2001 to 2006 and the third National Development Plan

(NDP 3) was implemented from 2007 to 2011. The fourth National Development Plan (NDP 4) was implemented from 2012/13 to 2016/17 and focused mainly on achieving high and sustained economic growth, employment creation, and reducing income inequality. This was followed by the fifth National Development Plan (NDP 5), which was implemented from 2017 to 2022 and aimed to improve the living conditions of all Namibians while contributing to the achievement of Vision 2030 (Government of the Republic of Namibia, 2004b). NDP 5 was structured around four main goals, namely: inclusive and sustainable economic growth, capable and healthy human resources, environmental sustainability and resilience, and good governance.

Within the NDP 5, there is also Thematic Area 4, which supports education for development. It is asserted that education is a fundamental right in Namibia, protected by the Constitution and the Basic Education Act 2020 (Act No. 3 of 2020) which mandate compulsory school attendance and marks a significant milestone. The Education for Development thematic area covers eight focus areas: integrated early childhood development (ECD), primary education; secondary education; life-long learning, technical, vocational training and skills development, higher education, integrated educational infrastructure, research, and science technology and innovation.

Currently, Namibia is implementing the sixth National Development Plan (NDP 6). NDP 6 has a section called Pillar 3: Environmental Sustainability. This pillar promotes the goals of achieving sustainable development through improved resilience, adaptability and the (green) growth opportunities.

From the above assertions, it can be inferred that NDP 6 supports climate change education by prioritizing environmental sustainability, climate resilience, and green growth across key sectors, including education, fisheries, agriculture, water, and tourism. This also means that the NDP 6 plan emphasises the need for an informed and skilled population capable of responding to climate risks and contributing to sustainable development pathways. This creates a clear mandate for integrating climate change concepts into formal education systems, including teacher training programmes.

Based on the above, it can be reasoned that NDP 6 provides a strong policy foundation for climate change education in particular and environmental sustainability in general. NDP 6 emphasizes the quality of curriculum, relevance, and responsiveness of education and training to national development priorities. Climate change education is emphasised, especially when integrated with marine science content aligns with this objective by equipping teachers with interdisciplinary knowledge and pedagogical skills to address real-world environmental challenges.

The Blue Economy Policy for Namibia

The draft Namibia's National Blue Economy Policy (2022-2031) focuses on the sustainable, integrated development of marine resources, balancing economic growth with environmental protection and social equity. The study by Machaka and Warikandwa (2025) indicates that the blue economy policy for Namibia is driven by the SDG 14 'Life below Water' one of 17 the UN Sustainable Development Goals for 2030. The gaols are focused on conserving and sustainably using oceans, seas, and marine resources. Key targets include reducing marine pollution, protecting ecosystems, addressing acidification, and ending overfishing to maintain healthy oceans.

The Blue Economy Policy for Namibia focusses on the development of blue fisheries (handled by the Ministry of Agriculture, Fisheries, Water and Land Reform, blue Tourism (Handled by the Ministry of Environment, Forestry and Tourism to promote sustainable tourism practices in aquatic ecosystems to ensure positive externalities for the environment, workers and local communities), blue Mining (managed by the Ministry of Mines and Energy for the exploration and extraction of various mineral and hydro-carbon resources), blue renewable energy (handed by the Ministry Mines and Energy for the generation and distribution of renewable energy such as wind, green hydrogen, hydrothermal, wave and tidal energy); blue maritime transport, ports and shipping (regulated by the Ministry of Works and Transport for movements of people and goods using the sea and inland waterways), blue water resources (managed by the Ministry of Agriculture, Fisheries, Water and Land Reform) and emerging or potential blue economy activities such as blue bio-prospecting and carbon Trading. There are also other emerging activities that are cross cutting, which will impact the blue space and the resources derived from the blue space.

The Namibia blue economy policy provides strategic direction on the principles of sustainable production, environmental protection, climate resilience, social of equity and Inclusivity, integrated governance, Regional and International collaboration, and also human capital development. This includes Investing in the human capital required to develop Namibia's Blue Economy.

Based on the preceding assertion, it can be reasoned that the Blue Economy provides a strong policy foundation for climate change education particularly by focussing on environmental sustainability in general and climate change education, particular. Climate change education is especially important when integrated with marine science content equipping teachers with interdisciplinary knowledge and pedagogical skills.

The National Policy on Climate Change

Namibian National Policy on Climate change (2011) Namibia strives for excellence in addressing climate change as a challenge by responding in a timely, effective and appropriate manner through exploring adaptation and mitigation approaches relevant to different sectors at the local, regional and national level in order to improve the quality of life of its citizens (Government of the Republic of Namibia, 2011). In particular the policy focused on Namibia's marine fisheries, which rely on the nutrient-rich upwelling of the Benguela Current system. The policy asserts that species are threatened by possible changes brought by climate change. When upwelling is suppressed by northerly or easterly winds, oxygen-poor water can accumulate near the seabed and suffocate marine life. Future changes in the distribution. and intensity of winds could therefore affect the fisheries sector

The National Policy on Climate Change promotes Education, Training, Capacity Building and Institutional strengthening. It is further explained in that document that:

> *"Education, training and capacity building should involve multiple stakeholders, including the government, NGOs, research institutions, local communities and international organisations (Government of the Republic of Namibia, 2011)*

As can be seen in the proceeding assertion the National Policy on Climate Change is increasingly supporting a cross-curricular approach to climate change education. This approach moves away from treating it solely as a science topic to integrating it across all subjects, including arts, humanities, and social studies, to foster a holistic, systems-thinking perspective. The goal of this approach is to equip teachers with the knowledge, skills, and values needed for teaching climate resilience and taking sustainable action.

The Influence of Educational Policies on Climate Change Education in Namibia

The 2030 Initiative on Education for Sustainable Development (ESD)

The ESD for 2030 road map outlines the pressing challenges facing the planet and emphasises the implementation of the *Education for Sustainable Development: Towards achieving the SDGs (ESD for 2030)* framework. The ESD for 2030 directly targets Quality Education, specifically Target 4.7 of the UN Sustainable Development Goals (SDG 4), which aims to ensure that by 2030, all learners acquire the knowledge and skills to promote sustainable development. It emphasises education for sustainable lifestyles, human rights, gender equality, peace, non-violence, global citizenship, and cultural diversity.

The ESD for 2030 framework focuses on five priority action areas

- **Advancing Policy:** Integrating ESD into international and national education and sustainable development policies.
- **Transforming Learning Environments:** Promoting a whole-institution approach where sustainability is integrated into teaching, management, and operations.
- **Building Capacities of Educators:** Empowering teachers and trainers to effectively deliver ESD.
- **Empowering and Mobilizing Youth:** Engaging young people as key actors in sustainable development.
- **Accelerating Local Level Actions:** Scaling up community-based sustainability initiatives.

As part of the "Education for Sustainable Development: Towards achieving the SDGs (ESD for 2030)" framework, UNESCO encourages every Member State to develop, implement, and monitor its own national strategy or "country initiative" to achieve the gaols set for 2030 (UNESCO, 2021). Thus, Namibia is currently in the advanced process of developing the ESD for 2030 country initiative. The goals of this initiative are to "reorient, integrate and upscale quality Environmental Education (EE) and Education for Sustainable Development (ESD) in environmental awareness, education and training systems, research and innovation systems, policies, programmes, and action for sustainable development (National EE/ESD Policy 2019). Moreover, the initiative is based on the National EE/ESD Policy and the National EE/ESD Strategy & Action Plan (2022-2026) was developed. This strategy has streamlined the policy strategies into five main priority areas as follows:

- EE/ESD Programmes, Training and Capacity Building
- Networking and Strengthening Partnerships
- Development and expansion of EE/ESD Curriculum and Learning Support Materials
- Developing Responsible Citizenship Action & Advocacy
- Promotion of Research, Innovation and Centres of Expertise

As seen above, the ESD for 203 initiative is being customised in Namibia. This framework aims to create an enabling environment for Namibia to incorporate climate change education and has potential to include marine science curricula and broader environmental studies programs.

The National Curriculum for Basic Education in Namibia

The National Curriculum for Basic Education builds on the experience and achievements of previous cycles of Namibian curricula and syllabuses since 1990. It is based on the Namibia Vision 2030, the Namibian Constitution and the Namibia Education Act. One of the goals of the National Curriculum for Basic Education in Namibia is to create an 'environmentally sustainable society'. Therefore, the National Curriculum for Basic Education in Namibia clearly articulates the following in that regard:

> *There is no atmospheric, land or water pollution from croplands, rangelands or mines, and minimal pollution from urban and industrial areas. Farms and natural ecosystems are productive and sustainable socially, economically and ecologically. There is high quality, low- impact tourism. The average family size is small, and there is food security (MoEAC, 2016).*

In addition, the National Curriculum for Basic Education incorporates cross-cutting themes like environment, sustainability, and climate change. This offers an opportunity to seamlessly integrate ocean and climate content into marine science in teacher education curriculum.

BARRIERS TO THE INTEGRATION OF MARINE SCIENCE CONTENT IN SECONDARY SCHOOL TEACHER TRAINING PROGRAMMES IN NAMIBIA

Based on the review of the key documents, such as the policy frameworks and institutional structures supporting climate change education in Namibia, the following barriers appears to affect the integration of marine science content in secondary school teacher training programmes in Namibia

Limited Teaching and Learning Resources

The review of teacher education curriculum seems to show that many institutions lack updated climate change teaching materials, laboratory equipment, and audio-visual resources necessary for modern science pedagogy.

Insufficient Educator Training

Teachers and lecturers lack adequate preparation to teach complex climate topics, limiting effective integration of climate change into marine sciences.

Complexity of Climate Concepts

Climate change science involves abstract concepts such as carbon cycles, feedback loops, and ocean-atmosphere interactions, which require strong foundational science knowledge (IPCC, 2021).

Curriculum Overload

Existing curricula are often dense, creating challenges for introducing new material without restructuring course content (MoEAC, 2016).

Limited Public and Community Awareness

Low climate literacy in some regions may weaken community-level support for school-based climate education initiatives (UNEP, 2021).

OPPORTUNITIES FOR INTEGRATING CLIMATE CHANGE EDUCATION IN MARINE SCIENCE CONTENT AMONG SECONDARY SCHOOL TEACHER TRAINING PROGRAMMES IN NAMIBIA

Based on the review of the key documents, such as the policy frameworks and institutional structures supporting climate change education in Namibia, the following opportunities exist for Namibian teacher education.

There is a Strong Policy and Institutional Support

Government commitment through national policies and international frameworks enables curriculum planners to introduce climate content into aquatic sciences (Government of the Republic of Namibia, 2011)

There is a Growing Interest in Marine and Environmental Studies

The expansion of marine science programmes and environmental education initiatives creates opportunities to mainstream climate issues across institutions. However, this is mostly evident only in coastal schools leaving behind schools inland.

There is an Availability of Local Scientific Data

Scientific research from the Benguela Current Commission, the Ministry of Agriculture, Fisheries, Water and Land Reform, and academic institutions provides localized datasets on ocean warming, biodiversity shifts, and fisheries variability (BCC, 2018). These datasets are valuable for supporting evidence-based fisheries management and improving stock assessments in Namibia. They can also be used to identify climate-driven changes in species distribution and ecosystem productivity. Furthermore, the available data can support marine spatial planning, climate change adaptation strategies, and long-term monitoring of marine ecosystem health. This ensures that policy and management decisions are based on reliable local scientific evidence rather than global estimates.

Climate Change Education is Highly Relevant to the Namibian Context

Climate change already affects Namibia's marine ecosystems, making the subject essential for learners studying aquaculture, aquatic ecology, and marine sciences (Jarre et al. 2013). Rising sea temperatures, ocean acidification, and shifting currents are altering the distribution of commercially important fish species, affecting both wild fisheries and aquaculture productivity (Young, 2025). Additionally, extreme weather events and changing rainfall patterns impact coastal habitats such as estuaries and mangroves, which serve as nursery grounds for many marine species (Young, 2025). Effective implementation of climate change education in Namibia requires overcoming resource limitations, enhancing teacher training, and improving public awareness. Integrating climate change knowledge into curricula equips learners and future professionals with the skills to monitor ecosystem shifts, develop adaptive management strategies, and contribute to sustainable fisheries and aquaculture practices. Furthermore, awareness programs targeting coastal communities can foster resilience by promoting sustainable resource use and informed decision-making in the face of climate-driven changes.

CONCLUSIONS AND WAY FORWARD

There is a significant and growing scope for integrating Climate Change Education into Marine Science Content among Secondary School Teacher Training Programmes in Namibia. This is driven by the country's high vulnerability to climate impacts and a national commitment to sustainable development. The government through policies such as the National Blue Economy (2024), the National Policy on

Climate Change (2011) and the National Environmental Education and Education for Sustainable Development Policy (2019), captured in the Namibia's ESD strategy for 2030 is actively working on a template to integrate climate awareness into formal and non-formal education, including teacher education.

It is important to promote Environmental Education programmes and awareness to empower preservice teachers following a transformative learning approach. The promotion of Environmental Education and awareness is fundamental to building climate-resilient societies. Climate change education fosters responsible environmental behaviour, enhances understanding of marine and freshwater systems, and encourages community participation in conservation. Awareness initiatives such as marine science outreach programmes, coastal clean-ups, school ocean clubs, and public climate dialogues can strengthen teacher education programmes and support the sustainable use of Namibia's ocean resources. Secondly, the empowerment of teachers with climate knowledge not only supports adaptation and mitigation efforts but also ensures the long-term sustainability of Namibia's aquatic ecosystems.

- Strengthen curriculum design to include practical, ocean-focused climate content
- Improve teacher training through workshops, in-service programmes, and collaboration with marine research institutions
- Utilize local datasets and case studies from the Benguela Current system
- Establish partnerships with government agencies, NGOs, and communities
- Promote active learning approaches such as fieldwork, citizen science, and project-based learning

This approach must include transformative learning to enable pre-service teachers to acquire pedagogical practices that encourage them to apply learner agency, ethical reasoning, and climate action.

SUMMARY

The focus of this paper is to report the barriers and opportunities for integrating climate change education in marine science content among secondary school teacher training programmes in Namibia. This chapter addresses this topic by discussing the theoretical framework, which is based on Transformative Learning Theory. This learning theory, is important because it focusses on how those in a learning environment can critically examine and change their deeply held assumptions, beliefs, and worldviews. The next section is about climate change and its effects on the Namibian ocean space. The discussion is significant as it relates to consequences that relate to impacts on food security in Namibia, fish processing and value chain, and damage to coastal Infrastructure. The chapter also presents the policy frameworks and institutional structures supporting climate change education in Namibia and the national policies that support climate change education in Namibia. This discussion also includes the influence of educational policies on climate change education in Namibia. The policies are important because they foster a cohesive environment for accomplishing educational opportunities among teacher education institution so that they understand what is expected of them. In addition, the chapter presents the barriers to the integration of marine science content in secondary school teacher training programmes and opportunities for integrating them into marine science content. The last section of this paper is about the conclusions and the way forward, which presents a variety of measures that must be undertaken to effectively address climate change education in marine science content among secondary school teacher training programmes in Namibia.

REFERENCES

Barnard, P. (2014). *Benguela Current Large Marine Ecosystem: Integrative Assessment of Climate Variability, Ecosystem Impacts, and Management*. BCC.

Blue Marine Foundation. (2024). Namibia's Ocean literacy Toolkit. 11 November 2024. https://www.bluemarinefoundation.com/education-resources/namibias-ocean-literacy-toolkit/

Boyer, D. C., Boyer, H. J., Fossen, I., & Kreiner, A. (2001). Changes in the abundance of the northern Benguela sardine stock during the decade 1990–2000. *Fisheries Research*, *58*(1), 79–94.

FAO [Food and Agriculture Organisation]. (2020). *The State of World Fisheries and Aquaculture 2020: Sustainability in Action.* Rome: FAO.

Foundation Tara Ocean Org. (n.d.) *Why is the Ocean a vital ally in the fight against climate change*? Available from https://fondationtaraocean.org/en/scientific-news/why-is-the-ocean-a-vital-ally-in-the-fight-against-climate-change/

Government of the Republic of Namibia. (2004a). *Namibia Vison 2030*. Office of the President.

Government of the Republic of Namibia. (2004b). *Namibia Development Plan 6*. Office of the President.

Government of the Republic of Namibia. (2011). *National Climate Change Policy for Namibia*. Ministry of Environment and Tourism.

Hutchings, L., van der Lingen, C. D., Shannon, L. J., & Crawford, R. J. (2006). The Benguela ecosystem: An ecosystem of four components. *Progress in Oceanography*, *7*(2–4), 1–48.

IPCC [Intergovernmental Panel on Climate Change]. (2019). *Special Report on the Ocean and Cryosphere in a Changing Climate (SROCC).* Geneva: Intergovernmental Panel on Climate Change.

IPCC [Intergovernmental Panel on Climate Change]. (2021). *AR6 Climate Change 2021: The Physical Science Basis.* Geneva: Intergovernmental Panel on Climate Change.

Jarre, A., Hutchings, L., & Kirkman, S. P. (2013). Climate effects on the Namibian marine ecosystem: Natural variability and human impacts. *Environmental Development*, *7*, 95–118.

Kanyimba, A., Nantinda, L., & Namutenya, M. (2014). The Scope of Climate Change Education in the Namibian Formal and Non-formal Education Systems. In Ed Nhamo,G & Shava, S. *Climate Change Education in the SADC School Curriculum 102-123*. Pretoria: Africa Institute of South Africa.

Machaka, C. L. A., Warikandwa, T. V. (2025). "Towards a Sustainable Blue Economy for Namibia and Its Coastal Communities: A Contemporary Socio-Legal Perspective" *PER / PELJ* 2025(28) - DOI http://dx.doi.org/DOI: 10.17159/1727-3781/2025/v28i0a17363

Mezirow, J. (1997). Transformative learning: Theory to practice. *New Directions for Adult and Continuing Education*, *74*(1), 5–12. DOI: 10.1002/ace.7401

MoEAC [Ministry of Education, Arts and Culture]. (2016). *National Curriculum for Basic Education*. Government of Namibia, Windhoek. https://www.nied.edu.na/assets/documents/05Policies/NationalCurriculumGuide/National_Curriculum_Basic_Education_2016.pdf

Moltke, A., & Blythe, J. L. (2020). Vulnerability of small-scale fishers to climate change in Namibia. *Marine Policy*, *118*, 104040.

BCC [Benguela Current Commission]. (2018). *State of the Benguela Current Large Marine Ecosystem Report*. Swakopmund: BCC Secretariat.

Spear, D. (2018). *Vulnerability and responses to climate change in drylands: The case of Namibia*. International Development Research Centre.

Streck, C., Keenlyside, P., & von Unger, M. (2016). The Paris Agreement: A New Beginning. *Journal for European Environmental & Planning Law, 13* (2016) 3-29. Available form https://climatefocus.com/wp-content/uploads/2022/06/The-Paris-Agreement-A-New-Beginning.pdf

Timmermann, A., & Friedrich, T. (2016). Ocean circulation and climate change in the Benguela region. *Nature Geoscience*, *9*(10), 718–724.

UNEP [United Nations Environment Programme]. (2021). *Climate Action Note – Climate Change Impacts in Africa*. UNEP, Nairobi.

UNESCO. (2021). *Education for Sustainable Development: A Roadmap*. UNESCO.

UNFCCC [United Nations Framework Convention on Climate Change]. (1992). *United Nations Framework Convention on Climate Change*. United Nations, New York.

UNFCCC [United Nations Framework Convention on Climate Change]. (2016). *The Paris Agreement*. United Nations, New York. https://unfccc.int/sites/default/files/resource/parisagreement_publication.pdf

World Bank. (2022). *Namibia Climate and Development Report*. World Bank.

Young, L. (2025). *How Does Climate Change Affect Fish?* Institute of Environmental Research. https://iere.org/how-does-climate-change-affect-fish/

Chapter 10

Climate Change Education: An Integrated Approach

Eurika Jansen van Vuuren
http://orcid.org/0000-0003-0906-4986
University of Mpumalanga, South Africa

ABSTRACT

This chapter explores Environmental Change Education (ECE) in the Foundation Phase, framed within the context of escalating global ecological crises and the urgent need for transformative educational responses. It positions environmental degradation—manifested in climate change, biodiversity loss, pollution, and resource scarcity—as both a planetary emergency and a profound educational challenge. It furthermore argues that these crises are largely rooted in unsustainable human practices characteristic of the Anthropocene, and therefore require a fundamental shift in human behaviour that must begin in early childhood education. The chapter conceptualises environmental change not only as an ecological issue but also as a matter of children's rights, highlighting the disproportionate vulnerability of young people to environmental risks. ECE is a multidimensional process that not only transmits scientific knowledge but also fosters environmental awareness and ethical responsibility, critical thinking, and active citizenship. The chapter aligns with the Curriculum and Assessment Policy Statement (CAPS), particularly in the Life Skills curriculum. Although CAPS promotes holistic development and integrated, interdisciplinary learning, teaching is knowledge-driven rather than transformative. The chapter argues that early, integrated, and creative pedagogical approaches are essential for fostering meaningful environmental understanding and long-term behavioural change. These approaches include interdisciplinary teaching and Arts-integrated teaching with the Creative and Performing Arts positioned as

DOI: 10.4018/979-8-3373-6975-4.ch010

a powerful pedagogical vehicle. Arts-based approaches—encompassing music, drama, visual arts, and movement—enable experiential, affective, and imaginative engagement with environmental concepts, making abstract ideas accessible and memorable for young learners. As far as methodology is concerned, the researcher used a qualitative research design within a pragmatic paradigm, which prioritises practical, contextually relevant knowledge. Document analysis and literature review are used to examine curriculum policy, international practices, and pedagogical strategies related to ECE and Arts integration. To support the study's arguments, integrated lesson ideas are provided that show how environmental themes can be infused through Arts-based approaches, which help narrow the gap between theory and practice. The Integrated Curriculum Model (ICM) serves as a theoretical framework, providing a structured lens for organising content, understanding, and thematic integration. When aligned with the Foundation Phase, the ICM emphasises conceptual depth and the development of higher-order thinking while using overarching themes such as environmental sustainability. Along with ICM, other theoretical perspectives—including constructivism, socio-cultural theory, experiential learning, and critical pedagogy—also support the use of Arts integration for fostering active, collaborative, and reflective learning. Various models of curriculum integration are suggested, but particularly the synergy model, which can be meaningfully implemented while maintaining disciplinary integrity. Despite a supportive curriculum, impeding challenges include limited teacher preparation, insufficient resources, and a lack of methodological guidance for inquiry-based and integrated teaching. In countries such as Finland and Singapore, the effectiveness of inquiry-based and interdisciplinary approaches is a good example for the South African context. The critical role of teacher education in addressing challenges, such as the development of disciplinary, pedagogical, and integrative knowledge among pre-service teachers, is emphasised in the chapter. Arts-integrated pedagogy can address many of the identified challenges by enhancing learner engagement, supporting diverse learning styles, and fostering emotional and ethical connections to environmental issues. Indigenous Knowledge enriches ECE by promoting cultural relevance and strengthening learners' sense of place and environmental identity. Arts integration is presented as a resource-conscious approach that is ideal for some South African schools with limited resources. In conclusion, the chapter argues that integrated Arts-based Environmental Change Education offers a transformative approach to teaching and learning in the Foundation Phase. It assists with the holistic development of learners to function optimally as critically engaged citizens. The study provides practical insights for teachers and policymakers who are seeking to strengthen environmental education in South Africa.

INTRODUCTION

"Adults keep saying we owe it to the young people, to give them hope… I want you to act as if the house is on fire, because it is."

— Greta Thunberg (Tannock, 2021)

These words were spoken by environmental activist Greta Thunberg, who compared the world's environmental crisis to a house on fire. The urgency captured in this statement reflects the broader global recognition that the world is facing an environmental crisis of unprecedented scale. Climate change, biodiversity loss, water scarcity, pollution, and extreme weather events signal that the "house" of our planetary home is indeed under threat. While it may not be possible to extinguish the fire entirely, there is an urgent need to contain and manage it responsibly. Prideaux, Thompson & Pabel (2020) propose a "flattening of the curve" approach to global warming, advocating for sustainable mitigation strategies rather than abrupt, unrealistic reversals.

The current crisis, however, is not merely part of natural climatic cycles that have occurred over millennia. It is largely the consequence of human activity during the Anthropocene—an era characterised by the extensive exploitation and misuse of natural resources. Unsustainable consumption patterns and environmental degradation necessitate a profound transformation in human behaviour. Such transformation cannot begin in adulthood; it must be cultivated from early childhood so that responsible and sustainable living becomes habitual rather than corrective.

Environmental Crisis as a Children's Rights Issue

The urgency of environmental education is not confined to environmental policy discourse; it has become a child-centred concern. In its report, "The Climate Crisis is a Children's Rights Crisis!" (2019), UNICEF highlights the disproportionate impact of climate change on children. Young people are increasingly exposed to air pollution, water scarcity, food insecurity, and climate-related diseases such as malaria and dengue fever. These realities position climate change not only as an ecological issue, but as a social justice and human rights concern.

Climate change education has been conceptualised in diverse ways. Stevenson et al. (2017, p. 2) define it as "learning in the face of risk, uncertainty and rapid change." This definition underscores both the unpredictability of environmental futures and the necessity of preparing learners to navigate complex, rapidly shifting ecological realities.

Environmental Change Education (ECE) in early schooling, therefore, extends beyond the transmission of scientific facts. It involves cultivating environmental awareness, ethical values, critical attitudes, and responsible action (Tilbury, 1995;

UNESCO, 2017). The task is not merely to inform children about environmental degradation, but to empower them to live in ways that sustain ecological balance.

Environmental Change Education in the South African Context

In South Africa, environmental themes are embedded within the Life Skills curriculum as prescribed by the Department of Basic Education in the Curriculum and Assessment Policy Statement (CAPS) (DBE, 2011). CAPS foregrounds holistic development—encompassing social, emotional, intellectual, and physical growth—which aligns closely with the principles of sustainability education. The curriculum recognises interconnectedness and promotes responsible citizenship as foundational educational aims.

Despite the inclusion of environmental content within CAPS, its classroom enactment often remains knowledge-driven rather than transformative. Young learners may acquire terminology related to pollution, recycling, or climate change without internalising sustainable behaviours. The challenge, therefore, lies not in the absence of curriculum content but in the pedagogical approaches through which it is delivered.

The Need for Early, Integrated, and Creative Pedagogy

Children do not necessarily acquire environmentally responsible behaviours through observation at home. If sustainable living is to become second nature, it must be intentionally cultivated in early childhood education. This requires teachers who are competent and confident in facilitating environmental awareness, compassion, and critical thinking in developmentally appropriate ways.

CAPS (DBE, 2011) explicitly recommend integrated teaching in the Foundation Phase (FP). Teaching cannot occur in isolated subject silos when real-world problems—such as climate change—are inherently interdisciplinary. Environmental responsibility must permeate Life Skills (including Natural Sciences, Social Sciences, Technology, Music, Art, Drama, Physical Education, and Creative Movement) as well as Languages and Mathematics.

Integrated teaching reflects the interconnected nature of ecological systems: human and non-human life forms exist in mutual dependence. Accordingly, educational practice should mirror this integration.

Towards Transformative Environmental Change Education

Although CAPS includes essential environmental themes, these are not always presented in ways that stimulate sustained interest or long-term retention. If knowledge alone were sufficient, the current environmental crisis would arguably not have reached its present intensity. Learners must be emotionally engaged and intellectually challenged to develop meaningful environmental commitment. Integrated pedagogy—particularly through the Creative and Performing Arts—offers a promising avenue for achieving this goal.

The Arts provide experiential, affective, and imaginative entry points into environmental learning. Children learn best through play, and the Creative and Performing Arts are an embodiment of playing (Alves-Oliveira, Arriaga, Xavier, Hoffman, & Paiva, 2022). Through music, drama, visual arts, and movement, abstract environmental concepts become embodied experiences. Play-based activities must vary between classroom and outdoor engagement, as outdoor learning enhances students' environmental literacy and fosters a deeper connection to nature (Roussou et al., 2025). Outdoor activities support creativity, sensory engagement, and hands-on experiences that foster curiosity and problem-solving. The outdoors can provide natural materials like soil, plants, and water for experimentation, fostering environmental awareness and care. Being outside also means that learners experience seasonal changes that deepen their understanding of nature (Kiviranta, Lindfors, Rönkkö & Luukka, 2024). Access to outdoor spaces and materials for hands-on activities can vary, especially in under-resourced contexts. However, scalable and flexible gamified approaches, such as storytelling or role-playing, can effectively mitigate these barriers by requiring minimal resources while maintaining educational value. Educator training emerges as a critical factor, with research emphasising the need for experiential learning and ongoing support to integrate STEAM and theatrical methods effectively.

Such approaches foster deeper involvement in learning and enhance retention by linking cognition, emotion, and creativity.

Chapter Focus

This chapter, therefore, explores Environmental Change Education in the Foundation Phase within the context of global ecological urgency and South African curriculum policy. It argues that integrated teaching, and specifically Arts-integrated pedagogy, offers a transformative approach to fostering environmental awareness and responsible action among young learners. By moving beyond fragmented instruction towards creative, interdisciplinary engagement, Environmental Change Education can meaningfully shape a generation capable of acting responsibly in a world that urgently requires change.

A substantial body of scholarship has examined pedagogical approaches to environmental education within the Senior Phase and Further Education and Training (FET) bands. In contrast, research focusing on environmental education in the Foundation Phase (FP) remains comparatively underdeveloped. Moreover, only a limited number of studies have investigated the role of the Creative and Performing Arts in supporting knowledge retention within this context. This study, therefore, foregrounds the Arts as an integrative pedagogical tool in the Foundation Phase, aiming to enhance both teaching and learning processes.

The primary objective of this research was to explore Environmental Change Education through Arts integration as a potential strategy for strengthening educational impact and long-term retention. In doing so, the study seeks to cultivate environmental awareness among future citizens by fostering an understanding of the urgent ecological challenges facing the planet and encouraging responsible stewardship to mitigate the adverse effects of human activity.

The central research question guiding this inquiry is: How can Arts integration enhance Environmental Change Education?

The study is further structured around the following sub-questions:

- In what ways does the South African school curriculum support Environmental Change Education in the Foundation Phase?
- How is Environmental Change Education conceptualised and implemented in other contexts?
- How can pre-service teachers be effectively prepared and empowered to facilitate Environmental Change Education with competence and confidence?
- How can integrated Arts pedagogy practically strengthen Environmental Change Education in classroom settings?

By addressing these questions, the study aims to generate insights that may inform educators, teacher educators, and the Department of Education, thereby enhancing Environmental Change Education in the Foundation Phase.

Key Terms

Key concepts underpinning this chapter include the term Arts, which is understood to encompass the full spectrum of foundational art forms represented within the formal schooling system. This includes both the Creative Arts (Visual Arts) and the Performing Arts (Dance, Drama, and Music), as offered in the Foundation Phase (FP). In this study, the Arts are conceptualised not as isolated disciplines but as interconnected modalities of expression and meaning-making that can contribute holistically to teaching and learning.

A further central concept is integration, a term that is subject to multiple interpretations and applications. While integration may occur within a single subject through the incorporation of diverse skills, themes, or competencies, the term in this study refers specifically to interdisciplinary integration across subject areas. In this context, integration involves the deliberate interweaving of Mathematics, Home Language, English First Additional Language, and the various components of Life Skills, as previously outlined. Through such an approach, subjects are not treated as discrete or compartmentalised entities; rather, they are presented as mutually reinforcing domains of knowledge and practice. Lessons are therefore structured to incorporate multiple subject areas simultaneously, fostering a cohesive and holistic learning experience. The particular form of integration is determined by the relative emphasis or proportional representation of each participating subject within a given lesson.

Building on the conceptual and contextual foundations outlined in the introduction, the study now turns to how these ideas are systematically investigated. Having established the urgency of Environmental Change Education (ECE), its positioning within the South African curriculum, and the potential of Arts-integrated pedagogy to foster meaningful engagement and retention, it becomes necessary to adopt a methodological approach capable of both interrogating existing frameworks and generating practical, classroom-relevant insights. The research questions that guide this inquiry require an approach that is interpretive yet application-oriented, one that not only analyses policy and literature but also translates these findings into pedagogical possibilities. Accordingly, the following methodology section outlines the research design, strategies, and analytical processes employed to explore how Arts integration can enhance Environmental Change Education in the Foundation Phase.

METHODOLOGY

Research Design and Paradigm

This study employed a qualitative research design situated within a pragmatic paradigm. Pragmatism prioritises practical solutions to educational challenges and values knowledge that is useful in real-world contexts. As suggested by Dube, Nkomo, & Apadile-Thokweng (2024, p. 1001), it is 'a philosophy that strives to unveil practical knowledge.' In alignment with this orientation, the study sought not only to interrogate theoretical perspectives on Environmental Change Education (ECE) in the Foundation Phase (FP), but also to generate a practical, classroom-based example of integrated Arts teaching.

The qualitative design enabled an interpretive exploration of curriculum policy, scholarly literature, and pedagogical approaches related to Environmental Change Education in South African primary schools.

Research Strategy

The study utilised two primary research strategies: document analysis and literature review. These approaches were selected to enable an in-depth examination of policy frameworks, curriculum documents, and scholarly discourse relevant to environmental education and Arts integration.

Document analysis focused primarily on national curriculum policy documents that guide Foundation Phase teaching in South Africa. Central to this analysis was the Department of Basic Education's Curriculum and Assessment Policy Statement (CAPS) for the Foundation Phase.

The document analysis aimed to identify how Environmental Change Education is embedded within the Life Skills curriculum, to examine the extent to which interdisciplinary integration is encouraged and finally to determine how Arts education is positioned within the FP curriculum framework.

Policy documents were analysed thematically to extract references to environmental sustainability, integration, holistic development, and learner agency.

A systematic literature search was conducted to explore approaches to Environmental Change Education in international contexts, theoretical perspectives on integrated and interdisciplinary teaching, the role of Arts integration in enhancing conceptual understanding and learner engagement, and strategies for empowering pre-service teachers to implement environmental education effectively.

Peer-reviewed journal articles, policy reports, and seminal theoretical texts were consulted. The literature review enabled a comparative analysis of South African practices and global trends in environmental education.

Alignment of Research Questions and Tools

Table 1 presents the alignment between the study's sub-questions and the research tools employed.

Table 1. Alignment of sub-questions and research tools

Research Question	Research Tool
How does the South African school curriculum support Environmental Change Education in South Africa's FP?	Document analysis; literature search
How is Environmental Change Education approached in other countries?	Literature search
How should pre-service teachers be empowered to efficiently and confidently facilitate Environmental Change Education?	Literature search Observation
How can integrated Arts teaching augment Environmental Change Education in practice?	Document analysis; literature search; observation

This alignment ensured methodological coherence and strengthened the study's internal validity by directly linking research questions to appropriate data sources.

Data Analysis

Data generated from document analysis and literature review were analysed using thematic analysis. Recurring themes were identified, coded, and categorised into the following conceptual strands: Curriculum support for environmental education, interdisciplinary and integrated teaching approaches, teacher empowerment and professional preparation, and Arts-based pedagogical strategies. These themes informed both the theoretical framing of the article and the development of a practical teaching example.

Development of the Integrated Teaching Example

In keeping with the pragmatic paradigm, the study culminated in the design of integrated Environmental Change Education lesson ideas that incorporate Arts-based pedagogies. The lesson ideas demonstrate how environmental themes can be approached interdisciplinarily to enhance young learners' understanding of the impact of their daily actions on the environment.

The integrated approach illustrates how environmental concepts can be explored through Visual Arts, Drama, Music, and Movement, how higher-order thinking and creativity can be cultivated through thematic integration and how learners can develop both conceptual understanding and environmental responsibility.

The inclusion of this practical example strengthens the study's applied relevance and bridges the gap between policy analysis and classroom implementation.

Ethical Considerations

As the study relied exclusively on publicly available documents, researcher observation over a long period, and existing literature, formal ethical clearance was not required. Academic integrity was ensured through accurate referencing and adherence to scholarly standards.

Methodological Rigour

Credibility was enhanced through triangulation between policy documents and scholarly literature. Transparency was maintained by clearly aligning research questions with research tools and analytic procedures. The pragmatic orientation ensured that findings are contextually grounded and pedagogically applicable within the South African Foundation Phase setting.

In summary, this qualitative, pragmatically oriented study combined document analysis and literature review to investigate Environmental Change Education in the Foundation Phase and to propose an integrated Arts-based pedagogical approach that is both theoretically grounded and practically implementable.

THEORETICAL FRAMEWORK

Primary Theoretical Lens: The Integrated Curriculum Model

This study is situated within a theoretically grounded framework anchored in the Integrated Curriculum Model (ICM) developed by VanTassel-Baska (1986). Originally conceptualised within the field of gifted education at the Centre for Gifted Education, the ICM provides a robust theoretical structure for aligning depth of content, higher-order cognition, and interdisciplinary thematic integration within curriculum design.

Although initially developed for learners with advanced cognitive abilities, the model's structural emphasis on conceptual depth, complexity, and coherence makes it adaptable to heterogeneous classrooms. In this study, the ICM functions not merely as a curriculum design model but as the primary theoretical lens through which Arts-integrated Environmental Change Education (ECE) in the South African Foundation Phase is structured and interpreted. Extending the ICM Beyond Gifted Education

The purpose of this study is not to propose a new conceptual model. Rather, it extends and applies the ICM within a specific pedagogical and socio-educational context.

Integrated teaching draws on multiple disciplines in purposeful, interconnected ways to deepen learners' understanding. Instead of presenting subject areas as isolated bodies of knowledge, integration foregrounds conceptual coherence and thematic relationships across disciplines.

Within this paradigm, the ICM organises the curriculum into three interrelated facets: Advanced Content, Higher-Order Thinking Processes, and Overarching Themes and Issues. Together, these facets provide the structural architecture through which Arts-integrated Environmental Change Education is conceptualised in this study.

In the original ICM, *advanced content* refers to discipline-based knowledge that extends beyond grade-level expectations in depth and complexity. Within the Foundation Phase context, however, this does not imply acceleration of content but rather conceptual richness and meaningful engagement with ideas.

When applied to Environmental Change Education, learners engage deeply with age-appropriate environmental concepts such as water conservation, biodiversity, and pollution. Arts integration allows abstract environmental ideas to be explored through multiple representational modes, including visual expression, music, drama, and movement.

In this way, advanced content becomes conceptually layered rather than cognitively exclusive, ensuring that environmental ideas remain accessible to learners across diverse ability levels while still promoting meaningful depth of understanding.

The second facet of the ICM emphasises *higher-order thinking* processes such as analysis, evaluation, synthesis, and creative problem-solving. These processes can be meaningfully cultivated even in the Foundation Phase when learning experiences are developmentally appropriate and engaging.

Arts-based learning offers particularly powerful opportunities to develop these cognitive processes. For example, dramatic role-play can be used to explore environmental dilemmas and ethical decision-making. Recycled art projects can stimulate creative problem-solving, such as when learners collect litter from the school environment and transform these objects into three-dimensional "litter monsters", accompanied by environmental awareness messages displayed around the school. Collaborative music creation, such as writing a class rap about littering, allows learners to synthesise environmental knowledge and values in creative ways. Visual contrasts, such as drawing parallel scenes of polluted versus sustainable environments, encourage learners to analyse the effects of human behaviour on nature. Through such activities, Arts integration becomes the pedagogical vehicle through which higher-order cognition is scaffolded. The creative process itself inherently requires interpretation, reflection, and decision-making.

The third facet of the ICM foregrounds *interdisciplinary themes* that transcend traditional subject boundaries. Environmental change is a theme that intersects scientific understanding, ethical considerations, social responsibility, and community life.

Within this framework, environmental sustainability functions as the organising interdisciplinary issue. The Arts serve as integrative bridges linking subjects such as Life Skills, Natural Sciences, Social Sciences, Mathematics and Literacy. Through this approach, learners explore environmental change not merely as academic content but as a societal and moral concern. This thematic orientation aligns closely with holistic approaches to sustainability education and with the principles of Education for Sustainable Development.

Theoretical Foundations Supporting Arts Integration

While the ICM provides the structural framework for the study, several complementary learning theories offer epistemological justification for Arts-based pedagogy. Constructivist theory, as articulated by Jean Piaget (1952) and Lev Vygotsky (1978), posits that learners actively construct knowledge through interaction with their environment. Arts-based activities—such as role-play, visual representation, music composition, and movement—create multimodal opportunities for experiential meaning-making.

Similarly, John Dewey (1934) emphasises the role of aesthetic experience in deepening learning through emotion, imagination, and reflection. Environmental issues that may appear abstract to young learners become more tangible when explored through embodied and sensory engagement.

Vygotsky's emphasis on social mediation further highlights the importance of dialogue, collaboration, and shared meaning-making in learning. Integrated Arts teaching fosters such collaborative processes as learners collectively explore environmental concerns through creative expression.

Finally, Paulo Freire (1970) introduces the concept of conscientisation, whereby education cultivates critical awareness and social agency. When learners create environmental posters, advocacy performances, or recycled artworks, they move from passive recipients of knowledge to active participants in social and ecological transformation.

Models of Curriculum Integration

Theoretical understandings of curriculum integration further inform this framework. Deirdre Russell-Bowie (2009) identifies three principal models of integration. The Service Connection Model involves one dominant subject while another subject supports it in a limited way. For example, during a mathematics lesson on patterns, a teacher may clap a rhythmic sequence to illustrate a pattern. Although music is used as an example, the lesson remains focused on mathematical concepts rather than musical learning.

The Symmetric Correlation Model gives equal emphasis to two subjects while maintaining their disciplinary integrity. For instance, a lesson might begin by exploring musical rhythm patterns and then draw explicit parallels between rhythmic structures in music and mathematical patterns.

The Syntegration Model involves multiple subjects aligning around a shared overarching theme while remaining structurally distinct. In such an approach, an entire grade or school may work around a common theme such as patterns. Natural Sciences might explore patterns in flowers, Mathematics patterns in numbers, Social Sciences cultural patterns such as isiNdebele decorative art, while Visual Arts and Dance explore patterns in colour and movement.

Building on these models, Jansen van Vuuren (2022) proposes the Synergy Model, which advocates deliberate interdisciplinary interaction that enhances learning while maintaining disciplinary integrity. In this approach, integration can occur during a single lesson or across a series of lessons.

For example, while a Natural Sciences lesson may focus on patterns in nature, connections may also be made to mathematical shapes or to visual arts activities in which learners create patterned artworks. Each discipline contributes meaningfully to the learning process rather than functioning as an add-on.

This study's lesson ideas align most closely with a synergy-oriented approach, in which Arts integration is neither peripheral nor dominant but structurally embedded within thematic Environmental Change Education.

Conceptual Synthesis and Theoretical Proposition

This study adopts the Integrated Curriculum Model as its organising theoretical lens. Within this framework where environmental change functions as the overarching interdisciplinary theme and where Arts integration provides multimodal pathways for engaging conceptually rich environmental content.

This study in addition advances the proposition that when Environmental Change Education in the Foundation Phase is structured according to the three facets of the Integrated Curriculum Model—content depth, higher-order cognition, and overarching thematic integration—and enacted through Arts-based pedagogies, it enhances learners' conceptual understanding, critical engagement, emotional connection with the natural environment, and capacity for environmentally responsible action.

The next section focuses on the literature review, which is predominantly structured around the study's research questions, with a primary focus on the concept of integrated teaching.

LITERATURE REVIEW

The main question of the study asks, "How can integrated Arts teaching augment Environmental Change Education (ECE) in practice?" This question was answered by exploring documents to examine the influence of the CAPS curriculum on ECE, investigating how ECE is approached in other countries, considering best practices for pre-service teacher training, and finally, the use of integrated Arts to augment ECE in practice.

In What Ways Does the South African School Curriculum Support Environmental Change Education in the Foundation Phase?

The South African Curriculum and Assessment Policy Statement (CAPS) (CAPS, 2011) for Life Skills (LS) incorporates a broad range of topics that may be directly associated with environmental health and Environmental Change Education in the Foundation Phase (FP). Climate change education is particularly embedded in the Beginning Knowledge component of Life Skills, drawing conceptually from Social Sciences (History and Geography), Natural Sciences, and Technology. Within the Social Sciences strand, learners engage with key concepts such as conservation, cause and effect, place, adaptation, relationships and interdependence, diversity and individuality, and change. These concepts provide a foundational understanding of environmental systems and human–environment interactions. The Natural Sciences component introduces foundational disciplinary knowledge through core content areas, namely life and living; energy and change; matter and materials; and planet Earth and beyond. In addition to conceptual knowledge, learners are introduced to scientific process skills associated with inquiry, including observing, comparing, classifying, measuring, experimenting, and communicating. Complementing these are technological process skills, which encompass investigating, designing, making, evaluating, and communicating (CAPS, 2011, p. 8). Collectively, these skills align with the principles of Inquiry-Based Science Education (IBSE), which emphasises active knowledge construction through investigation and problem-solving.

Moreover, opportunities for integration extend beyond Life Skills. Content from Mathematics and the Languages can be meaningfully linked to environmental themes, provided that teachers possess the requisite pedagogical insight and creativity. However, despite the presence of inquiry-oriented skills in the curriculum, CAPS offers limited explicit guidance on the effective implementation of Inquiry-Based Science Education within the Natural Sciences component. Research by Fredericks, Condy, Phillips, Maree, and Chigona (2025) indicates that methodological support for IBSE remains insufficient. Similarly, Stears, James, and Beni (2019) emphasise that teachers require robust content knowledge to teach Natural Sciences effectively. Compounding these challenges is the limited preparation of pre-service teachers for integrated, inquiry-based approaches in which learners actively construct knowledge rather than passively receive information.

Inquiry-based science education (IBSE) in elementary schooling is widely recognised as a learner-centred pedagogical approach in which learners construct scientific understanding through active engagement with questions, investigations, and evidence-based reasoning. Rather than receiving scientific knowledge passively, learners participate in processes that mirror those of scientists, including observing phenomena, formulating questions, making predictions, conducting simple investigations, and communicating findings. In this approach, the teacher's role shifts from knowledge transmitter to facilitator who scaffolds inquiry through structured or guided activities appropriate for young learners' cognitive development. Recent scholarship emphasises that IBSE in primary classrooms promotes hands-on and minds-on learning, supports collaborative knowledge construction, and strengthens learners' ability to interpret evidence and draw conclusions from real-world contexts (Edillor, 2024; Ölçer, 2025). Furthermore, inquiry-based approaches have been shown to enhance conceptual understanding, increase learner motivation, and foster positive attitudes toward science by making learning more meaningful and relevant (Madlela & Ramnarain, 2024; Lupión-Cobos et al., 2026). However, the literature also highlights implementation challenges, particularly the need for adequate teacher preparation, appropriate scaffolding strategies, and sufficient instructional time to ensure that inquiry processes lead to accurate conceptual development rather than superficial exploration. Overall, IBSE is positioned as a foundational approach for developing scientific literacy in elementary education through active, experiential, and reflective learning experiences grounded in authentic scientific practice.

The Personal and Social Well-being (PSW) study area within Life Skills further supports Environmental Change Education by fostering social and emotional development, responsible decision-making, and ethical awareness. PSW addresses relationships with others and with the environment, as well as values and attitudes aligned with constitutional principles. In this way, learners are encouraged to make informed, morally responsible choices regarding their health and environmental stewardship (CAPS, 2011, p. 9).

How Is Environmental Change Education Approached in Other Countries?

Approaches to Environmental Change Education differ from country to country. In several developed education systems, Environmental Change Education is regarded as foundational to young learners' development. In countries such as Finland and Singapore, Natural Sciences education often incorporates inquiry-based learning. These approaches prioritise investigation, problem-solving, and learner agency in constructing scientific understanding (Kotsis, 2024).

Finland extends this model further by adopting an interdisciplinary approach that integrates Natural Sciences with disciplines such as mathematics and languages (Bürgin, 2021). Such an integrated approach is particularly relevant in the South African FP context, where Mathematics and Languages are prioritised due to persistent challenges in learner performance.

Similarly, in Lithuania, inquiry-based activities are employed to contextualise and deepen learners' understanding of climate change and healthy living (Lamanauskas, 2022). Inquiry-based strategies also contribute to literacy development, as learners engage in reading to source information for practical investigations. This reciprocal relationship between scientific inquiry and literacy development underscores the value of integrated pedagogical approaches.

How Should Pre-Service Teachers Be Empowered to Efficiently and Confidently Facilitate Environmental Change Education?

Improving the quality of education in South Africa is closely linked to the preparation of competent and confident newly qualified teachers. Taylor (2019, p. 247) argues that educational quality can be enhanced by ensuring that graduates exiting universities possess adequate knowledge, skills, and professional readiness. This imperative must be considered against the socio-educational backdrop in which many South African pre-service teachers originate from under-resourced schools where inquiry-based pedagogies were not part of their own learning experiences. Rural schooling contexts often face constraints related to teacher training and familiarity with contemporary pedagogical methodologies, as noted by Du Plessis and Mestry (2019). Consequently, many schools continue to rely on rote-learning practices. For pre-service teachers, this translates into the necessity of engaging with pedagogical approaches that may differ substantially from their prior schooling experiences to be ready to practice as teachers.

Additional challenges arise from limited exposure to Physical or Natural Sciences at the secondary school level, which may constrain students' familiarity with scientific modes of thinking. Furthermore, learners from lower socio-economic backgrounds frequently enter higher education institutions (HEIs) with insufficient disciplinary knowledge of environmental concepts. As Taylor (2019, p. 263) contends, disciplinary knowledge constitutes the foundation upon which effective pedagogy is built. This perspective is echoed by Ramsaroop and Peterson (2020), who note that first-year Bachelor of Education students often require substantial support in developing content knowledge.

Within the Foundation Phase, Natural Sciences lecturers are not always disciplinary specialists but rather academics who began as generalist educators with FP training. This may limit their capacity to model and implement Inquiry-Based Science Education effectively. The present study advocates for IBSE, specifically facilitated through Arts integration, as a means of strengthening both conceptual understanding and pedagogical practice.

To prepare pre-service teachers to teach Environmental Change Education effectively, teacher education programmes should prioritise the development of disciplinary knowledge, pedagogical knowledge, didactical knowledge and integrated teaching knowledge. Collectively, these knowledge domains can equip future teachers to implement integrated, inquiry-based Environmental Change Education with confidence and competence.

A study by Oliver-Barcelo, Ferrer-Ribot, and Jové (2024) shows that the pre-service training dilemma is also evident in other global contexts and show definite similarities with what is happening in South Africa. A study done in Arts education within early childhood education (ECE) teacher training programmes in Ibero-American universities found several reasons for ECE teacher training to not be successful, including,

- Insufficient training: ECE teacher training programmes lack adequate arts education to meet curriculum demands. Training is often superficial and brief, and it does not prepare pre-service teachers to implement arts education effectively.
- Limited Connection with Arts and Culture: Teachers often lack exposure to local art and cultural resources, which hinders their ability to integrate cultural elements into classrooms.
- Cross-Disciplinary Integration: Curricula emphasise cross-disciplinary approaches, but teacher training programmes fail to adequately prepare educators for this integration.
- Social-Educational Challenges: Arts are undervalued compared to other subjects like mathematics and language. Institutional support for arts education and ECE is insufficient, especially in rural and underserved areas.
- There is a lack of arts specialists in public schools, with generalist teachers often responsible for arts education. These teachers often become the mentors to pre-service teachers doing their teaching practice.

Afu, Ekwesaranna, Ohio and Nwagu (2025) add the absence of the use of digital tools as another impeding factor. When digital tools are available, they provide an alternative to physical resources and they facilitate collaborative learning. Furthermore, digital tools improve assessment.

Pre-service training is further hampered by barriers at the schools where students do their teaching practice. Recent scholarship highlights a persistent lack of quality resources as a significant barrier to effective environmental change education in primary schools. Studies conducted in Southern African contexts indicate that many schools lack sufficient instructional materials, limited access to environmental learning resources, and inadequate infrastructure, all of which constrain meaningful engagement with environmental topics (Moremong, 2024; Mashaba et al., 2022). These material shortages are closely linked to challenges in teacher capacity, as educators often lack both pedagogical training and structured curriculum support to effectively integrate environmental education into everyday teaching (Gule, 2024; Van Wyk, 2023). Consequently, environmental education is frequently treated as an "add-on" rather than a core component of the curriculum, reducing its depth and consistency in classroom practice. In addition, time constraints, heavy workloads, and large class sizes further limit the implementation of resource-intensive, experiential approaches such as outdoor learning and project-based activities, which are widely recognised as essential for developing environmental awareness (Gule, 2024). These challenges are compounded by systemic inequalities between schools, where those in under-resourced or rural contexts face disproportionately greater barriers to accessing quality materials and learning opportunities (Mashaba et al., 2022). International literature similarly underscores the marginalisation of environmental education in early schooling, noting that without targeted investment in resources and teacher support, its potential to foster meaningful environmental understanding and action remains limited (Sánchez et al., 2022; Oranga, Gisore, & Areba, 2023). Collectively, this body of research demonstrates that the absence of adequate resources is not an isolated issue but part of a broader structural challenge that continues to undermine the effectiveness of environmental change education in primary school contexts.

How Can Integrated Arts Pedagogy Practically Strengthen Environmental Change Education in Classroom Settings?

In addition, integrated Arts teaching can meaningfully augment Environmental Change Education (ECE) in the South African Foundation Phase by fostering holistic development, experiential learning, critical engagement, and environmentally responsible citizenship.

Research in environmental education underscores the importance of affective engagement in fostering pro-environmental behaviour (Palmer, 1998; Chawla, 2007). Knowledge alone rarely results in behavioural change. It is through the Arts that the affective domain is uniquely engaged. Drama develops empathy (e.g., role-playing endangered animals). Visual Arts encourage reflective interpretation (e.g., drawing polluted vs healthy ecosystems). Music and movement can reinforce environmental messages through rhythm and repetition. Such engagement aligns with the Life Skills aims of CAPS, which emphasise emotional and social development alongside cognitive learning (DBE, 2011). Integrated Arts teaching thus strengthens the dispositional dimension of environmental education.

Teaching about environmental change is contextually relevant in South African Schools since South African learners often experience environmental challenges directly—water scarcity, waste management issues, flooding, and land degradation.

This notion aligns with Gruenewald's (2003) research, which advocates that environmental education must be linked to learners' ecological contexts.

In the South African context, Indigenous Knowledge (IK) includes local understandings of biodiversity, seasonal cycles, land use, and conservation practices, which can enrich formal curricula (Shava, Krasny, Tidball & Zazu 2021). When integrating IK into the curriculum, it promotes respect for cultural diversity and heritage, strengthens learners' environmental identity and sense of place, encourages sustainable practices rooted in community traditions, and, importantly, bridges the gap between school knowledge and lived experience (Shava et al., 2021; Ng'asike, 2019).

Integrated Arts pedagogy supports the use of IK by encouraging learners to depict local environmental realities in drawings and storytelling. It incorporates indigenous ecological knowledge through traditional songs and narratives. It can furthermore facilitate community-oriented performances that address local environmental concerns. Such practices affirm cultural identity while promoting environmental stewardship.

CAPS encourages integration in the Foundation Phase, particularly within Life Skills. Integrated Arts teaching does exactly that, as it allows for interdisciplinary reinforcement, such as literacy development through environmental storytelling and scriptwriting. In addition, it could support numeracy integration through environmental data representation in visual projects and through learning about Natural Sciences concepts (e.g., water cycle, plant growth) explored through movement and dramatisation.

Eisner (2002) argues that the Arts expand the forms of representation through which learners make meaning. In this way, Arts integration enhances conceptual understanding while supporting diverse learning styles in multilingual classrooms.

In line with Eisner's (2002) research, more recent studies by Roussou et al. (2025) highlight the success of the transdisciplinary Science, Technology, Engineering, Arts and Mathematics (STEAM) framework and theatrical methods, which align with contemporary research emphasising experiential and transdisciplinary learning. The method also bridges cognitive, affective, and behavioural dimensions to foster a comprehensive educational experience. It deepens students' understanding of critical environmental issues while cultivating essential skills such as creativity, critical thinking, collaboration, and empathy. The synthesis of scientific inquiry with artistic expression creates engaging, relatable, and impactful learning opportunities, preparing students to address real-world environmental challenges.

Research in environmental education highlights the need for approaches that go beyond traditional cognitive knowledge transfer, which often lacks emotional engagement and personal relevance. Utilising the Arts as a teaching vehicle addresses these gaps by incorporating narrative-driven exercises, hands-on activities, and collaborative projects.

It is clear that simulation-based and drama-related pedagogies significantly support the development of multidimensional empathy in teacher education contexts (Levi-Keren et al., 2024). This support is also found in classrooms where it was noted that drama activities enabled learners to develop empathy, social awareness, and perspective-taking over time (Deliman, 2025)

The integration of STEAM methodologies in the teaching context resonates with research advocating for transdisciplinary education to develop holistic problem-solving skills. STEAM's emphasis on creativity and innovation complements the scientific rigour of environmental education, enabling students to approach ecological challenges from diverse perspectives. Moreover, the integrated design aligns with the concept of place-based education, which emphasises connecting learning to local contexts and real-world experiences.

Finally, cultural adaptability remains crucial for fostering inclusivity and engagement in STEAM education; incorporating local environmental issues, traditions, and narratives enhances the resonance with learners, fostering ownership and responsibility. This multifaceted approach ensures global applicability while addressing specific ecological, cultural, and evaluative dynamics within local communities.

Beyond early childhood classrooms, this framework could be adapted for teacher training programmes to equip educators with innovative strategies for incorporating transdisciplinary approaches into their curricula. Additionally, the scenario offers potential for community-based environmental initiatives, fostering intergenerational learning and public awareness about sustainability.

Environmental education in the early years should cultivate a sense of efficacy and responsibility (UNESCO, 2017). Arts-integrated environmental projects—such as school recycling campaigns, environmental get-togethers to clean up certain areas, and collaborative mural creation—provide learners with tangible activities that could foster an attitude of care for nature. It is through creative expression that learners develop problem-solving, collaboration, and communication skills, and an early awareness of the role one has to play in a community. Environmental education aligns with South Africa's broader educational commitment to democratic citizenship and social transformation (DBE, 2011).

Using integrated Arts teaching makes sense in the South African context since it is a resource-conscious pedagogy. In many South African Foundation Phase classrooms, material resources are limited. Integrated Arts teaching offers a resource-conscious approach, as it can use recycled materials for art-making, incorporate body percussion and voice in music activities, and use the surrounding outdoor spaces as experiential learning sites. Such approaches reflect sustainable practice while modelling environmental responsibility.

Integrated Arts teaching augments Environmental Change Education in the South African Foundation Phase by

1. Making abstract environmental concepts developmentally accessible.
2. Engaging the affective and moral dimensions of learning.
3. Supporting curriculum integration within CAPS.
4. Promoting contextual relevance and indigenous knowledge.
5. Encouraging learner agency and environmental citizenship.

By drawing on constructivist, socio-cultural, and critical pedagogical frameworks, Arts-integrated environmental education moves beyond transmission of knowledge towards transformative learning. In this way, it contributes not only to environmental awareness but to the formation of environmentally responsible, critically engaged young citizens.

Environmental Change Education can be enhanced through the use of integrated elements of 'play' as encompassed in the Arts, since FP children learn best through creative play. When pre-service teachers are well-versed in the use of different forms of integration/multidisciplinary approaches in their teaching, they will have the confidence to use them. In turn, learners will be more successful in understanding and retaining knowledge, enabling them to learn about issues impacting their future in depth.

The preceding literature review has highlighted both the possibilities and constraints associated with Environmental Change Education (ECE) in the Foundation Phase. While curriculum policy provides a supportive framework and international perspectives affirm the value of inquiry-based and integrated approaches, significant challenges remain in translating these principles into meaningful classroom practice. At the same time, the literature strongly positions Arts integration as a powerful pedagogical strategy capable of addressing many of these challenges by fostering engagement, contextual relevance, and deeper learning. However, the effectiveness of such an approach ultimately depends on how these theoretical insights are enacted in real teaching contexts. It is therefore necessary to move beyond conceptual discussion towards practical illustration. The following section responds to this need by presenting a set of integrated lesson ideas that demonstrate how Environmental Change Education can be implemented through Arts-based pedagogy in the Foundation Phase classroom.

LESSON IDEAS FOR INTEGRATED ENVIRONMENTAL CHANGE EDUCATION

As the core for these example lesson ideas, a story was used. Creative teachers might want to write their own story tailored to curriculum needs and their school context, but it is also possible to select an existing story aligned with the theme to use as the lesson's core.

As a starting point, the story in Figure 1. The story of Gadini the garden goat, continued to Figure 10, can be read to the learners using reading techniques such as predicting what the story is about and describing the pictures.

Figure 1. The story of Gadini the garden goat

Figure 2. Continued

Figure 3. Continued

Figure 4. Continued

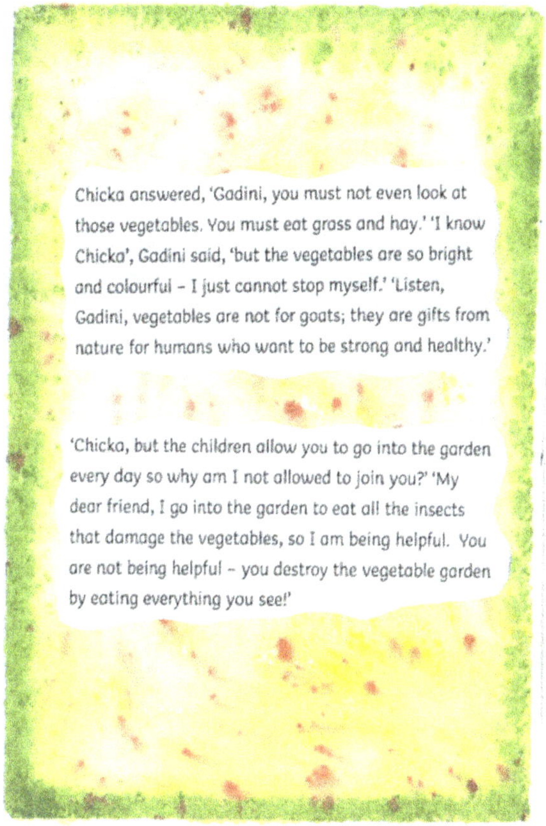

Chicka answered, 'Gadini, you must not even look at those vegetables. You must eat grass and hay.' 'I know Chicka', Gadini said, 'but the vegetables are so bright and colourful – I just cannot stop myself.' 'Listen, Gadini, vegetables are not for goats; they are gifts from nature for humans who want to be strong and healthy.'

'Chicka, but the children allow you to go into the garden every day so why am I not allowed to join you?' 'My dear friend, I go into the garden to eat all the insects that damage the vegetables, so I am being helpful. You are not being helpful – you destroy the vegetable garden by eating everything you see!'

Figure 5. Continued

Figure 6. Continued

Figure 7. Continued

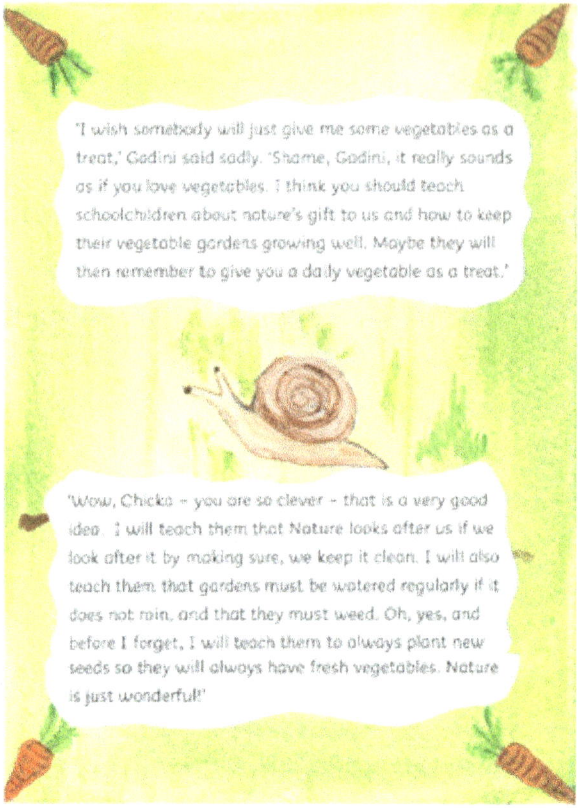

Figure 8. Continued

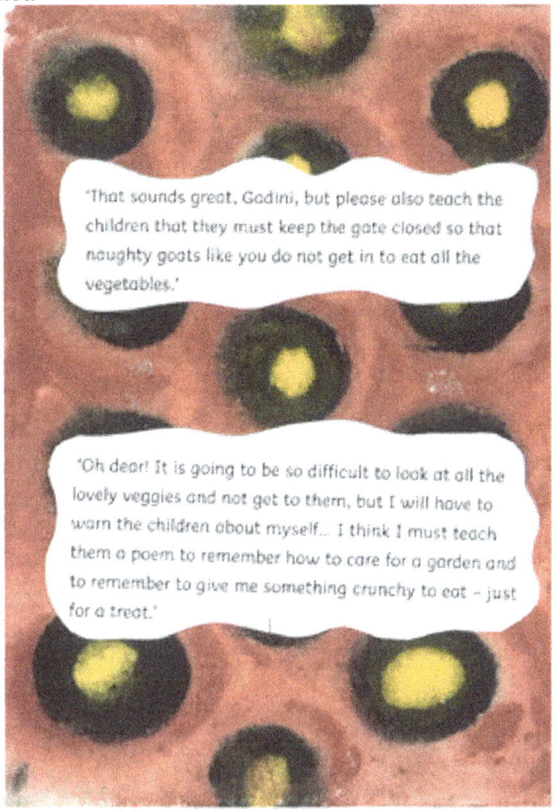

Figure 9. Continued

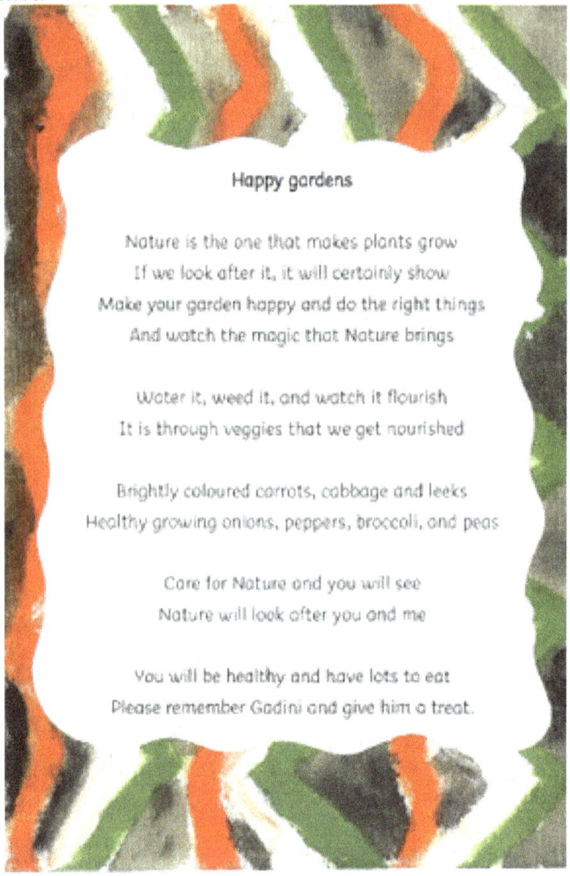

Figure 10. Continued

Language and Drama Ideas

Idea 1

Ask learners to find descriptive words in the story, for example juicy, crunchy, bright, colourful, and naughty. Learners should then say the words in a way that shows their meaning, using gestures or facial expressions to enhance the effect. For example, the word crunchy could be said with a rolling “r” sound to imitate the sound of a crunch. Discuss how the way something is said, as well as the gestures and facial expressions used, can enhance the meaning of a word.

Idea 2

Identify the rhyming words in the poem at the end of the story. Explain to learners what rhyming means, and then encourage them to think of means and words that rhyme with those used in the poem. For example, learners could add the word blow to rhyme with grow and show.

Idea 3

Use the word list at the back of the book to ensure that learners understand the meaning of each word. Play a game in which a word is explained without saying the actual word, and learners must guess what the word is. This will enhance retention of vocabulary.

Idea 4

Ask the following questions while reading the story, or set a short comprehension test that is appropriate for the learners' developmental level. For example:

Why is Gadini not allowed into the vegetable garden?
Name the vegetables that Gadini and Chicka saw.
What are the important things that Gadini must teach schoolchildren?
How must the children protect their garden from Gadini?
What is a treat?
What is your favourite treat?

You can also include multiple-choice questions or matching activities.

Idea 5

Give several learners the opportunity to play the roles of Gadini or Chicka. Learners can then act out the story of Gadini and Chicka.

Natural Science Ideas

Idea 1

Discuss the following questions:

What are worms?

How do worms change?

Study the life cycle of a silkworm. During silkworm season, a teacher could also keep silkworms in the classroom to expose learners to the different stages of their life cycle.

Idea 2

Ask learners to name the insects that appear in the story.
Discuss the following questions:

How do we know that a creature is an insect?
What characteristics do insects have?

Show pictures of different insects and ask learners to classify them according to their features.

Idea 3

Help learners grow their own lentil or bean plant in a small container in the classroom. Teach them what plants need in order to grow and encourage them to take care of their plants. Conduct a weekly inspection of the plants and ask learners to describe any changes they observe.

Art Ideas

Idea 1

Paint a butterfly. Learners paint one side of the butterfly using primary colours and then fold the paper in half while the paint is still wet so that the image transfers onto the other side. Use this activity to discuss symmetry in nature, such as in flowers, leaves, and animals.

Idea 2

Create giant insects from recycled materials that can be hung in the classroom.

Idea 3

Allow learners to collect litter from the school garden. They can then create a 3-D "litterbug" collage using the collected materials and other art materials such as crayons or paint. Crumpled paper, bottle tops, chip packets, and similar items can be glued onto the artwork.

Idea 4

Learners can draw each stage of the silkworm or bean life cycle whilst using a variety of media. It can even be done by shaping clay to represent each stage and presented on cardboard.

Mathematics Ideas

Idea 1

Discuss different shapes found in nature, especially those seen in the storybook, such as irregular shapes, circles, and ovals.

Idea 2

Explain what a graph is and show learners different types of graphs. Measure the height of the plant each week and record the results using a frequency plot or simple graph to show the plant's growth.

Idea 3

Use a picture containing many insects as a resource. Learners can create pictographs showing how many of each insect appear in the picture.

Music Ideas

Idea 1

When discussing littering in the poem, introduce the concept of a "litterbug." Explain what a rap is. Some learners can clap a rhythm while others perform the rap, and then they can switch roles. Learners can then be taught a short rap about being a litterbug, for example:

Hey all, listen, I have something to say
Don't let litterbugs get in your way
Litterbugs are ugly and certainly not cool
Do not let litterbugs rule!

Idea 2

Compose a simple melody for the rap and teach learners to sing it.

Idea 3

Teach learners existing children's songs that relate to aspects of the story. Focus on specific musical elements, such as dynamics (loud and soft) or tempo (fast and slow).

Many additional ideas can be included. However, it is important to ensure that any integrated activity is not simply an add-on but meaningfully teaches an aspect of the specific discipline.

CONCLUSION

The lesson ideas presented in this chapter demonstrate how Arts integration can enrich Environmental Change Education by connecting creative expression with concepts from multiple subject areas. Through language and drama activities, learners develop vocabulary, comprehension, and expressive communication skills. Natural Science activities stimulate curiosity about the natural world and provide opportunities for observation and inquiry. Visual art activities allow learners to explore creativity while developing fine motor skills and an appreciation for patterns and forms found in nature. Mathematics activities introduce important concepts such as shapes, measurement, and data representation, while music activities promote rhythm, listening skills, and collaborative participation. In this way, environmental themes become meaningful learning opportunities across the curriculum.

By integrating these disciplines through a shared theme or story, educators can create engaging learning experiences that support the holistic development of Foundation Phase learners. Such an approach enables learners to connect different areas of knowledge while encouraging creativity, critical thinking, and active participation. Importantly, however, integrated learning must be carefully planned so that each subject retains its educational focus and contributes meaningfully to the learning process.

As discussed throughout this chapter, the environmental challenges facing the world today demand an educational response that begins early and extends beyond the simple transmission of factual knowledge. The environmental crisis is not only an ecological concern but also a matter of social responsibility and children's rights. Within the South African context, the Curriculum and Assessment Policy Statement (CAPS) provides a valuable foundation by embedding environmental themes within the Life Skills curriculum and encouraging integrated teaching approaches in the Foundation Phase.

Nevertheless, the presence of environmental content within the curriculum does not automatically result in meaningful learning or behavioural change. The way in which Environmental Change Education is taught is equally significant. This chapter has therefore argued that integrated, creative pedagogies—particularly those drawing on the Creative and Performing Arts—can deepen learners' engagement with environmental concepts. Through music, drama, visual arts, and movement, complex environmental ideas can be transformed into concrete, experiential learning that resonates with young learners' natural inclination toward play and exploration.

The qualitative approach adopted in this study, combining document analysis and literature review, highlighted both the opportunities and the challenges associated with implementing Environmental Change Education in the Foundation Phase. While curriculum policy supports integration and holistic development, teachers require practical strategies and the confidence to translate these principles into effective classroom practice. Arts-integrated teaching offers one such strategy by linking environmental knowledge with creativity, emotional engagement, and critical thinking.

Ultimately, Environmental Change Education in the Foundation Phase should aim not only to inform learners about environmental problems but also to cultivate a sense of responsibility, care, and agency toward the natural world. When thoughtfully implemented, Arts integration can transform the classroom into a dynamic learning space where environmental understanding, creativity, and meaningful participation intersect. In doing so, education can play an important role in shaping environmentally conscious citizens who are better prepared to contribute to a more sustainable future.

REFERENCES

Afu, E. A., Ekwesaranna, F., Ohia, n., & Nwagu, E. K. (2025). Integrating digital technologies into environmental education curriculum: addressing pedagogical issues in the 21st century. Review of Education – [Special Edition] Volume 37, Issue 4 [2025}: 95-102 https://reviewofeducation.unn.edu.ng/

Alves-Oliveira, P., Arriaga, P., Xavier, C., Hoffman, G., & Paiva, A. (2022). Creativity landscapes: Systematic review spanning 70 years of creativity interventions for children. *The Journal of Creative Behavior*, *56*(1), 16–40. DOI: 10.1002/jocb.514

Bürgin, A. (2021). The European Commission. In Environmental Policy in the EU (pp. 93-109). Routledge. DOI: 10.4324/9780429402333-8

Chawla, L. (2007). Childhood experiences associated with care for the natural world. *Children, Youth and Environments*, *17*(4), 144–170. DOI: 10.1353/cye.2007.0010

Deliman, M. (2025). Drama-integrated dialogic pedagogy and the development of empathy and social awareness in classrooms. *Education Sciences*, *15*(6), 731. DOI: 10.3390/educsci15060731

Department of Basic Education (DBE). (2011). *Curriculum and Assessment Policy Statement (CAPS): Foundation Phase*. DBE.

Dewey, J. (1934). *Art as Experience*. Perigee.

Du Plessis, P., & Mestry, R. (2019). Teachers for rural schools–a challenge for South Africa. *South African Journal of Education*, *39*(Supplement 1), 39. DOI: 10.15700/saje.v39ns1a1774

Dube, B., Nkomo, D., & Apadile-Thokweng, M. (2024). Pragmatism: An essential philosophy for mixed methods research in education. *International Journal of Research and Innovation in Social Science*, *8*(3), 1001–1010. DOI: 10.47772/IJRISS.2024.803073

Edillor, J. A. A. (2024). Inquiry-based instruction in science among Grade 8 students. *International Journal of Educational Contemporary Explorations*, *1*(2), 129–137.

Eisner, E. W. (2002). *The Arts and the Creation of Mind*. Yale University Press.

Fredericks, E. J., Condy, J., Phillips, H. N., Maree, C., & Chigona, A. (2025). Novice teachers' professional discourse on teaching natural science in the Foundation Phase in the Western Cape, South Africa. *South African Journal of Education*, *45*(1), 1–12. https://www.ajol.info/index.php/saje/article/view/292174. DOI: 10.15700/saje.v45n1a2530

Freire, P. (1970). *Pedagogy of the Oppressed*. Continuum.

Gruenewald, D. A. (2003). The best of both worlds: A critical pedagogy of place. *Educational Researcher, 32*(4), 3–12. DOI: 10.3102/0013189X032004003

Gule, M. P. (2024). Strategies of implementing environmental education in primary schools (Doctoral dissertation, University of South Africa).

Jansen van Vuuren, E. (2022). Integrated Music Education in Primary School: A Position Paper. *Journal of Culture and Values in Education, 5*(1), 211–222. DOI: 10.46303/jcve.2022.16

Kiviranta, L., Lindfors, E., Rönkkö, M. L., & Luukka, E. (2024). Outdoor learning in early childhood education: Exploring benefits and challenges. *Educational Research, 66*(1), 102–119. DOI: 10.1080/00131881.2023.2285762

Kotsis, K. T. (2024). Integrating inquiry-based learning in the new Greek primary science curriculum. *European Journal of Education and Pedagogy, 5*(6), 63–71. DOI: 10.24018/ejedu.2024.5.6.899

Lamanauskas, V. (2022). Dvidešimt aštuntoji nacionalinė mokslinė-praktinė konferencija „Gamtamokslinis ugdymas bendrojo ugdymo mokykloje–2022": konferencijos apžvalga. Gamtamokslinis ugdymas, 19(2), 104-111.

Levi-Keren, M., Katz, R., & Shapiro, E. (2024). Simulation-based learning and empathy development in teacher education: A multidimensional approach. *Education Sciences, 14*(12), 1338. DOI: 10.3390/educsci14121338

Lupión-Cobos, T., García-Ruiz, C., & Toma, R. B. (2026). Preservice primary teachers' attitudes towards STEM education: A study developed through an integrated STEM training programme using inquiry-based approaches and engineering design practices. *Research in Science Education*. Advance online publication. DOI: 10.1007/s11165-026-10323-y

Madlela, B., & Ramnarain, U. (2024). *Utilising educational technologies to support inquiry-based learning in natural science*. International Journal of Emerging Multidisciplinary Educational Studies., DOI: 10.53378/ijemds.353093

Mashaba, E. K., Maile, S., & Manaka, M. J. (2022). Learners' knowledge of environmental education in selected primary schools of the Tshwane North District, Gauteng Province. *International Journal of Environmental Research and Public Health, 19*(23), 15552. DOI: 10.3390/ijerph192315552 PMID: 36497622

Moremong, T. C. (2024). The implementation challenges of sustainability education in primary schools (Master's thesis, University of Johannesburg).

Ng'asike, J. T. (2019). Indigenous knowledge practices for sustainable environmental education in early childhood development. *Journal of Outdoor and Environmental Education, 22*(2), 145–157.

Ölçer, S. (2025). Inquiry-based science teaching in primary school. *Journal of Child Development and Education, 1*(1), 55–66.

Oliver-Barcelo, M., Ferrer-Ribot, M., & Jové, G. (2024). Arts education in early childhood teacher training: An international analysis. *Teaching and Teacher Education, 148*, 104703. DOI: 10.1016/j.tate.2024.104703

Oranga, J., Gisore, B., & Areba, G. (2023). Barriers to transformative climate change education: Mitigation and resilience-building. *The International Journal of Social Sciences (Islamabad), 3*(3), 389–396.

Palmer, J. A. (1998). *Environmental Education in the 21st Century*. Routledge.

Piaget, J. (1952). Jean Piaget.

Prideaux, B., Thompson, M., & Pabel, A. (2020). Lessons from COVID-19 can prepare global tourism for the economic transformation needed to combat climate change. *Tourism Geographies, 22*(3), 667–678. DOI: 10.1080/14616688.2020.1762117

Ramsaroop, S., & Petersen, N. (2020). Portraits of primary school pre-service teachers at a South African university: Implications for nuanced student support. Journal of Education (University of KwaZulu-Natal), (81), 82-98.

Roussou, A. M., Argyrakou, C. C., & Milakis, E. D. (2025). Integrating STEAM and theatrical methods in early childhood environmental education: A framework for holistic learning. International Journal of Geography. *Geology and Environment, 7*(2), 19–42.

Russell-Bowie, D. (2009). Syntegration or disintegration? Models of integrating the arts across the primary curriculum. *International Journal of Education & the Arts, 10*(28), 1–23.

Sánchez, C.. (2022). Raising ecological awareness and digital literacy in primary school children through gamification. *International Journal of Environmental Research and Public Health, 19*(3). Advance online publication. DOI: 10.3390/ijerph19031829 PMID: 35162171

Shava, S., Krasny, M. E., Tidball, K. G., & Zazu, C. (2021). Agricultural knowledge in urban and resettled communities: Applications for environmental education. *Environmental Education Research, 27*(3), 393–411.

Stears, M., James, A. A., & Beni, S. (2019). Teaching science in the foundation phase: Where are the gaps and how are they accounted for? *South African Journal of Childhood Education*, *9*(1), 1–9.

Tannock, S. (2021). Pedagogies of Hope and Fear. In *Educating for Radical Social Transformation in the Climate Crisis* (pp. 87–117). Springer International Publishing. DOI: 10.1007/978-3-030-83000-7_4

Taylor, N. (2019). Inequalities in teacher knowledge in South Africa. In Spaull, N., & Jansen, J. D. (Eds.), *South African schooling: The enigma of inequality – A study of the present situation and future possibilities* (pp. 263–283). Springer. DOI: 10.1007/978-3-030-18811-5_14

Tilbury, D. (1995). Environmental education for sustainability: Defining the new focus of environmental education. *Environmental Education Research*, *1*(2), 195–212. DOI: 10.1080/1350462950010206

UNESCO. (2017). *Education for Sustainable Development Goals: Learning Objectives*. UNESCO.

UNICEF. (2019). The climate crisis is a children's rights crisis.

Van Wyk, M. M. (2023). Barriers to managing environmental education projects in primary schools. *Koers*, *85*(1). Advance online publication. DOI: 10.19108/KOERS.85.1.2535

VanTassel-Baska, J. (2023). Introduction to the integrated curriculum model. In *Content-based curriculum for advanced learners* (pp. 17–36). Routledge.

Vygotsky, L. S. (1978). *Mind in Society*. Harvard University Press.

Chapter 11
Climate Change Education Through Creative Arts:
Global Perspectives and the ICA1501 Module Case Study

Anriët Van Deventer
http://orcid.org/0000-0002-3358-0238
University of South Africa, South Africa

Georina Westraadt
http://orcid.org/0009-0006-3188-2079
Two Oceans Graduate Institution, South Africa

ABSTRACT

This chapter examines the integration of climate change education into higher education curricula through a case study of ICA1501, Introduction to Creative Arts, at a South African university. Situated in teacher education, the module advances a transformative pedagogy that integrates eco-pedagogy and arts-based learning to foster climate change literacy in Visual Arts and Drama. Informed by Freirean critical pedagogy, it frames climate change education as a global, holistic, justice-oriented approach encompassing social, ethical, cultural, and environmental dimensions. Arts-based practices, including visual production, performance, and reflective inquiry, render scientific knowledge affective and contextually meaningful. Climate change functions as an integrating curricular lens through eco-art, drama, and critical reflection. Positioned alongside regional models across five continents, ICA1501 demonstrates how climate change pedagogy can be adapted across contexts and disciplines to cultivate socially responsive, imaginative leaders for sustainability.

DOI: 10.4018/979-8-3373-6975-4.ch011

INTRODUCTION

The ICA1501 module is likened to international representations. In this chapter, the approach is qualitative and conceptual, drawing on existing scholarship on eco-pedagogy, arts-based learning, and climate change education to explain how climate education can be integrated into Creative Arts pedagogy. The ICA1501 module is used as an illustrative case to demonstrate how these theoretical ideas can be applied in teacher education. The chapter, therefore, connects theory, literature, and practice to conceptualise how Creative Arts pedagogy can support climate change literacy in higher education. For example, in the United States, preschoolers engage with sustainability in age-appropriate, playful ways through learners' Drama. In India, village puppetry groups dramatise floods, deforestation, and pollution to build local understanding and empathy. In Europe, at a Greek university, students create *trash art* mosaics from waste, promoting recycling and new attitudes toward consumption. In the Pacific, an exhibition of Chamorro and other Pacific-Islander art uses recycled materials and Drama, including storytelling, to reconnect island communities with their sacred land and water. These cases in North America, Europe, Asia, Latin America, Oceania, and Africa show that arts-integrated climate change education is a global trend that also resonates locally in South Africa. Throughout the chapter, we emphasise human, ethical and community dimensions of climate pedagogy. The ICA1501 Module and the other examples highlight *empathy, values, and collective action* rather than technology-driven solutions. We outline active teaching strategies, such as field-based projects, drama, artmaking, and discussion, together with diverse assessment modes, namely creative portfolios, community exhibitions and reflective essays that assess both knowledge and values. Finally, we discuss institutional support: UNESCO's Education for Sustainable Development goals urge that climate literacy be woven into all subjects, and universities must provide training, collaboration, and policy incentives for such integrated teaching. By comparing the ICA1501 Module to other representations, we show how creative climate pedagogy can be adapted across disciplines and cultures worldwide, empowering educators, lecturers, teachers, students and learners to imagine and create a more just and sustainable future.

PROBLEM STATEMENT

Despite the urgent need for climate change literacy to preserve the planet for future generations, traditional higher education curricula often overemphasise abstract scientific data and technical solutions. This data-centric approach frequently leaves learners feeling anxious, overwhelmed, or alienated, failing to foster the values and behavioural transformations necessary to address the crisis. Consequently, there is a critical need for transformative, arts-integrated pedagogies such as those used in the ICA1501 module that bridge the gap between scientific knowledge and socio-emotional engagement to cultivate socially responsive leaders.

BACKGROUND

Oziewicz (2023) defined climate change literacy as a dedicated consideration of the urgent *need to cultivate values and attitudes,* and of how we, as humans, should behave to preserve the earth for now and future generations. Climate change in education will lead to behavioural transformation among students in higher education (Cordero, Centeno, & Todd, 2020; Molthan-Hill, Worsfold, Nagy, Filho, & Mifsud, 2019) and subsequently, among learners in schools, they teach in this crucial form of literacy (Tytler & Freebody, 2023). Climate change education equips students and learners with the knowledge, skills, and values necessary to address the impacts of the climate crisis (Molthan-Hill et al., 2019; Powers, De Waters & Dhaniyala, 2021). It helps individuals understand the causes and consequences of climate change, preparing them to live with its effects and take appropriate actions to adopt more sustainable lifestyles.

The climate crisis refers to the urgent and wide-ranging problems caused by long-term changes in Earth's climate, especially those driven by human activities such as burning fossil fuels, deforestation, and industrial processes that increase greenhouse gas concentrations in the atmosphere (Intergovernmental Panel on Climate Change, 2018, 2021). These changes lead to global warming, extreme weather events, sea-level rise, biodiversity loss (Intergovernmental Panel on Climate Change, 2021; Pörtner, Scholes, Arneth, et al., 2023), food and water insecurity, health risks (Patz, 2014), economic disruption, displacement, and even conflict (Loucks, 2021).

More broadly, the term emphasises the severity and immediacy of climate change and its impacts on natural systems and human societies, signalling a decisive, potentially dangerous turning point (Armstrong-McKay, Staal, Abrams, et al., 2022) that requires urgent action to avoid catastrophic consequences (Kemp, Xu, Depledge, et al., 2022).

Scientific assessments, particularly those of the 2021 Intergovernmental Panel on Climate Change (IPCC), show that the climate crisis is unequivocally caused by human activities and is already affecting land, oceans, and the atmosphere worldwide, with worsening impacts expected if global warming is not rapidly limited.

Furthermore, the climate change crisis is not only a scientific and technological challenge but also a profound human one (Ford et al., 2018). Around the world, communities are feeling the effects in deeply personal ways – from drought-stricken farmlands in the Sahel, a semi-arid ecological belt in Africa that forms a transitional zone between the Sahara Desert to the north and the more humid savanna regions to the south (Traore & Owiyo, 2013), to flooded cities in Southeast Asia (Limthongsakul, Nitivattananon & Arifwidodo, 2017). Currently, large parts of the Northwestern United States are experiencing severe drought, with drought recovery probabilities 25-50% lower in recent decades (Williams et al., 2024).

Locally, in 2017, firestorms wiped out large parts of Knysna in the Southern Cape, South Africa (Forsyth, Le Maitre, Van den Dool, Walls & Pharoah, 2019). The Cape Province has experienced increasing fire frequency, with annual burnt area of fynbos rising significantly since 1980 (Kraaij, Baard, Cowling, Van Wilgen & Das, 2013). The Limpopo Province has experienced multiple extreme flood events linked to antecedent drought conditions (Franchi et al., 2024; Munzhedzi, Mugari, Nethengwe & Gumbo, 2024).

In such times, higher education carries a moral responsibility: to equip students not just with facts but also with creativity, empathy, and the ability to communicate and act for a sustainable future. For example, a recent analysis of Africa's warming climate notes that lecturers must prepare graduates who can *communicate, imagine, and act on environmental issues* (Filho, Sierra, Kalembo, et al., 2024; Manteaw, Boafo, Owusu, Enu & Amoah, 2025). Similarly, the United Nations Educational, Scientific and Cultural Organisation's (UNESCO) Education for Sustainable Development framework (ESD) emphasises that learning should equip students with the means to tackle climate change (United Nations Educational, Scientific and Cultural Organisation, 2020). The UNICEF Green Schools Initiative focuses on building climate-resilient schools and empowering learners as climate champions. It aims to integrate climate change education into school curricula, particularly in regions most affected by climate change (United Nations Children's Fund, 2024).

The chapter employs specific terminology, which is clarified as follows: *teachers and learners* are used in accordance with the definitions provided in the South African Curriculum and Assessment Policy Statement (CAPS). Within this policy framework, a *teacher* is a professionally qualified individual responsible for facilitating learning in a school context, while a *learner* means a pupil formally enrolled in a school and engaged in compulsory or elective education.

The term *educator* is applied more broadly as an inclusive, global descriptor for individuals involved in teaching, training, or facilitating learning across diverse contexts, including formal, non-formal, and informal settings.

In the context of the ICA1501 module, the terminology of *lecturers and students* is adopted, reflecting usage within a higher education institution. Here, a *lecturer* refers to an academic staff member responsible for teaching, assessment, and scholarly engagement, while a *student* denotes an individual formally registered for study at a tertiary institution.

Education for Sustainable Development Within the UNESCO Framework

Figure 1. Sunshine Seeds

Source: Shutterstock.com, UNESCO publication

Education for Sustainable Development (ESD), as articulated by UNESCO, positions education as a transformative, lifelong process that enables individuals and communities to participate meaningfully in creating environmentally sustainable, socially just, and economically viable futures. Rather than treating sustainability as a discrete curricular topic, ESD reframes learning itself as a vehicle for systemic change. It integrates ecological awareness, ethical responsibility, cultural sensitivity, and critical reflection into educational practice, thereby preparing learners to respond to complex global challenges such as climate change, biodiversity loss, inequality, and unsustainable patterns of production and consumption. Central to this orientation is the understanding that sustainable development requires not only scientific knowledge, but also shifts in values, behaviours and collective action.

UNESCO conceptualises ESD through interrelated cognitive, socio-emotional, and behavioural learning dimensions. The cognitive dimension develops a critical understanding of sustainability issues and their interconnected nature across environmental, social, cultural, and economic systems. The socio-emotional dimension nurtures empathy, solidarity, ethical awareness, and the capacity for collaborative problem-solving. The behavioural dimension emphasises agency, encouraging learners to translate knowledge and values into responsible decision-making and practical action. Together, these domains reflect a holistic pedagogy that moves beyond knowledge transmission toward transformative learning capable of reshaping individual and societal practices (UNESCO, 2020).

The current global implementation agenda, known as *ESD for 2030*, operationalises this vision through coordinated policy integration, the transformation of learning environments, strengthened educator capacity, youth empowerment, and locally grounded action (UNESCO, 2022). This framework aligns closely with Sustainable Development Goal 4.7, which calls for all learners to acquire the competencies necessary to promote sustainable development. Within higher education, ESD therefore functions not merely as content, but as an epistemological and pedagogical shift that reorients teaching, research, and community engagement toward sustainability-centred futures.

Traditional climate curricula often form parts of science and technology, carbon cycles, emission models, and renewable energy innovations. While these are important, an overemphasis on data can leave learners feeling anxious or alienated from the issue. As an educator observed, humans have a hard time hearing about climate change in abstract terms: it *feels too overwhelming, it feels too scary*. By contrast, the *Creative Arts,* namely Visual Arts, Drama, Music and Dance, can tap into emotion, values and culture, making the issue *tangible and hopeful*. In South Africa, according to the Department of Basic Education CAPS curriculum, Creative Arts is a compulsory subject in the Intermediate Phase (Grades 4–6) (Department of Basic Education, 2011) that integrates four art forms: Dance, Drama, Music, and Visual Arts. It aims to develop learners' creative expression, aesthetic awareness, critical thinking, and cultural understanding through practical and theoretical engagement with the different art forms in Creative Arts.

Artistic processes allow learners to *feel* climate change in their bodies and imaginations, not just their minds. For instance, in a New York learners' play about saving an urban park, young characters argue, strategise and even break into song, an approach that one drama-maker says helps kids process tough topics *without dwelling in despair.*

This chapter focuses on the module ICA1501, Introduction to Creative Arts (hereafter referred to as ICA1501), a course in South African teacher education within the Bachelor of Education (B.Ed.), Arts stream, which uses climate change as a central lens for learning to teach Creative Arts: Visual Arts and Drama. Rather than treating climate change as an add-on topic, ICA1501 integrates it across the module's goals and activities. Note that the Bachelor of Education is an undergraduate professional degree that prepares students for a career in teaching and educational practice. In the South African higher-education context, the B.Ed. is generally a four-year qualification aligned with teacher registration requirements through the South African Council for Educators (SACE). It serves as the foundational professional route into formal schooling.

We explore how this approach exemplifies two key theoretical traditions: *eco-pedagogy*, which is teaching that connects ecological and social understanding and *arts-based learning*, which refers to using creativity as a form of inquiry. Then we detail the pedagogical strategies, for example, field trips, drama enactments, and assessment methods such as portfolios and community projects, used in ICA1501, and discuss parallels with similar programs worldwide. Throughout, the emphasis is on the human, ethical, and community-oriented aspects of climate change education: building empathy, cultural relevance and a sense of responsibility, rather than simply delivering technical solutions. By highlighting ICA1501 and contrasting it with other regional models in North America, Asia, Europe, South America and Oceania, we show how climate change pedagogy can be adapted to diverse contexts and disciplines, ultimately preparing students worldwide to become imaginative and socially responsive actors for sustainability.

Visual creation and performance render scientific knowledge effective and contextually meaningful. Climate change serves as an integrating curricular lens through eco-art, drama, and critical reflection. Positioned alongside regional models across five continents, ICA1501 illustrates how climate change pedagogy can be adapted across contexts and disciplines to cultivate socially responsive, imaginative leaders for sustainability.

CONCEPTUAL FRAMEWORK

The chapter is conceptually informed by eco-pedagogy and arts-based learning as complementary pedagogical perspectives. Eco-pedagogy extends critical pedagogy into environmental contexts by emphasising ecological awareness, social justice, and critical engagement with environmental challenges. Arts-based learning positions creative processes such as visual production, performance, and storytelling as meaningful modes of inquiry that allow learners to explore complex social and environmental issues. Together, these perspectives provide a framework for understanding how Creative Arts education can support climate change literacy by engaging cognitive, emotional, and ethical dimensions of learning. Figure 1: Sunshine seeds supports the Education for Sustainable Development (ESD) framework. This visually anchors the *cognitive, socio-emotional, and behavioural* learning dimensions.

METHODOLOGICAL APPROACH

This chapter adopts a qualitative conceptual approach. Rather than presenting primary empirical data, the discussion develops through the analysis and synthesis of existing scholarship on eco-pedagogy, arts-based learning, and climate change education. Relevant literature is examined to identify theoretical perspectives and pedagogical approaches that support the integration of climate change education across disciplines. Through this interpretive engagement with the literature, the chapter explores how Creative Arts pedagogies can contribute to climate change literacy and sustainability education.

To illustrate how these theoretical ideas can be implemented in practice, the ICA1501 module (Introduction to Creative Arts) within a Bachelor of Education programme is presented as a case study. The module demonstrates how climate change themes can be integrated into Creative Arts pedagogy through visual and performative practices, reflective learning activities, and community engagement. The purpose of this illustration is not to present empirical findings, but to demonstrate how the principles of eco-pedagogy and arts-based learning can be applied in teacher education.

SCOPE AND LIMITATIONS

The chapter focuses on conceptual interpretation and pedagogical illustration rather than empirical evaluation. The ICA1501 module is therefore used as a contextual example to demonstrate possibilities for integrating climate change education into Creative Arts teacher education. While the discussion draws on international literature and global examples, it does not aim to provide a comprehensive empirical review of all arts-based climate education initiatives. Instead, the chapter offers a conceptual perspective highlighting the potential of Creative Arts pedagogy to support climate change literacy and sustainability-oriented teaching in higher education.

THEORETICAL FRAMEWORK: ECO-PEDAGOGY AND ARTS-BASED LEARNING

Eco-pedagogy and arts-based learning form the foundation of ICA1501's approach. Eco-pedagogy extends *Paulo Freire's critical pedagogy* into an environmental context. It urges learners to *read the world* through an ecological lens, critically examining how social, economic, and power structures impact the planet. In practice, this means connecting climate change science with issues of justice and equity. African students and learners stress that eco-pedagogy in their context *must* include indigenous knowledge and community wisdom. For example, Lotz-Sisitka, Wals, Kronlid and McGarry (2020) argue that sustainability education should value local practices and indigenous ecological perspectives alongside Western science. In other words, climate change education becomes a dialogue between global science and *lived African realities*. An African classroom might therefore discuss climate change not only in terms of rising CO_2 and the resulting rise in Earth's temperature, but also through cultural stories, traditional farming knowledge, and the legacy of colonial resource extraction. This holistic approach highlights that climate change is intertwined with history, politics and values, and is often called *climate justice*, so that learners recognise that marginalised communities worldwide often bear the harshest burdens.

Paulo Freire's Pedagogy

A widely accepted scholarly source that explicitly connects *Paulo Freire's pedagogy* to environmental justice and Eco pedagogy is Greg William Misiaszek's work (2023). His book on Freirean environmentalism explains that Eco pedagogy extends Freire's critical pedagogy into sustainability by linking social oppression, cultural inequality, and environmental destruction within a single socio-environmental framework (Freire, 1970). It argues that environmental justice is inseparable from broader struggles around globalisation, racism, gender, colonialism, and citizenship, and calls for critical, culturally grounded approaches to sustainability education.

Complementary research analysing Freire's vision of environmental education similarly highlights how key Freirean principles, praxis, dialogue, participation, contextualisation, and critical consciousness provide a theoretical bridge between critical pedagogy and environmental education, even though Freire himself was not primarily an environmental educator (Souza & Garcia, 2023; Misiaszek, 2023).

Complementing eco-pedagogy, arts-based learning treats artistic creation and performance as powerful tools for inquiry and meaning-making. Arts educator Elliot Eisner (2002) notes that the arts allow learners to explore complexity through metaphor, symbolism, and sensory experience. In the context of climate change education, the Creative Arts: Visual and Performing arts serve to humanise data and embody environmental issues. For instance, science alone can illustrate a melting glacier with graphs, but a drama scene in which a student personifies the glacier using themes like *"If I were a melting ice cap" or "I would say..." can* evoke empathy and urgency in a way that charts cannot. The ICA1501 Module in this chapter explains that, through the performance of stories (storytelling), students explore complexity, empathy, and meaning, thereby making climate change *tangible and emotionally resonant.*

Together, eco-pedagogy and arts-based methods engage both head and heart. Eco-pedagogy requires students to think critically about systems, power, and ethics, for example, examining why low-income communities are more affected by floods and disasters. Arts-based learning allows students to express feelings and imagine solutions creatively, for example, choreographing a dance about a forest regrowing after fire. This dual focus encourages *ethical dispositions*, such as care, empathy and a sense of agency alongside factual knowledge. In summary, climate change pedagogy in this framework is not just about transmitting scientific facts; it is a transformative process that fosters climate literacy grounded in values. A climate-literate graduate, in this view, is someone who understands climate change science and *values equity and justice* when seeking solutions. The Creative Arts is a *key-way* to cultivate those values, as one commentator observes: creative activities *humanise data, making environmental issues tangible and emotionally resonant.* By linking these theoretical strands, ICA1501 and similar programs aim to produce learners who not only know about climate change but are morally motivated to address it in their communities.

PEDAGOGICAL STRATEGIES FOR CREATIVE CLIMATE CHANGE EDUCATION

Implementing *eco-pedagogy and arts-based learning* requires *active, participatory teaching strategies.* Across cultures, climate change educators find that students learn best by doing, reflecting, and collaborating rather than by passively listening to lectures. Key methods include:

- *Experiential and place-based learning*: Students engage with real environments and communities. For example, ICA1501 students visit local sites such as streams, rubbish dumps and working farms to observe ecological issues firsthand, then interpret these observations through Creative Arts: Visual Arts or Drama. In other programs, learners might measure local temperature variations, document campus biodiversity, or help manage a school garden.

Fieldwork and service projects, for example, tree planting and community cleanup, make climate change concepts concrete and relevant. Such experience is core to learning: one study notes that making and performing climate change scenarios through Drama builds empathy and active agency. In line with the CAPS Creative Arts Topic 1 for *Performing Arts: Appreciate and reflect* (Department of Basic Education, 2011) Students can look at and discuss the dance drama *"Karatara"*, a Figure of 8 production choreographed by Shaun Oelf and Grant van Ster (Figure of 8, 2023), performed in the aftermath of the devastating fires which consumed large sections of Knysna in 2017.

- *Arts-based creation (Creative Arts: Visual and Performing Arts):* Students use creative media to explore climate change topics. Projects that will foster acute awareness of the environment and knowledge of how everyone can make a difference. In ICA1501, Visual Arts projects ask teacher-trainees to study eco-art by African and global artists, for instance, the work done by Congolese performance artist Junior Mungongu, who creates and wears costumes made from plastic bottles and lids picked up at dump sites and then creates their own. They might paint scenes of drought, build collages from recycled materials, or design community murals for water conservation. For a positive incline, especially as the focus is on training for Intermediate Phase teachers, picture-making projects of trees in forests, the floral kingdom, the wonders of nature, marine life, and birds and animals in the wild. While looking at photographs or real life, they become more aware, their eyes are opened, and their observation skills are honed; the desire to protect and conserve grows. Projects aim to focus students' attention on the importance of conserving the environment and cultural treasures (Westraadt, 2015). These activities turn abstract issues (such as air pollution or deforestation) into tangible experiences that engage the senses and foster meaning-making. Similarly, Drama (and movement) is used to embody climate change narratives. Techniques include role-playing elements of nature, students dramatise cause-and-effect, acting as plants growing, trees swaying, and fish swimming; creating mime sequences; exploring contrast and emotional expression; and using time-and-space to create weight, energy and force. They devise short plays like *"Switch off Something" and "The Day the Earth Spoke"* that reflect local environmental stories. These methods invite imagination and voice: students speak *from the perspective of nature* or co-write stories that connect personal experiences to global change. By integrating Creative Arts (Visual and Performing Arts), lecturers tap into learners' creativity and emotions, helping them internalise the moral urgency of climate change.

- *Reflective practice and dialogue:* Art and experience are paired with reflection. ICA1501 students keep journals linking their artwork or skits to questions of climate ethics. They discuss prompts like *"What would nature say to us? "What happens if the turtle cannot dive down for food because its stomach is filled with plastic?"* or debrief after a performance on how it made them feel. Such reflection helps solidify learning: studies show that when students write or talk about their creative work related to climate, their understanding and empathy deepen. Group dialogue in class or community workshops also allows the sharing of ideas and cultural perspectives. For example, after a play, young audience members in New York discussed sustainability ideas such as composting and public transit with the actors. Facilitated reflection ensures students connect the creative activities back to real-world concerns and values. In line with the Curriculum and Assessment Policy Statement (CAPS), Creative Arts, Topic 1 for *Visual Arts: Visual literacy* (Department of Basic Education, 2011), students look at and appreciate or analyse two-dimensional artworks by Noria Mabasa, *"The Flood";* Penny Siopis, *"Patience on a Monument";* and Helmut Starcke, *"Legend"*. CAPS is the national curriculum framework implemented by the Department of Basic and three-dimensional work made by South African artists, e.g., Education in South African schools. It provides detailed guidance on what teachers must teach and how assessment should occur in each subject and grade.
- *Interdisciplinary and collaborative teaching:* Effective climate change modules often involve lecturers from different fields working together. In ICA1501, the Creative Arts lecturers partnered with science and geography lecturers to ensure that climate change content is scientifically accurate and aligns with curricula across multiple subjects. This mirrors cross-campus, whole-institution approaches recommended by UNESCO, in which physics, history, literature, and the Creative Arts reinforce sustainability themes. Interdisciplinary projects might have students, for instance, collect local climate change data with science students and then represent findings through public art or dance. Collaboration also reflects the real-world reality that climate change solutions require contributions from many disciplines. Importantly, ICA1501 demonstrates to students-in-training that *all* teachers, not just science or geography educators, have a role: Creative Arts teachers can be *agents of change*, without becoming despondent, always bearing in mind that students are being trained to teach young learners. The message that there is still hope, that everything is not lost, and that everyone can make a difference should be conveyed by embedding environmental themes into their subjects. Faculty teams can co-design lessons, review each other's con-

tent, and share resources, such as lesson plans, images, and narratives, to ensure that climate change learning is coherent across courses.

- *Community engagement:* Many climate change-learning activities extend beyond the classroom. ICA1501, for example, includes public exhibitions of student art and performances held at schools or local venues. In these events, students present their climate change stories or murals to families, environmental groups, and local leaders, thus anchoring learning in a community context. Global cases follow suit: the Pacific art exhibit *"Ta Na'i Animu"* brings climate change-themed paintings and performances to Guam's public museum; the Kuikuro exchange project in Brazil involved indigenous artists documenting their culture and the threats posed by climate change to share with international audiences. At an open day at an Art Centre in the Western Cape, South Africa, parents and guests were invited to collaborate with the learners to celebrate World Environmental Education Day on 26 January 2026 in the making of insects and birds from cut up shampoo-, detergent- and other plastic containers, CD's, etc., joined with soft wire and using a glue gun. After the workshop, the assemblages were put up and exhibited in the foyer. Such community-based projects reinforce that climate change education is not insular; it invites dialogue with cultural traditions and seeks social change. They also help to validate student work, making climate change issues visible and actionable in society.
- *Scaffolded support and equity:* Because students enter with varied backgrounds, lecturers build in support. Workshops on basic art skills or climate change science concepts are provided so beginners can participate fully. Clear rubrics are developed so all students understand what is expected. Peer collaboration further balances skill gaps: in ICA1501, students with stronger art abilities mentor others during projects. This attention to equity is crucial: educators note that differences in ability or resources can pose challenges in creative climate change courses. Addressing this might mean using readily available materials, such as recycled materials, instead of expensive art supplies. The Visual Redress Chair Project illustrates how arts-based practices foster meaning-making through material-discursive intra-actions. Figure 2: The Visual Redress Chair Project shows how Creative Arts Education students at a university repurposed and transformed discarded campus chairs dumped under a stage in a hall into public artworks. addressing social and educational issues. The materiality of the broken chairs became an enabling constraint, shaping both concept and process (Meiring, 2025).

Figure 2. The Visual Redress Chair Project, South Africa

Source: Copyright the authors

In summary, creative climate change pedagogy employs *active, learner-centred methods* that situate climate change in learners' lives and cultures. These strategies are inherently *human and ethical*: they evoke feelings, encourage reflection on values, and root learning in community needs. Rather than relying on tech-gadgets or abstract simulations, the emphasis is on *people engaging with each other and the environment* through art. As one review notes, active Visual Arts and Drama (Creative Arts) projects *build empathy and ethical insight,* turning students into active 'agents of change' (Gaztambide-Fernández, 2013). Our global examples (below) will illustrate these strategies in action across different contexts.

GLOBAL CASE EXAMPLES

Climate-change Creative Arts education takes many forms around the world. Below, we highlight representative cases from various continents that demonstrate how different cultures apply creative pedagogy to address local realities of climate change.

To infuse our discussion with the African and Southern African Development Community (SADC) context, we focus on pedagogical artefacts and case studies from sources that ground climate change education in local realities.

Here are the key SADC focused examples and methods:

1. The ICA1501 Module (South Africa)

This is a core teacher-training module at a South African university that uses Visual Arts and Drama to teach climate awareness. It is explicitly designed to weave environmental themes into the national CAPS curriculum. Students create:

- Eco-Art Projects: Murals for water conservation or sculptures made from e-waste.
- Drama Enactments: Students role-play local environmental events, such as the Limpopo floods or Cape firestorms, personifying elements of nature to build empathy.

2. *The !Haos and Sitstories Projects*

These are collaborative, community-based projects in South Africa:

- *!Haos Project*: Student teachers used branches and family-linked materials to create woven installations that explored ecological *relationality*.
- *Sitstories*: A project at a community centre involving performance, music, and dance to celebrate local heritage and creativity as a response to environmental change.

3. *"Karatara"* Dance Drama

This performance, choreographed by Shaun Oelf and Grant van Ster, serves as a powerful pedagogical tool. It was created in the aftermath of the 2017 Knysna fires, allowing students to *appreciate and reflect* on the trauma of environmental disaster through professional SADC choreography. Derived from Khoekhoe origins, as is typical of many Western Cape place names, "Karatara" has no single agreed meaning but is generally interpreted as referring to a water- or river-associated place, likely linked to the nearby Karatara River.

4. The Green School South Africa (Paarl)

Located in the Western Cape, this school serves as a *green* institutional model. Its architecture is inspired by local Paarl Berg boulders, and it utilises an "*iRESPECT*" values framework (Integrity, Respect, Empathy, etc.) to integrate sustainability into every aspect of the curriculum.

5. Indigenous Knowledge and Traditional Arts

The sources suggest moving beyond Western technology by utilising traditional African art forms:

- Bogolanfini (Mud Cloth): Students are encouraged to research this traditional Malian art form as a sustainable way to develop colourants from vegetables and mud.
- Earthenware Clay: Tapping into indigenous knowledge for making and firing clay in wood fires, rather than relying on electric kilns.

6. Localised Scientific Context

Rather than focusing solely on global CO_2 levels, the sources emphasise African scientific realities:

- The Sahel belt's experience with drought.
- Invasive alien vegetation in South Africa, which fuels massive veld fires and destroys indigenous biodiversity. It also clogs up natural vleiland (wetland) and sucks up millions of litres of water in an already dry country. A *vleiland* is a low-lying wetland area or marshy piece of land, often waterlogged and rich in vegetation.

To raise awareness of the importance of propagating indigenous flora, picture-making projects focus on natural and endemic species, particularly those under threat.

- *Africa – ICA1501 (South Africa):* As described, ICA1501 is a required module in teacher training that uses Creative Arts: Visual Arts and Drama (for Grades 4–6) to teach climate change awareness (Gadotti, 2008). In South Africa, this course weaves environmental themes into art curriculum outcomes: students learn climate change science in its social and ecological context, practice creative expression on climate change topics, and develop strategies to teach these ideas in future classrooms (Gadotti, 2008). For example, student assignments include designing public art on waste management or creating recycled sculptures on water conservation (Orr, 1992). Drama lessons involve students acting out floods or droughts through role-play (some becoming rivers or factories) and freeze-frames contrasting life *before* and *after* climate change impacts (Pascoe, Leong, MacCallum, et al., 2017). A notable aspect is the integration of *place and reflection*: students visit local sites, e.g., a landfill or riverbank and then create art or performances based on those observations (Jickling & Wals, 2008). *"What do we plant?"* Alien vegetation plantations spread invasive plants that grow so densely that natural, indigenous plants become extinct. This affects the entire ecosystem, from insect life to birds and animals, to massive veld fires that destroy thousands of hectares of natural vegetation. The alien plants seed quickly and grow fast because there are no natural enemies. Consider the effect of forests planted with trees like pine, bluegum and black wattles, which burn easily due to the resin they contain. Their work is also shared with community stakeholders, such as murals in a village school or a climate change play performed for parents, ensuring cultural relevance. Assessments, namely creative portfolios and lesson-plan projects, emphasise both artistic growth and understanding of climate change (Eisner, 2002). Student feedback has shown deep engagement: by semester's end, many feel more confident discussing climate change issues and see art as a powerful way to inspire action (Eisner, 2002). In short, the ICA1501 case exemplifies an *eco-pedagogy-plus-arts approach*: learning is contextualised to African realities, e.g., by using local artists and Ubuntu values, and students become *storytellers* who can *inspire environmental awareness and sustainable action through the arts* (Freedman, 2003). A local example from another university with Bachelor of Education (B.Ed) students, which will be exhibited at the 2026 *"(Un)School"* community centre called *"Sitstories"*. This is a collaborative project which will take place in a community centre. It involves performance, music, dance, storytelling, exploring and celebrating people's stories, heritage and creativity. Another example is the *!Haos project* in which student teachers worked on a stick or branch as a departure, and twined, wove, and manipulated family-linked materials, reconfiguring understandings of relationality through material engagement. In the end, each student's project contributed to an installation, combined with text, that was exhibited at a large arts festival (Meiring, 2025). *!Haos* is a Khoekhoe term denoting a place associated with water or wetland conditions. Figure 3 and Figure 4 shows photographs taken at the *!Haos project* exhibition showcases the use of natural materials, such as sticks and branches.

Figure 3. The !Haos project, South Africa

Source: Copyright the authors

Figure 4. The !Haos project, South Africa

Source: Copyright the authors

In a different project in South Africa's Western Cape Province, close to the sea, learners from Art Centres in the province were encouraged to reduce, recycle, reuse, and create for an exhibition in the foyer of a public theatre complex. The public was encouraged to purchase artwork at the exhibits. The money collected was donated to the Two Oceans Aquarium turtle conservation centre. Figure 5: Suidoosterfees 2025 illustrates how child art exhibitions can support local conservation efforts, such as turtle protection, while encouraging the minimisation of plastic bag use, which ends up in the sea and threatens turtles' natural habitat. The Suidoosterfees is an annual arts festival in Cape Town that showcases theatre, music, and visual arts, with a focus on Afrikaans culture and inclusivity. Figure 6: Child art exhibition at Suidoosterfees 2025, South Africa below illustrates exhibition pieces.

Figure 5. Child art exhibition at Suidoosterfees 2025, South Africa.

Source: Copyright the authors, https://www.facebook.com/share/1C4uBuPq7X/

Figure 6. Child art exhibition at Suidoosterfees 2025, South Africa.

Source: Copyright the authors, https://www.facebook.com/share/1C4uBuPq7X/

Green School in South Africa educates children for sustainability in a progressive, learner-centered way. The purpose of the school is to educate for sustainability through community-integrated, entrepreneurial learning, immersed in a natural environment. The learning programme comprises learning objectives, 21st-century skills, and "*iRESPECT*" values, a framework of core behavioural principles, often used in educational, organisational, or leadership contexts where each letter in "*iRESPECT*" represents a specific value, such as *integrity, respect, empathy, service, professionalism, excellence, collaboration, and trust* that guides ethical conduct and responsible interaction. Learners are prepared to face the different challenges of a changing world, challenges unlike those faced by previous generations. In addition to learning how to spell and count, sustainability is integrated into the learning process. The campus is green, the habits are green and the curriculum is green. The school strives to balance a sense of play and wonder, social and emotional learning, and academic challenges. Learners discover themselves, their relationships, their strengths, their interests, their environment, and their world. The United Nations Sustainable Development Goals are introduced in the thematic learning, and a growing awareness of our planet's challenges and the interconnectedness of society, nature, and economics is cultivated. At Green School South Africa, art plays a crucial role in fostering sustainability and innovation. The school emphasises sustainable art education, focusing on eco-conscious material choices and innovative practices to minimise environmental impact. The Art Teacher shares her journey of integrating art into a sustainable environment, ensuring that art classes contribute positively to the school's commitment to sustainability. The program is designed to nurture creativity, individuality, and expression, with a focus on eco-conscious material choices and responsible disposal of artwork. The school's architecture and design also reflect a commitment to sustainability, creating a harmonious blend of nature-inspired elements.

Overall, Green School South Africa exemplifies how art can be integrated into a holistic educational approach that prioritises sustainability and creativity. The school's architecture draws inspiration from the natural landscape, echoing the organic shapes of the Paarl Berg, Paarl in the Western Cape and local boulders. Completed in February 2021, the first phase features clusters of buildings interconnected by landscaped walkways and verdant courtyards. These clusters house various facilities, including 16 classrooms spanning kindergarten to Grade 8, the versatile Sangkep multipurpose space, an administrative building, and the central *Heart of the School* (Green School South Africa GSSA).

- *Asia – Traditional Puppetry in India:* In Assam, Northeast India, a master puppeteer and teacher, Manoranjan Roy, has used *string puppet theatre* to teach learners about ecology and climate change. His troupe, Sagarika Puppet Drama, performs in schools and villages around Guwahati on topics such as deforestation, plastic pollution, and urban flooding (Bhattacharjee, 2026; UNESCO, 2020; UNESCO, 2022). Using the local Putala Naach style, stories are narrated through puppets of humans and animals. For instance, one script highlights how unchecked tree-cutting leads to *artificial floods* in the city by showing a sequence of scenes from sunny days to cloudbursts and the havoc they cause (UNESCO, 2020). Roy's background in education led him to use narrative and emotional appeal: before teaching preservation tips, he dramatises the *harmful consequences* of environmental damage in daily life (UNESCO, 2020). The puppets illustrate common man-vs-nature conflicts, such as wild elephants entering towns as forests shrink (UNESCO, 2020). After performances, students discuss solutions bridging the fantasy of the puppet show with practical action. This case shows how *folk arts* can bridge science and community tradition. It aligns with ICA1501's philosophy by making climate change issues concrete, for example, *the drawbacks of felling trees*, in a culturally familiar medium. The puppetry project has reached numerous schools, suggesting that even without high-tech resources, traditional Drama can instill climate change consciousness and positive attitudes among learners (UNESCO, 2020; UNESCO, 2022).

- *North America – Children's Theatre (USA):* In New York City, a *children's theatre* staged *"The Pocket Park Kids"*, a musical play about preserving a tiny urban park (Burton, Horowitz & Abeles, 1999). Though set in an unrelated plot (saving a local park), the story serves as an allegory for environmental action and the UN Sustainable Development Goals. Young characters with varied interests, like a community garden or a playground, etc., work out differences in a narrative that resonated with elementary students. Notably, the playwright designed the play to be engaging rather than preachy: she wanted sustainability lessons to emerge organically rather than *through overt teaching* (Robinson, 2012). After performances, actors' broken characters lead audience discussions on eco-actions, showing how theatre can spark conversation. Crucially, interviews revealed that theatre helped preschoolers and early learners grapple with climate change issues without feeling overwhelmed (Burton et al., 1999). An educational researcher commented that even small theatre activities like skits (parodies) and puppet shows. Elementary classes can serve as powerful entry points for climate change dialogue (Duncum, 2010). This US example underscores the emotional and developmental dimension: by couching climate change topics in a playful, imaginative narrative, children found it *accessible and inspiring*. As one co-creator noted, artists can address heavy topics *in ways that are uplifting and activating, rather than demoralising* (Monroe et al., 2019). Like ICA1501, the theatre project taught local activities, such as kids saving *their* park, and global thinking, tying each child's goal to a UN Goal (Tavin, 2003; Robinson, 2012). This model highlights how performance art can be integrated into early childhood or general curricula to foster early climate change awareness, literacy and motivation.

- *Europe – Trash Art (Greece):* In Europe, a novel program at the University of West Attica (Greece) illustrates arts-based climate change leadership. The Skyros Project (Part of an environmental education initiative) involves students creating large mosaics entirely from discarded waste – called *Trash Art* (GSGS, n.d.; Gude, 2007). Over the course of weeks, students collect materials such as bottle caps, bits of metal, etc. and compose images that address themes such as ocean pollution or biodiversity loss. This hands-on art process serves a dual purpose: it physically embodies recycling and circular-economy principles, and it prompts reflection on consumption. Researchers analysing the Skyros case found that participants' environmental attitudes improved significantly (Anderson, 2012). In interviews, students said the art project made them feel more responsible and hopeful about sustainability. Notably, the project did not rely on traditional lectures or tests; instead, it gauged impact through participants' own words and experiences. This Greek case parallels ICA1501's recycled-art activities (Orr, 1992). It shows that even at the tertiary level, creative expression can transform waste into a *meaningful statement* about climate change. The findings suggest that such art-based interventions can foster long-term ecological awareness among tertiary students (Anderson, 2012).
- *Latin America – Indigenous Exchange (Brazil):* In Brazil's Amazon, a cross-cultural arts project connected British theatre-makers with the Kuikuro indigenous community (Xingu). The People's Palace Projects organisation facilitated an exchange in which Kuikuro artists documented their culture through film and art, educating outsiders about the Amazonian climate change and cultural threats (Costandius, 2019). A Kuikuro filmmaker, Takumã Kuikuro, even spent time in London creating a short film, *"London as a Village"*, from an indigenous perspective. Over nearly a decade, this partnership has built trust and mutual learning around climate change justice: its stated goal is *to develop an understanding of indigenous practices and their connection to climate change justice and sustainability* (Desai & Hamlin, 2010). Though not a formal classroom, this is an educational initiative that uses storytelling, film, and art residencies to amplify *indigenous voices* on climate change issues. For instance, Kuikuro ceremonial art (the *Kuarup* dance) was documented and shared, linking cultural loss to environmental change. This project illustrates how climate change education can be *two-way*: educators learn from local knowledge while communities share their concerns. It also shows art acting as a *language* between cultures, similar to how ICA1501 uses art to link global science with local African realities (Kahn, 2010; Desai

& Hamlin, 2010). The Brazilian exchange underscores the ethics dimension of eco-pedagogy: it treats indigenous communities not merely as subjects of study, but as co-creators of knowledge and storytellers for change.

- *Oceania – Art-Based Climate Change Resilience (Australia/New Zealand):* A recent international effort between artists and climate change researchers developed the *Art-Based Climate Resilience Toolkit* for educators (Parsons & Pratt, 2024). Originating in Australia and New Zealand, this toolkit emphasises hope, empowerment, and collective action through Visual Arts and Drama. It includes activities such as creating clay *community-resilience figures, writing climate-change-justice* narratives, and performing interactive *Drama of the Oppressed* routines on adaptation (Chappell & Cahnmann-Taylor, 2013). For example, one exercise, *"United We Stand: Clay Figurines,"* has participants sculpt themselves or community members as guardians of nature, fostering a sense of agency. The toolkit's stated aim is to shift learners from anxiety to imagination by using art as a bridge between scientific data and personal stories (Parsons & Pratt, 2024); Chappell & Cahnmann-Taylor, 2013). This aligns with ICA1501's ethos: as ICA1501 students craft tangible art from abstract concepts, toolkit users produce visual or performative artefacts that symbolically convey climate change messages. An evaluation notes that such creative methods can transform people's environmental attitudes, making them more engaged (Chappell & Cahnmann-Taylor, 2013). While the toolkit is designed for broader community workshops, its core principles of inclusive storytelling and positive framing resonate with ICA1501's place-based Drama lessons. Together, these examples of Oceania reflect a global trend: educators are increasingly integrating the arts into climate change education to connect feelings with facts.

These diverse cases demonstrate the global relevance of arts-integrated climate change education. Despite cultural differences, common patterns emerge, situating learning in a local context, using creative expression to build empathy, and engaging communities outside school. For instance, both the Chamorro-Palau art exhibit in Guam (Sobel, 2004; Davis, 2010) and the Greek Trash Art project (GSGS, n.d.) required students to use recycled materials, demonstrating a universal emphasis on sustainability. When *"Day Zero"* loomed in the Western Cape Province of South Africa due to a severe drought, a school community manufactured large trees to adorn the school's entrance, as they could no longer water real trees. For this project, people from the surrounding community contributed. The stems were made entirely from large plastic containers, paint buckets, and plastic basins, and as the branches grew, caps, bottles, and a huge array of upcycled materials in the brightest of colours appeared. Similarly, the New York children's play and the African Drama lessons both used storytelling to frame climate change as a community story. In each region, arts-based pedagogy makes the technical *human* and honours cultural values, e.g., indigenous land stewardship in the Pacific, Ubuntu values in Africa (O'Neill, 1995). By comparing ICA1501's approach with these models, we see that *Creative Arts can be a unifying medium* for climate change pedagogy worldwide. Where science curricula might differ, art provides a *common language*: it allows students in South Africa, Kenya, Canada, India or Brazil to feel the urgency of climate change and imagine local solutions using their own cultural tools.

ASSESSMENT PRACTICES FOR CLIMATE CHANGE COMPETENCY

Assessing learning in creative climate change education is as much about process and values as about content. Traditional multiple-choice quizzes are inadequate to capture the depth of understanding and ethical growth we seek. Instead, educators use authentic assessment aligned with their active strategies. Key practices include:

- *Creative Portfolios:* Students compile visual artworks, photographs of installations, scripts or videos of performances, each accompanied by a reflective statement linking the piece to climate change themes (Boal, 1979; Eisner, 2002). For example, an ICA1501 portfolio might include a sculpture made from e-waste on pollution, with the student's explanation of the choices and climate change message. Portfolios allow lecturers to assess both artistic skill and the student's articulation of climate change concepts. They also become personal records of the student's growth in *creative climate change literacy – the ability to see, feel, and communicate environmental change* through art (Nicholson, 2005).
- *Lesson Plans and Teaching Tasks*: Since ICA1501 is a training course for future teachers, students create sample lesson plans that integrate Creative Arts with climate change science in an Intermediate Phase, Grade 4-6 classroom. In these assignments, students must demonstrate pedagogical understanding by describing activities e.g., a group mural on recycling, specific aims amongst which: to make informed and responsible decisions about their environment; to develop creative, expressive and innovative individuals; to develop skills such as awareness, problem-solving, interpersonal relations, leadership, decision-making (Department of Basic Education, 2011), and assessment tasks. Crafting such plans tests whether students can *translate* their own climate change learning into teachable units. It is vital that carefully considered, age-appropriate lesson topics serve as the basis for Creative Arts projects.

Recycling, upcycling and repurposing materials have been recommended for a long time. The CAPS document (Department of Basic Education, 2011), for instance, suggests that, for creating in 3D, the Intermediate Phase (Grade 4-6) recommendation is to use recyclable materials, such as cardboard, paper off-cuts, beads, sequins, ribbon, natural objects, and other suitable materials. Students are encouraged to look at and learn from examples of excellent practice for lesson ideas on the reuse and repurposing of materials and lesson topics for learners' art exhibitions, especially the work of experienced teachers at schools, art centres or Eisteddfod. In South Africa, an Eisteddfod is a structured cultural competition in which learners or community members participate in art disciplines such as Drama, Poetry, Music, Dance, and Visual Arts before an adjudicator. It serves as both a developmental platform and an evaluative space, offering participants formal feedback, certificates, and opportunities for artistic growth within school and community contexts. Many exhibitions are available for viewing on social media or websites if accessibility is a problem. For Drama, the recommendation is to use basic hand- and/or head-puppets made from found or recycled materials, inspired by an African story (traditional or contemporary) (Department of Basic Education, 2011). However, it is not only the use of materials that is important. What is vital is that, in climate change literacy education, the cause of climate change is conveyed to learners through the process of learning in and through the arts. Furthermore, *what can each person do to make a difference?* This approach has also been used in other programs: one case study had teacher candidates design climate change-related museum lessons, which were graded on how well they blended content accuracy with engagement (Boal, 1979). In general, asking trainees to teach what they learn is a powerful assessment because it requires deep processing and creativity.

- *Research Essays and Reflection Papers:* Written tasks let students analyse broader issues. For instance, ICA1501 students write essays on the role of art in environmental awareness, using local examples like African eco-artists (Eisner, 2002). These essays are evaluated on their understanding of climate change justice, their critique of art's impact, and their integration of course content. Reflective journals are also graded: lecturers look for evidence that students connect their feelings and values to the subject matter. The Creative Arts curriculum, CAPS, that ICA1501 students will eventually be expected to teach in schools, suggests assessment tasks which bear evidence of learners being able to appreciate and reflect as well as observe and discuss. For example, a good journal entry might describe how writing a poem about deforestation prompted the writer to consider their own water use at home. Another example of reflection is to require students to review and discuss selected environmental films, of which there are many documentaries, movies and animated films, and to consider the relevance to their own situation. Cinematography has a profound emotive impact. Examples with positive messages, such as *"2040"* (Damon, 2019), or animations like *"Princess Mononoke"* (Miyazaki, 1997), *"WALL-E"* (Stanton, 2008), or *"The Lorax"* (Renaud, 2012), are recommended. At the tertiary level, the work of David Attenborough cannot be left out, especially *"Ocean"* (Attenborough, n.d.), which concludes with significant notes of hope. It brings good news about experiments underway to restore the waters and prevent disaster. A qualitative assessment of reflective essays and film discussions emphasises growth in ethical awareness, rather than just factual recall (Bolton, 1979). To further their research, ICA1501 students study local visual art, for instance, at the Green School and other examples where climate themes are incorporated into the artwork and possibly connect with arts educators to obtain information on the programmes followed there. Using this information, they compile research reports on the findings. Their research establishes whether the examples of excellent practice they researched can be applied in the environments they work in.

- *Group Projects and Community Engagement*: Collaborative works like a student group painting a climate change mural for a local school, or a short film they made with their smartphones, following their collaboratively written script, or an animation or comic strips, with their own stories and scripts, with group members contributing to the drawings. If these are shown at a primary school or to youth groups, students work on a positive, hopeful message, which can be assessed for creativity, social relevance, and teamwork (Greene, 1995; Eisner, 2002). Grading criteria might include questions like: *Did the project accurately represent a climate change issue? Did it engage local audiences? Did students demonstrate critical thinking about solutions?* This mirrors service-learning evaluation: students receive feedback not only from lecturers but also from community partners.
- *Rubrics Blending Knowledge and Values:* Across all tasks, lecturers use holistic rubrics that combine cognitive and affective criteria (Bolton, 1979). For example, an arts project rubric might assess scientific accuracy, e.g., the correct cause-and-effect of climate change, alongside emotional depth or evidence of justice-oriented reflection, e.g., how well the piece conveys empathy for vulnerable communities. This ensures that students are credited for insights and caring, not only for technical skills. Research suggests that these combined rubrics help capture dispositions such as *environmental stewardship and collaboration* (Bolton, 1979).

- *Feedback and Iteration:* Formative assessment through peer review and reflection is also key. In ICA1501, students share drafts of art projects, scripts or skits and give each other feedback on clarity and impact. Lecturers might hold class evaluations of climate change exhibits. The goal is ongoing improvement: if a student's puppet shows on water scarcity initially misses the connection to local culture, peers can suggest adding a regional myth or language element. This mirrors best practices in Creative Arts education, where iterative refinement is expected. The creative climate change toolkit from Oceania, for example, recommends group discussions after each art activity to reinforce learning and prepare for subsequent tasks (Hickey-Moody, 2013; Chappell & Cahnmann-Taylor, 2013).

These varied assessment practices align closely with the goals of climate change in education: they value *engagement, empathy and action*. A study of one teacher-prep program found that when students create art and narratives about climate change, they become not just informed but also empowered and engaged (Bolton, 1979; Eisner, 2002). In ICA1501, for example, after compiling their portfolios and teaching plans, many students reported feeling more confident and motivated to incorporate climate change into their future classrooms (Eisner, 2002). Assessment is thus not an afterthought but an integral pedagogical tool: it reflects the principle that climate change education should *produce climate-literate graduates who can translate learning into action* (Heathcote & Bolton, 1995).

INSTITUTIONAL AND POLICY IMPLICATIONS

Embedding creative climate change pedagogy at scale requires support beyond individual classrooms. Institutional policies and leadership are critical to sustain these approaches. Key implications include:

- *Whole-Institution Commitment:* UNESCO and other bodies stress that climate change education must be cross-cutting, not siloed in a few courses (Filho, Shiel, Paço, et al., 2019; UNESCO, n.d.). This means universities should incorporate sustainability outcomes into all programs. For example, ICA1501 practitioners aligned their art curriculum with science, language and life-skills standards (Wagner, 1976; Pink, 2006), and similar alignment should be encouraged across campuses. Some universities now create sustainability offices or networks to coordinate such efforts (Filho et al., 2019). These entities can foster collaboration, e.g., by bringing together arts education faculty with peers from engineering and the humanities so that curricula reinforce one another. Funding for interdisciplinary projects, seed grants for climate change-art initiatives, and recognition (such as teaching awards for sustainability innovation) send a strong signal that the arts have a valued role in the climate change agenda.

- *Faculty Development and Communities of Practice:* Lecturers often need training to teach climate change topics, especially if they come from non-science backgrounds. ICA1501 overcame this by holding workshops on climate change science and artistic techniques for the students (Pink, 2006; Neelands & Goode, 2015). At the institutional level, universities should offer professional development for faculty across disciplines. For instance, training sessions on eco-pedagogy can empower literature or philosophy professors to introduce climate change literature, while the art faculty can learn basic environmental science to support their teaching. Communities of practice (regular faculty learning groups) allow lecturers to share success stories and resources. As one sustainability expert advises, student teachers must view the environment *as a character in our classroom* but to do so, they need examples and support on how to do it effectively (Van Deventer, 2026). Administrative backing is vital: without it, an enthusiastic faculty may feel isolated.
- *Curriculum and Accreditation Standards:* Policymakers can catalyse change by embedding climate change literacy into accreditation criteria. Several governments and accrediting bodies now encourage or require sustainability learning outcomes for graduates. For example, Article 12 of the Paris Agreement explicitly calls on nations to *enhance climate change education, public awareness and participation* (UNESCO, n.d.). In practice, some education systems have already mandated environmental education hours or competencies in teacher standards. The Department of Basic Education, Republic of South Africa published an Education for Sustainable Development (ESD) strategy stating that, by 2030, education for sustainable development, involving an inter-disciplinary, integrated and active approach to learning, is a vital element of all levels and programmes of the education and training system, creating and maintaining environmentally literate and active citizens, ensuring that all South Africans, present and future, enjoy a decent quality of life through the sustainable use of resources. Connecting climate change objectives to official standards ensures longevity: climate change education then becomes part of *what we teach*, not merely what we *hope to teach.* Research conducted by Matsepe and Maluleke (2024) concluded that much remains to be done to integrate climate change education into current CAPS curricula. The study recommends ongoing awareness campaigns to raise teachers' understanding of the importance of climate change education. At the

26th Conference of the Parties (COP26) at the United Nations Framework Convention on Climate Change (UNFCCC), education ministers committed to integrating climate change education across all educational institutions. However, a UNESCO study found that fewer than half of the countries surveyed included climate change in their educational policies, highlighting significant gaps that need to be addressed.

- *Physical and Financial Resources:* Sustainability initiatives often also involve campus operations, green buildings, gardens, recycling, etcetera, which can be integrated with teaching. ICA1501's module suggests such links: students cultivating a campus garden, a lucrative business, and hands-on biology and agriculture could inform an art project on plants, flowers, insects, birds, growth, and care (Van Deventer, 2026). Providing spaces such as university gardens, nature trails, or art spaces that use upcycled and found materials encourages collaboration between living and learning. Budgeting for art supplies, especially recycled or donated materials, is also crucial. Challenges have been noted in resource-strapped environments; one case study noted that uneven access to art supplies or internet tools can hinder climate change-art lessons (Shephard, 2008). Institutions can mitigate this by sharing supplies across departments, applying for grants aimed at climate change innovation, or tapping into open educational resources, online art portfolios, and free climate change data to keep costs low. Experimentation and innovation can be encouraged to create in 2D, for which students can use opened-up cardboard boxes as a background for picture-making, newspaper or magazine and to create in 3D, use earthenware clay, tapping into indigenous knowledge on how to make and fire earthenware in a wood fire, without the need for electric kilns. Repurposed packaging materials, egg boxes, sugar packets, or potato pockets can be converted into plants, animals, or insects. They can make their own paper with teabags and other recycled stuff. Weaving can be done with plastic bags cut into strips. Colourants can be developed from vegetables, tea, or mud, and research mud cloth printing (Bogolanfini), a traditional art form of Mali.
- *Global and National Networks:* Climate change education benefits from shared knowledge across institutions and borders. International networks, such as UNESCO's Higher Education sustainability initiatives or the International Teaching Artist Collaborative, compile case studies and best practices (Lozano, Lukman, Lozano, Huisingh, & Lambrechts, 2013; Rowe, 2007). The creators of ICA1501 drew on such literature in designing the course. To spread successful models, conferences and online platforms should highlight arts-based climate change pedagogy. Similarly, national education ministries can promote pilot programs: for example, one article argues that

teacher training institutions must incorporate climate change into all subject methods, not just science or social sciences (geography), a view clearly embodied by ICA1501; Filho et al., 2019; Trott et al., 2020 and Van Deventer, 2026.

Lastly, the policy environment must align with the pedagogical vision. Sustainability cannot rely solely on passionate individual lecturers and teachers. Global agreements like the Paris Accord and UNESCO's 2030 Agenda frame education as a *root of change*, urging that no field be exempt from the climate change conversation (UNESCO, n.d.; Sipos, Battisti, & Grimm, 2008). Practical governance actions, policy mandates, curricula reform, funding and institutional leadership training, resources will determine whether creative climate change modules flourish or falter. The evidence suggests a positive trend: numerous recent reports call on universities to become beacons of *lifelong learning* for sustainability (Warburton, 2003). If fulfilled, such commitments would make Creative Arts teacher training modules like ICA1501 the norm rather than the exception, embedding arts-based climate change literacy in teacher education worldwide.

CONCLUSION

Climate change education must be as much about *the heart as the head*. The ICA1501 Module case study in this chapter, along with our African and international examples, illustrates a powerful truth: when students engage with climate change through Creative Arts and community, they don't just learn about climate change, they learn through it (Inwood, 2010). By dramatising or dancing the story of a river, painting the plight of a forest, a family hiking in nature, carrying our shopping in our own bags (not plastic), cycling instead of driving, or performing as a flood survivor, a tour guide taking a group on a hike along the coast, learners experience environmental change in a deeply personal way. This process builds empathy, ethical awareness, and a sense of action, creative climate change literacies that purely technical instruction rarely achieves (Inwood, 2010; McNaughton, 2012).

Moreover, arts-integrated pedagogy is adaptable across cultures. In each continent we examined, teachers find locally relevant narratives and artistic forms from Indian puppetry to Brazilian indigenous film to Pacific Island exhibits to carry climate change education. Yet the underlying approach is coherent: combine ecological understanding with Drama and storytelling, action and art-making, and assess through authentic creative work. This method aligns with UNESCO's vision of education for sustainability and justice (Kwauk, 2020; Stevenson, Brody, Dillon,

& Wals, 2013), *proving that the stories we tell about our relationship with the planet* indeed shape how we live on it (Kagan & Kirchberg, 2016).

The comparison of ICA1501 with other representations shows that while context differs, the *human factor* remains central. Climate change pedagogy across disciplines must address values, emotions, and community ties. For instance, even a high-tech country like Australia now recognises the need to pair data with *hope, empowerment, and collective action* through art (Curtis, Reid & Ballard, 2012). Conversely, in resource-limited settings, traditional arts are used to powerful effect, as in India and remote Pacific islands. The lesson is that all educators in the arts, sciences, and humanities have a role in climate change education. Teacher-educators and lecturers should be empowered to incorporate climate change lenses into every subject; similarly, engineers, doctors, artists, and historians should see climate change awareness as integral to their fields.

To support this shift, institutions must sustain interdisciplinary innovation and policy backing. The journey is challenging (teacher training, resources, and curriculum time are constraints; Shephard, 2008; Reid & Petocz, 2006), but cases like ICA1501 demonstrate that it can be done within existing structures. The reward is graduates who are not only informed about climate change but also imaginative, conscientious storytellers ready to inspire collective action in their classrooms and communities (McNaughton, 2012).

In closing, we echo a call from the Creative Arts-and-climate change community: before we act to solve climate change problems, we must engage in the creative work of *imagining* a better world (Blackwood, 2024). By learning about climate change through Creative Arts, reflection, and community dialogue, students develop the moral and emotional resolve needed for the long struggle ahead. As one ICA1501 lecturer put it, teachers have *a moral obligation to create an environment that shapes the character of our classroom* (Van Deventer, 2026). By making climate change issues central to Creative Arts as to any other subject, lecturer/educator-advocates can nurture hope, responsibility, and a mindful, positive changed mindset in students and learners worldwide, one artwork *"Art for art's sake"*, one play, one lesson at a time (Inwood, 2010; UNESCO, n.d.; Winner, Goldstein & Vincent-Lancrin, 2013).

AI Disclosure Statement: Artificial intelligence tools were utilised to assist with paraphrasing, language refinement and the verification of selected resources.

REFERENCES

Anderson, A. (2012). Climate change education for mitigation and adaptation. *Journal of Education for Sustainable Development*, *6*(2), 191–206. DOI: 10.1177/0973408212475199

Armstrong-McKay, D. I., Staal, A., Abrams, J., Winkelmann, R., Sakschewski, B., Loriani, S., Fetzer, I., Cornell, S. E., Rockström, J., & Lenton, T. M. (2022). Exceeding 1.5°C global warming could trigger multiple climate tipping points. *Science*, *377*(6611), eabn7950. Advance online publication. DOI: 10.1126/science.abn7950 PMID: 36074831

Attenborough, D. (n.d.). David Attenborough documentaries. Timeout. https://www.timeout.com/film/the-15-best-david-attenborough-documentaries-to-stream

Bhattacharjee, M. (2026) Striking the Ecological Chord with Puppetry. The Environmental Issues. Fall/winter issue no. 52. https://www.unima-usa.org/ecological-chord

Blackwood, K. A., & Communications, S. (2024). On climate change, artists 'imagine the world otherwise'. Latin American Studies. https://as.cornell.edu/news/climate-change-artists-imagine-world-otherwise

Boal, A. (1979). *Theatre of the oppressed*. Pluto Press.

Bolton, G. (1979). *Towards a theory of drama in education*. Longman.

Burton, J. M., Horowitz, R., & Abeles, H. (1999). Learning in and through the arts: Curriculum implications. In Fiske, E. B. (Ed.), *Champions of change: The impact of the arts on learning* (pp. 35–46). Arts Education Partnership.

Chappell, S. V., & Cahnmann-Taylor, M. (2013). No child left with crayons: The imperative of arts-based education and research with language "minority" and other minoritized communities. *Review of Research in Education*, *37*(1), 243–268. DOI: 10.3102/0091732X12461615

Cordero, E. C., Centeno, D., & Todd, A. M. (2020). The role of climate change education on individual lifetime carbon emissions. *PLoS One*, *15*(2), e0206266. Advance online publication. DOI: 10.1371/journal.pone.0206266 PMID: 32017773

Costandius, E. (2019). *Engaging the curriculum in visual communication design: A social justice perspective*. Sun Press.

Curtis, D. J., Reid, N., & Ballard, G. (2012). Communicating ecology through art: What scientists think. *Ecology and Society*, *17*(2), 3. Advance online publication. DOI: 10.5751/ES-04670-170203

Damon, D. (Director). (2019). 2040 [Film]. Good Thing Productions.

Davis, J. H. (2010). *Why our schools need the arts.* Teachers College Press.

Department of Basic Education. (2011). Curriculum and Assessment Policy Statement: Creative Arts, Intermediate Phase Grades 4–6. Pretoria: Department of Basic Education. https://www.education.gov.za/Curriculum/CurriculumAssessmentPolicy Statements%28CAPS%29/CAPSIntermediate.aspx

Department of Basic Education South Africa. (n.d.). Addressing the impact of climate change in schools. https://www.education.gov.za/ArchivedDocuments/ArchivedArticles/Addressing-the-impact-of-climate-change-in-schools-0524.aspx

Desai, D., & Hamlin, J. (2010). Visual culture and the (re)production of the social. In Costantino, T., & White, B. (Eds.), *Essays on aesthetic education for the 21st century* (pp. 189–208). Sense Publishers.

Duncum, P. (2010). Seven principles for visual culture education. *Art Education, 63*(1), 6–10. DOI: 10.1080/00043125.2010.11519047

Eisner, E. W. (2002). *The arts and the creation of mind.* Yale University Press.

Figure of 8. (2023). Karatara [Video]. YouTube. https://www.youtube.com/watch ?v=zlcLtjr14wo

Filho, W. L., Shiel, C., Paço, A., Mifsud, M., Ávila, L. V., Brandli, L. L., Molthan-Hill, P., Pace, P., Azeiteiro, U. M., Vargas, V. R., & Caeiro, S. (2019). Sustainable development goals and sustainability teaching at universities: Falling behind or getting ahead of the pack? *Journal of Cleaner Production*, *232*, 285–294. DOI: 10.1016/j.jclepro.2019.05.309

Filho, W. L., Sierra, J., Kalembo, F., Ayal, D. Y., Matandirotya, N., Pereira Amaroda Costa, C. I. V., Sow, B. L., Aabeyir, R., Mawanda, J., Zhou, L., & Baleh, D. (2024). The role of African universities in handling climate change. *Environmental Sciences Europe*, *36*(1), 72. Advance online publication. DOI: 10.1186/s12302-024-00933-6

Ford, J. D., Sherman, M., Berrang-Ford, L., Llanos, A., Carcamo, C., Harper, S., Lwasa, S., Namanya, D., Marcello, T., Maillet, M., & Edge, V. (2018). Preparing for the health impacts of climate change in Indigenous communities. The role of community-based adaptation. *Global Environmental Change*, *49*, 129–139. Advance online publication. DOI: 10.1016/j.gloenvcha.2018.02.006

Forsyth, G., Le Maitre, D., Van den Dool, R., Walls, R., & Pharoah, R. (2019). The Knysna fires of 2017: Learning from this disaster. Disaster Risk Reduction Centre, University of Cape Town. https://riskreductionafrica.org/assets/files/Knynsa%20 Fires%20Report%202019.pdf

Franchi, F., Mustafa, S., Ariztegui, D., Chirindja, F., Di Capua, A., Hussey, S., Loizeau, J.-L., Maselli, V., Matanó, A., Olabode, O., Pasqualotto, F., Sengwei, W., Tirivarombo, S., Van Loon, A. F., & Comte, J.-C. (2024). Prolonged drought periods over the last four decades increase flood intensity in southern Africa. *The Science of the Total Environment*, *923*, 171489. Advance online publication. DOI: 10.1016/j.scitotenv.2024.171489 PMID: 38453074

Freedman, K. (2003). *Teaching visual culture: Curriculum, aesthetics, and the social life of art*. Teachers College Press.

Freire, P. (1970). Pedagogy of the oppressed. *Continuum.*

Gadotti, M. (2008). Education for sustainability: A critical contribution to the Decade of Education for Sustainable Development. Green Theory & Praxis. *The Journal of Ecopedagogy*, *4*(1), 15–64. DOI: 10.3903/gtp.2008.1.2

Gaztambide-Fernández, R. A. (2013). Why the arts don't do anything: Toward a new vision for cultural production in education. *Harvard Educational Review*, *83*(1), 211–236. DOI: 10.17763/haer.83.1.a78q39699078ju20

Green School South Africa GSSA. (n.d.). https://greenschoolsa.co.za/programme/

Greene, M. (1995). *Releasing the imagination: Essays on education, the arts, and social change*. Jossey-Bass.

GreenSchool South Africa. (n.d.). iRESPECT values [Values framework]. Green-School South Africa. https://greenschoolsa.co.za/wp-content/uploads/2020/05/ LAUNCH_eiRESPECTvalues-1-min.pdf

Gude, O. (2007). Principles of possibility: Considerations for a 21st-century art & culture curriculum. *Art Education*, *60*(1), 6–17. DOI: 10.1080/00043125.2007.11651621

Heathcote, D., & Bolton, G. (1995). *Drama for learning: Dorothy Heathcote's mantle of the expert approach to education*. Heinemann.

Hickey-Moody, A. (2013). *Youth, arts and education: Reassembling subjectivity through affect*. Routledge. DOI: 10.4324/9780203855829

Intergovernmental Panel on Climate Change. (2018). *Global warming of 1.5°C. An IPCC Special Report on the impacts of global warming of 1.5°C above pre-industrial levels and related global greenhouse gas emission pathways*. Cambridge University Press.

Intergovernmental Panel on Climate Change. (2021). Summary for policymakers. In *Climate Change 2021: The Physical Science Basis. Contribution of Working Group I to the Sixth Assessment Report of the Intergovernmental Panel on Climate Change*. Cambridge University Press.

Intergovernmental Panel on Climate Change (IPCC). (2021). *Climate change 2021: The physical science basis*. Cambridge University Press.

Inwood, H. J. (2010). Mapping ecoliteracy. *Canadian Journal of Environmental Education, 15*, 86–98.

Jickling, B., & Wals, A. E. J. (2008). Globalization and environmental education: Looking beyond sustainable development. *Journal of Curriculum Studies, 40*(1), 1–21. DOI: 10.1080/00220270701684667

Kagan, S., & Kirchberg, V. (2016). Music and sustainability: Organizational cultures towards creative resilience–A review. *Journal of Cleaner Production, 135*, 1487–1502. DOI: 10.1016/j.jclepro.2016.05.044

Kahn, R. (2010). *Critical pedagogy, ecoliteracy, and planetary crisis: The ecopedagogy movement*. Peter Lang.

Kemp, L., Xu, C. W. Q., Depledge, J., Ebi, K. L., & Gibbins, G.. (2022). Climate Endgame: Exploring catastrophic climate change scenarios. *Proceedings of the National Academy of Sciences*. DOI: 10.1073/pnas.2108146119

Kraaij, T., Baard, J. A., Cowling, R. M., van Wilgen, B. W., & Das, S. (2013). Historical fire regimes in a poorly understood, fire-prone ecosystem: Eastern coastal fynbos. *International Journal of Wildland Fire, 22*(3), 277–287. DOI: 10.1071/WF11163

Kwauk, C. (2020). *Roadmaps to quality education in a time of climate change*. Brookings Institution.

Limthongsakul, S., Nitivattananon, V., & Arifwidodo, S. D. (2017). Localized flooding and autonomous adaptation in peri-urban Bangkok. *Environment and Urbanization, 29*(1), 51–68. Advance online publication. DOI: 10.1177/0956247816683854

Lotz-Sisitka, H., Wals, A. E. J., Kronlid, D., & McGarry, D. (2020). Transformative, transgressive social learning: Rethinking higher education pedagogy in times of systemic global dysfunction. *Current Opinion in Environmental Sustainability, 16*, 73–80. DOI: 10.1016/j.cosust.2015.07.018

Loucks, D. P. (2021). Impacts of climate change on economies, ecosystems, energy, environments, and human equity: A systems perspective. In *Impacts of Climate Change on Food Security in Small Island Developing States* (pp. 359–394). Elsevier., DOI: 10.1016/B978-0-12-822373-4.00016-1

Lozano, R., Lukman, R., Lozano, F. J., Huisingh, D., & Lambrechts, W. (2013). Declarations for sustainability in higher education: Becoming better leaders, through addressing the university system. *Journal of Cleaner Production, 48*, 10–19. DOI: 10.1016/j.jclepro.2011.10.006

Manteaw, B. O., Boafo, Y. A., Owusu, E. H., Enu, K. B., & Amoah, A. B. (2025). Transgressive pedagogies in a climate-constrained Africa: Grounding climate change education and sustainability learning in local communities. *Frontiers in Education, 10*, 1519914. Advance online publication. DOI: 10.3389/feduc.2025.1519914

McNaughton, M. J. (2012). Implementing education for sustainable development: The potential use of time-honoured pedagogical practice from the progressive era of education. *Journal of Sustainability Education, 3*, 1–18.

Meiring, K. (2025). Stay with the trouble: Entangled relations–A tribute to Professor Elmarie Costandius's embodied arts-based practice. *Critical Studies in Teaching and Learning, 13*(SI1), 44–68. DOI: 10.14426/cristal.v13iSI1.2753

Misiaszek, G. W. (2023). *Freire and environmentalism: Ecopedagogy*. Bloomsbury Academic. DOI: 10.5040/9781350292130

Miyazaki, H. (Director). (1997). Princess Mononoke [Film]. Studio Ghibli.

Molthan-Hill, P., Worsfold, N., Nagy, G. J., Filho, W. L., & Mifsud, M. (2019). Climate change education for universities: A conceptual framework from an international study. *Journal of Cleaner Production, 226*, 1092–1101. DOI: 10.1016/j.jclepro.2019.04.053

Monroe, M. C., Plate, R. R., Oxarart, A., Bowers, A., & Chaves, W. A. (2019). Identifying effective climate change education strategies: A systematic review of the research. *Environmental Education Research, 25*(6), 791–812. DOI: 10.1080/13504622.2017.1360842

Munzhedzi, L., Mugari, E., Nethengwe, N. S., & Gumbo, A. D. (2024). Trends in extreme rainfall and their relationship to flooding episodes in Vhembe district, South Africa. *Environmental Research Communications*, *6*(9), 095016. Advance online publication. DOI: 10.1088/2515-7620/ad7702

Neelands, J., & Goode, T. (2015). *Structuring drama work: 100 key conventions for theatre and drama* (3rd ed.). Cambridge University Press.

Nicholson, H. (2005). *Applied drama: The gift of theatre*. Palgrave Macmillan. DOI: 10.1007/978-0-230-20469-0

O'Neill, C. (1995). *Drama worlds: A framework for process drama*. Heinemann.

Orr, D. W. (1992). *Ecological literacy: Education and the transition to a postmodern world*. State University of New York Press.

Oziewicz, M. (2023). Climate literacy through speculative fiction. [PMLA]. *Publications of the Modern Language Association of America*, *138*(2), 415–422. DOI: 10.1632/S0030812923000068

Parsons, M., & Pratt, S. (2024). Resilience Toolkit: Arts-based activities for changing climates. https://weadapt.org/wp-content/uploads/2024/02/climate-resilience-toolkit-CC.pdf

Pascoe, R., Leong, S., MacCallum, J., Mackinlay, E., Marsh, K., Smith, B., Church, T., & Winterton, A. (2017). *The role of arts in education for building 21st century competencies: National Exceptional Teachers for Disadvantaged Schools (NETDS) Program*. Australian Government Department of Education and Training.

Patz, J. A. (2014). Climate variability and change: Food, water, and societal impacts. In *Environmental Public Health* (pp. 281–319). Springer., DOI: 10.1007/978-1-4614-8417-2_12

Pink, D. H. (2006). *A whole new mind: Why right-brainers will rule the future*. Riverhead Books.

Pörtner, H.-O., Scholes, R. J., Arneth, A., Barnes, D. K. A., Burrows, M. T., Cheung, W. W. L., Cheng, L., Djalante, R., Eakin, H. C., Frölicher, T. L., Gattuso, J.-P., Hashimoto, S., Hua, F., Huq, S., Insarov, G. E., Iversen, T., Leadley, P., Levin, L. A., Lutz, W., & Mintenbeck, K. (2023). Overcoming the coupled climate and biodiversity crises and their societal impacts. *Science*, *380*(6653), eabl4881. Advance online publication. DOI: 10.1126/science.abl4881 PMID: 37079687

Powers, S. E., De Waters, J., & Dhaniyala, S. (2021). Climate literacy—Imperative competencies for tomorrow's engineers. *Sustainability (Basel), 13*(17), 9684. Advance online publication. DOI: 10.3390/su13179684

Reid, A., & Petocz, P. (2006). University lecturers' understanding of sustainability. *Higher Education, 51*(1), 105–123. DOI: 10.1007/s10734-004-6379-4

Renaud, C. (Director). (2012). The Lorax [Film]. Universal Pictures.

Robinson, K. (2012). *Out of our minds: Learning to be creative* (2nd ed.). Capstone Publishing Ltd., DOI: 10.1002/9780857086549

Rowe, D. (2007). Education for a sustainable future. *Science, 317*(5836), 323–324. DOI: 10.1126/science.1143552 PMID: 17641184

Shephard, K. (2008). Higher education for sustainability: Seeking affective learning outcomes. *International Journal of Sustainability in Higher Education, 9*(1), 87–98. DOI: 10.1108/14676370810842201

Sipos, Y., Battisti, B., & Grimm, K. (2008). Achieving transformative sustainability learning: Engaging head, hands and heart. *International Journal of Sustainability in Higher Education, 9*(1), 68–86. DOI: 10.1108/14676370810842193

Sobel, D. (2004). *Place-based education: Connecting classrooms and communities*. The Orion Society.

Souza, K. S., & Garcia, P. H. M. (2023). Environmental education from the perspective of Paulo Freire: A critical analysis. *Fórum Ambiental da Alta Paulista, 19*(5). Advance online publication. DOI: 10.17271/1980082719520234192

Stanton, A. (Director). (2008). WALL-E [Film]. Pixar Animation Studios.

Stevenson, R. B., Brody, M., Dillon, J., & Wals, A. E. J. (Eds.). (2013). *International handbook of research on environmental education*. Routledge. DOI: 10.4324/9780203813331

Suidoosterfees. (2025). Reduce, recycle, re-use, create [Facebook post]. https://www.facebook.com/share/1C4uBuPq7X/

Sustainability initiatives that are developed in Greece GSGS, (n.d.). Skyros project. https://observatory.sustainable-greece.com/en/practice/skyros-project.2103.html

Tavin, K. (2003). Wrestling with angels, searching for ghosts: Toward a critical pedagogy of visual culture. *Studies in Art Education, 44*(3), 197–213. DOI: 10.1080/00393541.2003.11651739

Traore, S., & Owiyo, T. (2013). Dirty droughts causing loss and damage in Northern Burkina Faso. *International Journal of Global Warming*, *5*(4), 498. Advance online publication. DOI: 10.1504/IJGW.2013.057288

Trott, C. D., Even, T. L., & Frame, S. M. (2020). Merging the arts and sciences for collaborative sustainability action: A methodological framework. *Sustainability Science*, *15*(4), 1067–1085. DOI: 10.1007/s11625-020-00798-7

Tytler, R., & Freebody, P. (2023). How can we improve climate change education in schools? Start with the teachers. World Economic Forum. https://www.weforum.org/stories/2023/07/climate-change-education-schools-teachers/

UNESCO. (2020). *Education for sustainable development: A roadmap*. UNESCO.

UNESCO. (2022). *Education for sustainable development for 2030: Progress and global action*. UNESCO.

UNESCO. (n.d.). Education for sustainable development [Campaign poster]. https://www.unesco.org/en/climate-change/education

United Nations Children's Fund. (2024). The Green School Initiative (GSI): UNICEF approach to building resilience and accelerating climate action for children in and through education in West and Central Africa. UNICEF Regional Office for West and Central Africa. https://www.unicef.org/wca/media/10196/file/UNICEF-WCAR-Green-School-Initiative-June-2024-EN.pdf.pdf

United Nations Educational, Scientific and Cultural Organization. (2020). Education for sustainable development: A roadmap. UNESCO. https://unesdoc.unesco.org/ark:/48223/pf0000374802

Van Deventer, A. (2026). *ICA1501 Introduction to Creative Arts. Module framework and content*. Unisa.

Van Ster, G., & Oelf, S. (2023). Karatara. Figure of 8. https://www.youtube.com/watch?v=zlcLtjr14wo

Wagner, B. J. (1976). *Dorothy Heathcote: Drama as a learning medium*. Hutchinson.

Warburton, K. (2003). Deep learning and education for sustainability. *International Journal of Sustainability in Higher Education*, *4*(1), 44–56. DOI: 10.1108/14676370310455332

Westraadt, G. (2015). Visual Art Education in the Primary School. Mowbray: Future Managers.

Williams, E., Abatzoglou, J. T., Hegewisch, K. C., & Williams, P. (2024). Anthropogenic climate change has reduced drought recovery probabilities across the western US. *Communications Earth & Environment*, *5*(1), 425. Advance online publication. DOI: 10.1038/s43247-024-01640-z

Winner, E., Goldstein, T. R., & Vincent-Lancrin, S. (2013). Art for art's sake? The impact of arts education. OECD Publishing. Updated February 2026. DOI: 10.1787/9789264180789-en

Chapter 12
Teachers' Use of Information and Communication Technologies and Artificial Intelligence in Climate Change Lessons

Debbie Ann Sanders
http://orcid.org/0000-0002-4978-5158
University of South Africa, South Africa

Shirley Mukhari
University of South Africa, South Africa

ABSTRACT

Climate change is widely recognised as one of the most demanding global challenges of the 21st century, affecting environmental sustainability, social equity, economic growth and human wellbeing. Teaching can play a pivotal role by supporting learners to develop the knowledge, skills, values and attitudes needed to understand climate change and respond to its impacts. However, teaching it is challenging, as the topic is complex, interdisciplinary, value-laden and emotionally demanding. This chapter examines how information and communication technologies (ICTs), including artificial intelligence (AI) tools, can enhance climate change teaching. It highlights how ICTs and AI can improve lessons while recognising pedagogical, ethical and practical challenges. Their effectiveness depends on pedagogically

DOI: 10.4018/979-8-3373-6975-4.ch012

informed, context-sensitive and carefully planned implementation. The chapter also discusses ongoing professional development, ethics and alignment between curriculum, policy and ICT integration.

INTRODUCTION: TEACHING AND LEARNING ABOUT CLIMATE CHANGE IN A SOUTH AFRICAN CONTEXT

Climate change is widely recognised as one of the most demanding global challenges of the 21st century, for the significant implications it holds for environmental sustainability, social justice, economic development, food security, water availability, health and human wellbeing. In South Africa, these effects are particularly evident due to existing socioeconomic inequalities, high levels of vulnerability and uneven access to resources amongst the citizenry. The education sector in particular is well positioned to support and enhance the younger generation's understanding of climate change, as well as their ability to respond meaningfully to its consequences (United Nations Educational, Scientific and Cultural Organization [UNESCO], 2024; 2025).

In the school curriculum, lessons on climate change play a critical role in equipping learners with the knowledge, skills, values and attitudes required to engage with local and global environmental issues. Admittedly, teaching topics related to climate change can present significant pedagogical challenges (Intergovernmental Panel on Climate Change [IPCC], 2023; UNESCO, 2024). Since the topic is complex and informed by multiple disciplines it raises ethical and value considerations, and can place emotional demands on teachers and learners alike. In this chapter, the argument is made that information and communication technologies (ICTs), including emerging artificial intelligence (AI) tools, can serve as powerful enablers of efforts aimed at raising awareness of climate change.

The use of ICT tools, which are increasingly promoted as pedagogical enablers to support teaching and learning, have notable advantages. One such advantage is that digital technologies offer access to real-time climate data, interactive simulations, global perspectives and innovative learning opportunities that would be difficult to achieve through traditional teaching methodologies alone (Olaniyi et al., 2025; Phiri et al., 2025). Despite this, the integration of ICTs and AI raises important pedagogical, ethical and contextual concerns, particularly in under-resourced school environments (Khoza & van der Walt, 2025; Oubibi et al., 2024).

This chapter examines not only the use of ICTs and AI in climate change lessons, but also the importance of aligning those lessons with the South African Curriculum and Assessment Policy Statements (CAPS) (Department of Basic Education [DBE], 2011). In particular, the focus is on primary and secondary schools in South Africa. The social constructivism and the Diffusion of innovation (DoI) theories are used as guiding frameworks throughout this chapter. Although ICTs and AI can have valuable benefits for climate change-related lessons, their actual effectiveness depends on pedagogically sound, context-sensitive, principled and reflective implementation. Finally, the proposal put forward here, is that teachers should remain central to learning, and that the gamut of available technologies must be seen as a support for, rather than a replacement of, teaching expertise.

The chapter commences with an overview of the theoretical foundations of social constructivism and DoI in Section 2. Section 3 offers an analysis of the role of ICTs and AI in climate change teaching and learning, while Section 4 examines the alignment of these tools with the CAPS curriculum, with particular emphasis on lesson design and assessment practices. Section 5 discusses the key factors that influence teachers' adoption of ICTs and AI in their lessons on climate change, while Section 6 discusses the importance of ethical and professional considerations associated with the use of technologies in classroom contexts. Finally, Section 7 synthesises the main insights and identifies implications for future practice and research.

CONCEPTUAL FRAMEWORK

To guide the discussion on integrating ICT and AI in teaching and learning, this chapter adopts a conceptual framework that illustrates how technology, pedagogy, and learner engagement interact. The framework (Figure 1 below) positions AI and ICT tools as enablers of inquiry-based learning, differentiated instruction, assessment and data visualisation. It also recognises the critical role of the teacher in mediating learning, scaffolding understanding and ensuring meaningful social interaction.

The framework also highlights potential risks, such as over-reliance on ICTs and AI or limited access to digital resources, and the importance of aligning classroom practices with educational policies. By situating these elements together, the framework provides a roadmap for debating both the opportunities and challenges of ICT and AI-supported climate change lessons.

This conceptual framework guides this chapter and it shows the linkage of the practical applications, pedagogies and critical considerations discussed in the sections that follow.

Figure 1. Conceptual Framework for AI and ICT Integration in Teaching and Learning

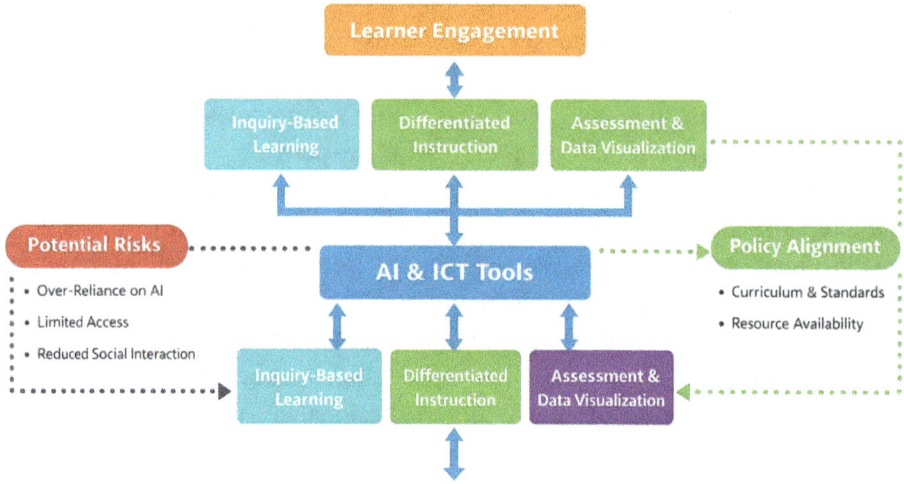

THEORETICAL FRAMEWORK: SOCIAL CONSTRUCTIVISM AND DIGITAL LEARNING

Social constructivism

Founder

Lev Vygotsky (1896–1934), who is known as the father of social constructivism, conceived of a sociocultural theory which emphasises that learning is a socially mediated and culturally situated process (Vygotsky, 1978). He argued that cognitive development occurs because of our interaction with others and through our engagement with cultural tools – in particular, language. According to this theory, learning is mediated by social relationships, historical context and shared meaning-making practices. Social constructivism thus positions learners as active participants who construct understanding through interaction, dialogue and experience, rather than through knowledge being transmitted from the teacher to them.

Principles

Social constructivism is based on a number of interrelated principles, which are briefly discussed below.

Mediation

Mediation is the process through which individuals transfer knowledge from their social environment to themselves, by means of cultural tools, signs and symbols (Vygotsky, 1978). Mediation provides teachers with a framework to develop an understanding of how learners learn within social and cultural settings. Language is notably a major mediation tool for cognition and communication, meaning teachers can effectively employ language, symbols and digital technologies to support learners' cognitive development – as Vygotsky (1978) notes, all these tools can assist users in making sense of abstract concepts and complex phenomena.

Teachers can use ICT tools such as interactive graphs, climate simulations or AI-driven visualisations as mediating tools to help learners make sense of abstract concepts conveyed in the course of a lesson dealing with climate change.

Language as a Primary Mediation Tool

Language plays a crucial role in the development of higher mental functions (Vygotsky, 1962), serving as the main instrument that elevates the so-called "inner speech" monologue. These are the silent, internal voices that are used to think, plan and solve problems. Collaborative exercises and conversations in the language classroom are essential for acquiring academic language proficiency skills (Vygotsky, 1978).

Teachers can group their learners and ask them to discuss relevant climate-related concepts. These discussions can also occur via online platforms where learners use language to mediate understanding, share perspectives and construct knowledge collaboratively. Such discourse promotes critical thinking and problem solving, thereby helping to achieve the lesson outcomes.

Cultural Tools and Symbols (Semiotics)

Vygotsky's sociocultural theory emphasises that cultural tools and signals (semiotics) mediate human behaviour at both the social and the individual levels. Examples of semiotic tools include counting systems, mnemonic devices, algebraic symbol systems, artwork, writing, diagrams and maps. These semiotic tools serve as aids for future endeavours in autonomous problem solving, as well as instruments for the co-construction of knowledge (Vygotsky, 1978).

More Knowledgeable Other

Skilled people with an excellent understanding of how to perform different tasks are referred to as "More Knowledgeable Others" (MKOs). They are able to scaffold learning for learners who are deemed less knowledgeable. Teachers or peers can act as MKOs and assist learners in solving tasks that are slightly above their cognitive level (Vygotsky, 1978). In technology-enhanced classrooms, AI tools may function as supplementary MKOs by providing prompts, feedback or modelling processes, while the teacher remains the primary mediator of learning.

Zone of Proximal Development

The Zone of Proximal Development (ZPD) is indicative of the difference between the learners' actual developmental level (as determined by independent problem solving) and their potential developmental level (as determined by problem solving under the guidance of, or in collaboration with, an MKO). Learners participate in tasks with the assistance of teachers or peers who have more advanced knowledge, in order to enhance their understanding. Development in the ZPD thus occurs through interaction and guidance (Vygotsky, 1978).

Actual Development Level

The Actual Development Level (ADL) is the phase in learners' development when they are mature enough to solve problems on their own. It indicates what a learner can do independently, without assistance. It is essential for teachers to assess learners' ADLs to determine their linguistic abilities at that time (Vygotsky, 1978). For instance, before introducing an AI climate model, the teacher might first assess learners' ability to interpret graphs or calculate averages using spreadsheet software.

Potential Development Level

The Potential Development Level (PDL) is the level learners can attain with the assistance of teachers or more knowledgeable peers. Tasks in the PDL can be achieved with guidance, but not yet independently. Understanding the PDL enables teachers to scaffold instruction and promote their learners' development (Vygotsky, 1978).

Teachers can design lessons that target learners' PDL by incorporating ICTs or AI tools, with guidance. For instance, learners might initially need help analysing large climate datasets, but with teacher or AI support they can develop the skills to complete the task independently.

Scaffolding

Scaffolding refers to the temporary support which MKOs provide to help learners perform tasks that are beyond their current abilities. Support is steadily withdrawn as learners achieve competence in what is being taught. Scaffolding allows learners to develop skills they would not be able to acquire independently (Wood et al., 1976). In lessons based on climate change, teachers could successfully use scaffolding to provide their learners with step-by-step guidance so they can understand and interpret climate simulations or datasets. Such support is then gradually reduced or even withdrawn, as learners become increasingly confident and competent.

Internalisation

Internalisation is the process through which social interactions are transformed into internal cognitive processes. It enables learners to absorb external, cultural and social activities as internal thought processes. Through internalisation, learners develop higher cognitive functions by using language and other symbolic tools (Vygotsky, 1978). Internalisation helps them reflect on those concepts associated with climate change, which were discussed in class. For instance, after exploring global temperature data using AI tools, learners might summarise trends in their own words, and relate these to local environmental impacts.

Culture as a Determinant of Learning

Culture is one of the central principles of sociocultural learning theory. Learners construct meaning by applying their past knowledge and experiences (Vygotsky, 1978). Notably, they acquire knowledge at different levels and in different ways because their backgrounds, upbringing and social experiences differ. In addition, they have varying rates of access to resources.

Collaborative Learning and Cooperative Activity

Learners acquire new skills by participating in cooperative activities. Interacting and speaking with peers function as mediating tools that enable them to cooperatively develop understanding and negotiate meaning. Cooperative learning thus facilitates the joint construction of knowledge – as Vygotsky (1978) points out, knowledge is not acquired in isolation.

Guided Participation

Guided participation, facilitated by an MKO, emphasises that cognitive development occurs within a social context, where engaging in supervised social activities with peers fosters learners' cognitive growth. While guided participation builds on and extends Vygotsky's (1978) notion of the ZPD, group work must be carefully structured, with clear guidance on using digital simulations or AI tools. For example, a teacher might guide learners through a step-by-step analysis of a climate scenario, providing prompts and feedback as they gradually take the lead in the investigation.

Advantages

One significant advantage of social constructivism is its strong alignment with learner-centred and inquiry-based pedagogies. The theory promotes active participation, collaboration and critical thinking. These learning processes are crucial for helping learners engage with complex, interdisciplinary topics such as climate change. According to Lenkauskaitė et al. (2020), another key advantage is that social constructivism values learners' prior knowledge, lived experiences and cultural contexts, making that approach particularly relevant in highly diverse South African classrooms. Furthermore, the theory supports the meaningful integration of ICTs and AI as mediating tools that encourage interaction, exploration and conceptual understanding, rather than promoting the passive consumption of information.

Critique

Despite its strengths, social constructivism also has certain limitations. Implementing related approaches can be challenging in school contexts characterised by large class sizes, limited resources or uneven access to digital technologies. Accordingly, teachers may find it difficult to implement such an approach. The theory also places significant demands on teachers as they need to be skilful at scaffolding their learners, managing collaboration and ensuring conceptual depth. As Dewodo et al. (2025) explain, without adequate guidance, learners may engage superficially with the learning content and struggle to construct a coherent understanding of key concepts. These critiques emphasise the importance of a well-structured pedagogical design and teacher support when implementing social constructivist principles, particularly in ICT- and AI-enhanced learning classrooms.

Application to Climate Change, ICT, and AI

By adopting a social constructivist approach to lessons about climate change, teachers can successfully integrate ICTs and AI to support teamwork, effective problem solving and in-depth discussions on complex environmental subjects. These tools can also promote interactive learner engagement with real-world data: for instance, using digital simulations, virtual labs, data visualisation tools and interactive modelling environments, teachers can effectively support learners in exploring cause-and-effect relationships, testing hypotheses, and visualising abstract concepts such as greenhouse gas emissions, feedback loops and long-term temperature changes.

Besides supporting collaborative learning, social constructivism can also be valuable in lessons focusing on climate change as it recognises that knowledge is constructed through social interaction and shared values. As Olaniyi et al. (2025) point out, climate change is not only a scientific issue but also a social and ethical one. This requires learners to consider different perspectives and experiences. Through classroom discussions, reflective exercises and guided dialogue, learners can explore how knowledge of climate change is developed and understood. This approach encourages them to think critically about information, question misunderstandings and engage responsibly with environmental issues. As a result, social constructivism supports the development of an informed, thoughtful, environmentally responsible learner cohort. If integrated successfully, ICTs and AI can promote social constructivist learning by encouraging discussions of shared knowledge, as well as collaborative learning. When teachers use relevant digital tools it becomes evident that the technology enhances critical thinking and ethical engagement, rather than having learners passively interact with content.

Limitations of Social Constructivism in AI-Mediated Learning Environments

Although social constructivism emphasises learning through interaction, dialogue and cultural tools, the increasing use of AI in classrooms introduces challenges that the theory did not anticipate. First, AI systems can sometimes act like a *More Knowledgeable Other (MKO)* by providing information or guidance. However, AI generates responses based on patterns in data rather than real-life experience. This may blur the authenticity of the MKO role and shift authority away from human teachers and peers to technological systems.

Second, when learners rely on AI tools or automated feedback, they may interact less with classmates. Since peer discussion and collaboration are central to social constructivism, reduced interaction can limit opportunities for learners to negotiate meaning and learn from one another.

Third, AI tools often are dependant dominant global knowledge sources, which may emphasise English-language materials or Western perspectives. Without careful guidance from teachers, this could limit the inclusion of local knowledge, indigenous perspectives and multilingual resources.

Finally, AI tools can provide scaffolding for learners, but over-reliance on these tools may prevent learners from developing their own metacognitive skills, such as planning, monitoring and evaluating their learning. As a result, learners may become dependent on technology rather than developing independent learning strategies and critical thinking skills.

From the above, it is clear that teachers should continue to play a central role in guiding learning. AI should be used to stimulate discussion, comparison and critical thinking rather than replacing interaction between teachers and learners. Teachers should encourage learners to explain their thinking, discuss ideas with peers, and draw on their diverse knowledge to ensure that learning remains collaborative and meaningful.

Diffusion of Innovation Theory

Founder

In addition to social constructivism, this chapter is guided by Everett Rogers' Diffusion of innovation (DoI) theory, which was first developed in 1962. This theory explains how new ideas and technologies are communicated and adopted over time within a specific social system. The point of departure is that individuals progress through stages of knowledge, persuasion, decision, implementation and confirmation, when faced with a novel technological development.

In this chapter, the DoI framework is used to examine how ICT tools can enhance lessons focusing on climate change. It is also implemented to help explain how factors such as the availability of digital resources, the professional development of teachers, technical support and the school culture influence this process. DoI is widely used in educational technology research, as it offers a structured way of understanding why some teachers, schools or systems adopt digital tools more readily than others.

Principles

Rogers (1962; 2003) discusses five main principles of DoI which determine the likelihood of effective technology use. First is *relative advantage*, which refers to the extent to which a technological tool is regarded as better/more effective than existing teaching methodologies. Teachers frequently perceive ICT as a means of increasing their learners' engagement in a lesson, improving the quality of the lesson and ensuring the successful achievement of the learning outcomes. Second is *compatibility*, which is defined as how well ICT aligns with existing values, teaching approaches and curriculum expectations. Teachers are more likely to adopt relevant technologies that complement CAPS-aligned lesson plans. Third, *complexity* describes the perceived difficulty of using a particular technology. When digital tools are challenging to navigate or require advanced skills, teachers tend to be reluctant to incorporate them in their lessons. Fourth, *trialability* indicates the extent to which teachers can experiment with ICT tools before fully implementing them. The benefit lies in them trying out tools in a low-risk environment during practical teaching experiences. Finally, *observability* refers to how visible or evident the positive outcomes of ICT use is to others. If teachers use ICT successfully in their climate change lessons, for instance, other teachers would be more likely to adopt similar practices.

DoI (Rogers, 1962; 2003) also distinguishes between five adopter categories: innovators, early adopters, early majority, late majority and laggards. Innovators confidently experiment with new ICT and AI tools as soon as these become available, while early adopters may be more willing to experiment with advanced digital technologies (such as AI tools) before carefully selecting and adopting them, and acting as role models for others in the process. Early majority adopters are those who adopt ICT and AI tools after observing their successful use by their peers. Those in the late majority category are hesitant to adopt new technologies and prefer using more familiar digital resources; only eventually adopting technology due to pressure or by necessity. Finally, laggards resist change and adopt ICTs and AI only when there are no alternatives.

These indicators help to explain the progress and methodological approach of ICT and AI integration in lessons, including those on climate change.

Advantages

DoI (Rogers, 1962; 2003) provides a valuable, well-established framework for understanding how new ICTs are adopted and implemented. It assists in identifying which factors support or impede adoption, for instance, teacher beliefs, the school context and infrastructural conditions. This framework allows for targeted interventions, mentorship and professional development to support ICT integration. Additionally, DoI clarifies the variations in ICT and AI uptake and adoption among teachers in different school contexts.

The adopter categories can help to explain the differences in teachers' readiness and confidence when it comes to integrating ICTs and AI into their climate change lessons. Innovators and early adopters (younger teachers, for example) might experiment with advanced digital applications such as AI tools, while the early and late majority (older teachers) might prefer more familiar digital resources such as PowerPoint presentations, educational videos and online worksheets. This understanding is important, as it allows professional development initiatives to be tailored to each teacher's level of experience, resource availability and educational institutional support.

Critique

Although the DoI (Rogers, 1962; 2003) framework is useful for teaching lessons about climate change, it has certain limitations. The theory focuses on the adoption process, rather than the quality or impact of technology use. The linear progressions that occur throughout the adoption stages may not reflect the iterative and context-dependent nature of teaching practice. Moreover, the framework does not explicitly make provision for various sociocultural factors, learner diversity or even the ethical concerns associated with technology implementation.

Although the DoI's categorisation of adopters assists in explaining the differences in teachers' readiness to use ICTs and AI, there are certain limitations. It assumes that teachers progress logically through the various categories but, in reality, adoption is often nonlinear and is influenced by multiple contextual factors. For example, a teacher classified as an early adopter might successfully integrate AI-based climate change simulations in one school but find it difficult in another because of a lack of technology, poor infrastructure or insufficient support.

Additionally, the DoI framework does not consider how social influence, mentorship or collaboration can support or prevent adoption across categories. Older teachers in the early or late majority stages may be reluctant to try these new tools, even when the benefits are clear. At the same time, innovators might encounter resistance from colleagues, or constraints within educational institutions. These limitations are particularly relevant to climate change lessons which require teamwork, peer modelling and shared experimentation, rather than isolated adoption, to ensure successful implementation and, ultimately, comprehension on the part of the learner.

Application of DoI to Climate Change, ICT, and AI

In this chapter, DoI is applied as an analytical lens for interpreting teachers' actual use of ICTs and AI in climate change lessons. Besides focusing on whether technology is adopted, DoI is used to examine precisely how teachers make sense of digital and AI tools in relation to supporting the attainment of lesson outcomes, meeting learner needs and accommodating classroom realities. Most lessons on climate change involve complex concepts, abstract processes and global–local connections, which require teachers to carefully plan and select the most effective teaching methodologies. DoI shows how teachers can move from awareness of ICT and AI tools to meaningful classroom use that supports learning about diverse climate change issues.

The theory is also used as a guiding framework to explore how factors such as access to reliable devices, internet connectivity, support from educational institutions and professional development affect whether teachers are able to progress through the various stages of adoption. It also influences teachers' decisions about integrating ICTs and AI into their lessons on Earth's changing climate. DoI therefore provides a useful way of interpreting differences in technology use across diverse school environments, without attributing these differences solely to teacher motivation.

Furthermore, DoI is applied to inform strategies for supporting teachers' integration of different technologies in their lessons. By acknowledging that teachers are at different stages of adoption, and belong to different adopter categories, the theory accommodates the design of differentiated support initiatives. These may include mentorship, peer collaboration and targeted professional development that aligns with teachers' readiness levels. In this way, DoI sheds light on current practices, while also identifying ways in which to strengthen teachers' technology implementation in their classrooms.

ICTS AND AI IN CLIMATE CHANGE TEACHING AND LEARNING

ICTs play an increasingly important part in supporting climate change lessons, by enabling teachers to present complex environmental concepts in more accessible and engaging ways. Most teachers find it difficult to explain abstract processes, long-term trends and global–local interactions, as required by the curriculum, using textbooks alone. This is particularly prevalent when topics related to climate change are taught, because such concepts are often abstract, multi-layered and difficult to grasp. ICTs can effectively enhance traditional teaching approaches by incorporating multimedia, interactive resources, simulations and real-world data into lessons. For instance, research conducted by Olaniyi et al. (2025) found that ICT tools have been used successfully in lessons on climate change, to ensure that knowledge is more easily accessible and connected to actual community contexts. In addition, interactive scientific simulations help to improve learners' understanding of dynamic processes, by offering visualisations of complex natural and man-made phenomena (Kundariati et al., 2025). Other recent research Chasokela & Hlongwane, 2025; Tisogluet al., 2025) have shown that ICTs can effectively enhance traditional teaching approaches by incorporating multimedia, interactive resources, simulations and real-world data into lessons. They have demonstrated that ICT-supported lessons and interactive digital tools can significantly enhance learner engagement and conceptual understanding when integrated with appropriate pedagogy.

Types of ICT Tools Used in Climate Change Lessons

A wide range of ICT tools are commonly used in climate change lessons in South African schools. These include presentation software such as PowerPoint, which can help teachers to structure lessons and display key concepts visually. Educational videos and documentaries are also beneficial, as they help learners visualise the various climate change topics that are the focus of their lesson. Online simulations and virtual laboratories allow learners to explore cause-and-effect relationships, such as how global climate change influences local weather patterns and extreme natural events. In addition, data visualisation tools and interactive maps allow learners to interpret climate data, weather patterns and environmental changes over time.

Teachers find such ICTs particularly useful in educational settings where access to physical resources (such as laboratory equipment or fieldwork opportunities) is limited. Clearly, ICTs offer different platforms for supporting experiential and inquiry-based learning.

Pedagogical Purposes of ICT Use in Climate Change Education

Climate change lessons that incorporate ICTs can enhance learner engagement, improve their conceptual understanding and promote higher-order thinking skills. When used effectively, digital tools can encourage learners to explore real-world problems, analyse evidence and propose solutions to environmental challenges. ICTs also support inquiry- and problem-based learning approaches, thereby developing learners' critical thinking skills. This aligns particularly well with the interdisciplinary nature of lessons on climate change. Furthermore, ICTs allow teachers to differentiate teaching and learning, since these tools can provide multiple representations of content. This is particularly important in diverse classrooms where learners' language backgrounds, learning needs and levels of prior knowledge differ.

Understanding AI in School Education Contexts

In the school environment, AI includes tools such as adaptive learning platforms, intelligent tutoring systems and generative AI tools (e.g., ChatGPT). These innovations can serve as excellent support tools that teachers might use to provide feedback, suggest additional learning resources and adapt content to each individual learner's needs (Nonkula, 2025). AI tools can also assist teachers with their lesson planning, assessment design and the actual content being taught.

AI tools can support climate change lessons by allowing learners to analyse large datasets, explore climate models and engage with scenario-based learning. For example, teachers might use AI simulations to help learners visualise various long-term climate trends and interpret specific environmental data (Mokoena & Seeletse, 2025; Nonkula, 2025). In addition, adaptive tools can be used to scaffold learners who have difficulty grasping scientific relationships. To maintain academic integrity teachers must, however, ensure that AI tools are used to support learning processes, rather than to produce completed learner work.

Despite the above advantages, AI must be viewed as a support tool for teaching and learning, not a replacement for teachers. As Nikitina and Ishchenko (2024) point out, teachers remain central to the learning process by facilitating discussions and dealing with the ethical and emotional aspects of climate change education. Clearly, AI tools are most effective when integrated into lessons in pedagogically informed and reflective ways.

Although ICT and AI offer many benefits for climate change lessons, it is important to note the risk of learners becoming too dependent on AI tools. If learners become reliant on AI-generated answers, explanations or summaries, they may not fully engage with the learning material. Teachers, therefore, need to guide learners to think critically and discuss the actual information they obtain. In practical climate change lessons, AI must support learning activities such as inquiry, discussion and data analysis. AI must not be seen as a tool that replaces the teacher's role. Careful planning and teacher training in this regard are therefore necessary to ensure that technology strengthens learning rather than impeding learners' understanding.

ICTs and AI Within a Social Constructivist Framework

From a social constructivist perspective, ICTs can be used effectively as mediating tools in climate change lessons to facilitate the development of shared meaning-making and collaborative learning. Teachers can include digital discussion platforms, group-based research tasks, and collaborative online activities in their lessons to help learners achieve the desired outcomes. In that way, knowledge is constructed collectively, through dialogue and interaction. Rather than transmitting information themselves (teacher-talk), teachers can use ICTs to facilitate discussion, scaffold learning and guide learners towards a deeper understanding of relevant climate-related topics.

ICTs and AI Within the DoI Framework

According to DoI (Rogers, 1962; 2003) theory, teachers will integrate ICTs into climate change lessons if the proposed tools are seen to offer a relative advantage over traditional teaching methodologies and are easy to use. ICT uptake will also improve if such tools are compatible with curriculum outcomes and assessment criteria, and teachers can observe how these tools are used in real classroom contexts before adopting them in their own teaching environments. Most teachers will find it easier to use different technologies in their lessons if they have opportunities to test these in settings where the risk level is relatively low, prior to opting to adopt them. These five components (relative advantage, compatibility, complexity, trialability and observability) help to explain why adoption varies across schools and teachers.

Emerging Classroom Practices in ICT- and AI-Supported Climate Change Education

Emerging research and pilot projects on climate change teaching and learning within South African schools and similar educational settings have demonstrated promising results where teachers successfully used ICTs and AI-driven simulations, interactive data visualisations and collaborative digital platforms to ensure that their learners actively engage with climate change concepts (Olaniyi et al., 2025; Mokoena & Seeletse, 2025). For instance, case studies reveal how teachers at times have used low-bandwidth tools and offline resources to help them overcome infrastructural constraints. These practical applications validate the theoretical models of social constructivism and Diffusion of Innovation chosen to guide this chapter and also provide practical ideas for teachers seeking to replicate such approaches.

Practical examples further illustrate how ICT and AI can support climate change lessons. For example, if teachers utilise AI-based simulations, learners will be able to explore different climate scenarios by adjusting variables such as carbon emissions or deforestation levels and actually observing possible environmental outcomes. Similarly, ICT-based assessments can include interactive quizzes, online collaborative tasks or digital portfolios that allow teachers to monitor learners' progress and provide immediate feedback. Teachers should also use real climate datasets, such as temperature records or rainfall trends, which learners can analyse using digital visualisation tools. Working with such authentic climate data will improve learners' analytical skills and also help them understand the real-world relevance of climate change concepts.

Potential Risks of Over-Reliance on ICTs and AI in Education

While ICTs and AI offer novel and enticing ways to support teaching and learning, over-reliance on these tools can create risks. Learners may become dependent on ICTs and AI for answers, rather than developing critical thinking, problem-solving and independent learning skills. AI systems can also produce inaccurate or biased information, which learners may accept if they lack the skills to evaluate it.

Additionally, as mentioned, unequal access to technology can exacerbate existing educational inequalities. Schools with limited technological devices, connectivity and limited opportunities for teacher training may find it difficult to use ICTs and AI effectively. For these reasons, ICTs and AI should be used as supporting tools, complementing teacher instruction, discussions and collaborative learning, rather than replacing them.

Conflicts Between Policy and Classroom Practice

Educational policies often promote and insist on the integration of AI and other digital tools, but the reality is that practical challenges in day-to-day classrooms often make it difficult, if not impossible, for teachers to successfully implement the policy expectations. This gap between policy and what actually occurs in an actual classroom setting highlights the necessity of context-aware strategies. Teachers need to be empowered to adapt policy recommendations to fit their local resources and learner needs. Successful implementation of technology-enhanced learning, therefore, requires policy support but more importantly, adequate infrastructure, professional development and ongoing support to ensure that ICTs and AI tools enhance rather than complicate learning (Abedi, 2024).

CLIMATE CHANGE EDUCATION WITHIN THE CURRICULUM AND ASSESSMENT POLICY STATEMENT

In South Africa, CAPS (DBE, 2011) constitutes the guiding framework for teaching and learning in schools from grades R–12. CAPS (DBE, 2011) stipulates that climate change topics should be taught across subjects such as Natural Sciences, Social Sciences, Geography, Life Sciences and Environmental Studies. At the primary school level learners are taught foundational concepts, such as understanding weather patterns, the use of natural resources, the human impact on the environment and basic sustainability practices. These lessons, taught in the early phases of schooling, aim to build learners' awareness and develop their curiosity about the natural world. High school learners are taught concepts associated with climate systems, environmental change and human–environment interactions in much more detail, particularly in Geography and Life Sciences. This progression is intended to gradually develop learners' scientific literacy and critical thinking skills, while the cross-disciplinary approach reflects the complex and interconnected nature of climate change by fostering a holistic understanding of critical environmental, social and economic issues among learners.

Although CAPS (DBE, 2011) offers numerous opportunities for lessons on the topic of climate change, research (Hebe, 2024; Matsepe & Maluleke, 2024) has revealed that the actual integration of climate change into the curriculum is often sporadic and limited. Related topics are mainly taught as isolated subjects, rather than being integral to other related subjects, as stipulated by CAPS. For example, important concepts related to climate change, such as "greenhouse gases", may be introduced in Natural Sciences but not integrated clearly into, or taught in, Geography lessons on weather and climate systems. This fragmentation can make it difficult for learners to develop comprehensive knowledge and skills related to climate change, or fail to connect scientific knowledge with ethical, social and economic considerations. It can also complicate teachers' task of developing learners' climate literacy systematically, and relating scientific concepts to social, economic and ethical dimensions.

A further professional consideration is teacher agency. While AI can support teaching, it cannot definitively replace teachers' professional judgement. Teachers are central to the teaching and learning process; they are the ones who must decide what is required of the curriculum. Teachers have to ensure that technology supports learning, rather than dictates it. Importantly, however, an overreliance on AI may weaken critical engagement and reduce teachers' control over what knowledge is imparted to their learners (Mogoale et al., 2025; UNESCO, 2025).

Curriculum and Policy Context: DBE, CAPS, and Digital Ethics

The integration of ICTs and AI tools in school-based climate change lessons should be understood within the policy and curriculum framework of the South African Department of Basic Education (DBE). The DBE's ICT in Education Policy emphasises the ethical, structured and equitable use of digital technologies to enhance teaching, learning and assessment while protecting learners' rights and well-being (DBE, 2018). Even though this policy was developed before the widespread use of ICTs and AI tools, its core principles, such as digital literacy, teacher professional development and responsible technology use, remain relevant. More recent DBE assessment guidelines reiterate the importance of assessment integrity, learner accountability and fair assessment practices in environments which are digitally driven (DBE, 2024). Together with international guidance on ethical AI use, these frameworks ensure that teachers do not compromise curriculum standards or academic integrity.

The content of climate change lessons is explicitly embedded within the CAPS curriculum, particularly in the Further Education and Training (FET) phase. In Geography (Grades 10–12), for example, climate change is covered through topics such as climatology, global atmospheric processes, human-environment interaction, environmental sustainability and the impacts of climate change at global and local scales (DBE, 2018). In order to successfully achieve the lesson outcomes, learners are expected to interpret climate data, analyse spatial patterns, and evaluate the consequences of human activities on climate systems. Similarly, in Life Sciences (Grades 10–12), climate change topics include human impact on the environment, biodiversity loss, ecosystem disruption and sustainable resource use, where learners explore and reflect on the influence of human behaviour on ecosystems (DBE, 2011b).

At the General Education and Training (GET) level, climate change content is more specifically integrated across subjects such as Natural Sciences and Social Sciences. At this level, learners explore foundational concepts related to weather and climate, conservation, responsible resource use, and human interaction with the environment. These topics lay the foundation for more explicit climate change learning in the FET phase. This highlights CAPS' intention to develop climate literacy gradually, moving from basic environmental awareness to more complex scientific, ethical and socio-economic analyses.

Understanding where climate change content is embedded within CAPS is essential for teachers when integrating ICTs and AI into their lessons. It ensures that digital tools are not used as add-ons but are intentionally and systematically aligned with prescribed curriculum content, assessment standards and policy expectations. When teachers align ICTs and AI tools with DBE policies and CAPS, it ensures effective, ethical and professional teaching and learning practices.

Aligning ICTs and AI Use With CAPS Outcomes

A practical method for integrating ICTs and AI into lessons is the backward design model developed by Wiggins and McTighe (1998). In accordance with this model, teachers must first identify desired CAPS-aligned lessons, then decide what acceptable evidence will demonstrate those results (thus, how the learners will be assessed), and finally plan targeted ICTs and AI-supported teaching and learning activities. For example, if one of the outcomes of a Grade 10 Geography lesson requires learners to interpret spatial climate data, teachers can select GIS mapping tools or AI-driven climate simulations that directly show these spatial patterns in temperature, rainfall or extreme weather events.

Overall, backward design (Wiggins & McTighe, 1998) ensures that ICTs and AI are integrated meaningfully, with the focus on the lesson outcomes and assessment criteria. It also leads to purposeful pedagogy. This is particularly important in climate change lessons, as these topics involve complex content that requires carefully structured learning experiences to ensure deep understanding, critical thinking and informed decision-making.

Designing Inquiry-Based and Constructivist Lessons

Modern-day education requires teachers to use inquiry-based methodologies that encourage learners to ask questions, explore problems and construct their own knowledge. This approach helps young minds to grasp abstract concepts such as global warming, sea-level rise and extreme weather temperatures. ICTs and AI can effectively develop learners' inquiry-related skills and competencies by providing access to real-time data, interactive maps and simulations that make climate processes easier to visualise and understand.

AI tools can also assist learners in attaining the envisaged lesson outcomes. These applications can suggest probing questions and provide individualised feedback which is personalised to each learner's specific response. Collaborative digital platforms such as Google Workspace Tools, Google Docs and Google Sheets allow learners to create relevant data charts, summaries and reflective notes collaboratively, promoting topical discussions and mutual understanding. Evidently, ICTs and AI can effectively aid in the development of skills related to planning investigations, analysing information and drawing evidence-based conclusions, all of which are required by CAPS (DBE, 2011).

Differentiating Instruction Using ICTs and AI

ICTs and AI tools support the use of differentiated teaching and learning methods in lessons dealing with climate change. These applications offer teachers and learners more relevant ways of accessing learning material and engaging with the content. Learners can, for example, use videos, diagrams, simulations or simplified texts, which can be adapted depending on their individual learning needs and language proficiency. Additional explanations or specific activities can be included for learners who require extra support in mastering complex concepts, such as climate systems or weather patterns. Differentiation is particularly significant in South African classrooms where learners often come from diverse cultural backgrounds, have varying levels of language proficiency, and learn and progress at different paces (Molotsi, 2022).

Integrating ICTs and AI Into Assessment Practices

It is essential for teachers to incorporate interdisciplinary approaches when assessing their learners on climate change topics. For instance, to complete an assessment, learners may be required to use AI simulations to predict the impacts of climate change on local water resources, and to integrate concepts from Life Sciences related to biodiversity and ecosystem health. If teachers align both the outcomes and the assessment criteria of their lessons with ICT-related activities they can better measure learners' understanding and engagement, as well as their ability to apply knowledge across disciplines and engage meaningfully with complex, real-world issues.

Teachers could ask learners to complete project-based assessments, do data-interpretation exercises and write reflective journals. Learners might be asked to integrate various ICTs and AI tools, such as collaborating on virtual platforms, visualising complex climate data through simulations and interactive models, and exploring cause-and-effect relationships more concretely. Such activities can enhance their understanding of climate change-related terms, while developing their digital literacy and problem-solving skills. ICTs and AI can also be successfully used to scaffold learners' knowledge by suggesting hypotheses, guiding data analyses or generating personalised prompts for reflection. This allows learners to engage critically with concepts associated with climate change, giving the teacher additional time to focus on higher-order facilitation and ethical discussions.

ICTs and AI can assist teachers in evaluating the higher-order skills of their learners. As an example, teachers could design online quizzes, data-interpretation tasks, digital posters or even short reflective responses related to climate change themes. These tools make it easier to carefully monitor each learner's understanding of the lesson content, identify gaps, and offer constructive and timely feedback (de Beer & Lautenbach, 2024; Matlala et al., 2025). Teachers can also design assessments featuring scenario-based questions, where learners have to analyse various challenges related to climate change, such as water scarcity and how global warming affects water supply and agriculture in their specific area of South Africa. These concerns are central to lessons on climate. This approach aligns with CAPS, which promotes learners' problem-solving and critical thinking skills, and advocates the application of knowledge to real-world contexts (DBE, 2011; 2024).

FACTORS INFLUENCING TEACHERS' ADOPTION OF ICTS AND AI IN CLIMATE CHANGE LESSONS

Teachers' adoption of ICTs and AI in their lessons could be influenced by several interrelated factors which might occur at the individual, institutional and systemic levels. Policy frameworks (DBE, 2018; UNESCO, 2018) often assume that if teachers have access to ICTs they will naturally adopt them, leading to enhanced teaching practices. However, research consistently demonstrates that adoption is neither automatic nor uniform (Matlala et al., 2025; Mnisi et al., 2024). The social constructivist and DoI theories provide a useful lens for analysing the factors prompting teachers to refrain from adopting innovations in their lessons, particularly in relation to climate change. These theories highlight how perceptions, contextual conditions and social dynamics influence teachers' decisions to adopt or resist new technologies. Some of these factors are discussed below.

Teachers' Pedagogical Beliefs

Teachers' beliefs about teaching, learning, ICTs and AI strongly influence whether or not they will use technology to enhance their climate change lessons. Those who value learner-centred, inquiry-based approaches are more likely to perceive ICTs and AI as valuable and relevant for achieving lesson outcomes. By contrast, teachers who prioritise the delivery of lesson content or who doubt the usefulness of digital technologies may resist adoption, particularly if they feel that technology threatens or undermines their authority.

Infrastructure, Access, and Digital Inequality

Researchers such as Mnisi et al. (2024) and Mwapwele et al. (2019) argue that the most crucial factor in teachers' successful uptake of ICTs and AI is access to reliable, up-to-date infrastructure. Uninterrupted electricity supply, internet connectivity and functional devices are essential for meaningful and successful technology integration (Mnisi et al., 2024; Mwapwele et al., 2019). In South Africa, persistent inequalities among schools mean that access to these resources varies widely.

Under-resourced rural schools in particular have to deal with intermittent internet connectivity, limited access to functional digital devices and outdated technological infrastructure. In such contexts, teachers are limited in terms of their ability to integrate ICTs to support the successful attainment of the outcomes of climate change-related lessons (Ahiaku et al., 2025; Chisango & Marongwe, 2025).

Following the social constructivist theory (Vygotsky, 1978) adopted for this research, it is clear that ICTs and AI function as mediating tools that help to support interaction, dialogue and guided learning. When access to these tools is limited it is difficult for teachers to develop collaborative and learner-centred lessons, which reduces opportunities for learners to engage meaningfully with the concepts taught.

From a DoI (Rogers, 1962; 2003) framework perspective, when infrastructure is limited it reduces the perceived relative advantage of ICTs and AI. If technologies are unreliable or difficult to access, teachers may perceive them as being disruptive, not beneficial, and thus prefer not to use them. When teaching aspects of climate change this situation may become particularly challenging, as related topics often rely on up-to-date data, multimedia resources and interactive tools. In such scenarios, teachers might revert to traditional teaching methods – not because of resistance to innovation on their part, but because of actual pragmatic and logistical constraints.

Professional Development and Teacher Preparedness

Research by several scholars (Mpungose, 2020; Msila, 2015; Padayachee & Mbatha, 2022; Nyathi & Joseph, 2024) consistently shows that teachers' willingness and ability to adopt digital technologies are closely linked to the quality and relevance of the professional development opportunities available to them (Mpungose, 2020; Msila, 2015; Nyathi & Joseph, 2024). In the South African context, professional development initiatives all too often focus on basic technical skills only, rather than providing sufficient guidance on digital pedagogy, the ethical use of AI and subject-specific applications for teaching climate change-related concepts. Without sufficient support, teachers might lack confidence in selecting appropriate tools, designing meaningful learning activities or managing those classrooms with available state-of-the-art technology (Mpungose, 2020; Nyathi & Joseph, 2024).

From a social constructivist perspective, professional development should be collaborative, reflective and practice-based. Undoubtedly, teachers benefit most when engaging in shared learning experiences and peer collaboration with professional learning communities which allow them to co-construct knowledge with their fellow teachers about how ICTs and AI can support inquiry-based and interdisciplinary lessons on climate change (Msila, 2015). Such approaches enable teachers to experiment with digital tools, reflect on their classroom experiences and adapt their teaching methodologies in ways that are meaningful within their unique school contexts.

From a DoI perspective, professional development is crucial for enhancing teachers' perceptions of the relative advantage, compatibility and trialability of ICTs and AI. When professional development training focuses explicitly on showing how digital tools can be linked to curriculum outcomes and assessment practices, teachers are more likely to view these technologies as useful rather than cumbersome, and to adopt them for their lessons (Mpungose, 2020; Nyathi & Joseph, 2024). Ongoing support, mentoring and opportunities to observe successful practice further enhance adoption by helping to reduce uncertainty, as teachers perceive that there are risks involved in ICT and AI uptake. Clearly, limited professional development can constrain teachers' perceptions of the relative advantages of ICTs and AI, as well as their self-efficacy in using these tools and their willingness to adopt them in their lessons.

Overall, professional development that aligns with both social constructivism (Vygotsky, 1978) and the DoI theory (Rogers, 1962; 2003) chosen to guide this research can help teachers to successfully implement new technologies. It may also assist them in using these tools confidently and in developing crucial skills such as critical thinking, collaboration, and meaningful engagement with relevant climate change issues.

Lesson Outcomes and Assessment Criteria

Alignment with existing lesson outcomes and assessment criteria is another critical factor influencing adoption. Teachers are more likely to integrate ICTs and AI once they realise that these tools align with the CAPS outcomes and assessment criteria. As mentioned, climate change concepts are embedded across multiple subjects, yet they are often treated as an add-on rather than a central focus. For this reason, teachers might struggle to justify the time and effort required to integrate the latest technologies.

Assessment pressures can further constrain adoption. When assessments are prioritised, teachers may decide not to implement new ICT tools and pedagogical approaches, instead spending all their available time preparing their learners for upcoming assessments. This highlights the importance of aligning assessment practices with pedagogical goals that value inquiry, critical thinking and systems understanding.

School Management Support

The school's management and culture significantly influence teachers' willingness to adopt ICTs and AI. Supportive leadership that values innovation, provides access to resources and encourages experimentation creates conditions that are conducive to adoption. By contrast, schools with rigid structures, limited ICT support or management who responds indifferently to failure, may discourage teachers from experimenting with novel technologies.

From a social constructivist viewpoint, it is clear that the social context of the school significantly affects teachers' ICTs uptake. In schools with a culture that supports shared responsibility, teachers can collaboratively explore ICTs and AI tools, and reflect on both successes and failures. Communities of practice can also be developed to support shared learning, thereby reducing teachers' feeling of isolation and creating a culture of reflective experimentation.

One of the key principles of DoI is observability, which is also a vital driver of adoption. When teachers observe colleagues successfully using ICTs and AI in their climate change lessons, they might view these tools as effective and worth using. Peer modelling, mentorship and collaborative planning therefore play important roles in establishing innovative ICT practices within schools.

ETHICAL AND PROFESSIONAL CONSIDERATIONS

It is important for teachers and school management to realise that the use of ICTs and AI in education presents significant ethical and professional challenges (Mogoale et al., 2025; UNESCO, 2025). Notably, one of the main concerns is data privacy, as AI-based applications can collect and store sensitive learner information. This means that teachers and schools must be very careful about ensuring that personal data are protected and used responsibly, and that legal and institutional requirements are adhered to, as Mogoale et al. (2025) caution.

Another concern relates to algorithmic bias. It is possible for AI systems to reflect or reinforce existing social, cultural and gender biases. For that reason, teachers must carefully and critically consider AI-generated feedback and recommendations, rather than accepting them uncritically (Chinta et al., 2024). Closely related to this is the issue of transparency. Learners should first be taught the basics of how AI tools work, what data are used, and how decisions or feedback are generated, particularly when AI influences assessment or learning pathways (UNESCO, 2025).

Ethical integration requires teachers to prioritise the emotional wellbeing of their learners, particularly in lessons on climate change: if they use digital resources to show videos or simulations of environmental disasters, they must be aware that while such images can increase awareness, they may also cause anxiety or feelings of helplessness in sensitive learners. Teachers must therefore assume a mediating role by supporting reflection, encouraging discussion and carefully guiding their learners towards constructive and informative responses to climate action.

When teachers incorporate AI into climate change lessons they must use professional judgement and critically evaluate any AI-generated outputs. It is essential that these digital tools enhance learners' critical thinking skills, rather than simply letting them passively listen to the teacher impart the required knowledge. Ultimately, clear ethical guidelines require targeted teacher training and institutional support, to ensure that AI is used responsibly and equitably to enhance lessons (Mokoena & Seeletse, 2025; UNESCO, 2025).

Research in South African educational contexts shows that AI tools can unintentionally exacerbate existing inequalities, unless ethical concerns are taken into account. A study by Mokoena and Seeletse (2025) found that the use of AI in under-resourced and rural schools revealed unequal access to digital infrastructure, risks to data privacy and algorithmic bias, all of which can limit the meaningful and fair use of AI in classrooms. Thus, meticulous planning is essential for making sure that AI does not benefit already advantaged learners, while marginalising others.

Ethical concerns might influence teachers' decisions about whether to use AI in their climate change lessons when it comes to data privacy, misinformation, bias and learners' over-reliance on AI-generated content. Where such issues are pertinent, teachers might become overly cautious. These concerns might arise in lessons dealing with climate change, as the complex nature of the topics taught increases the risk of misinformation or misrepresentation. Teachers might refuse to use AI tools to achieve lesson outcomes unless clear ethical guidelines and school support are in place.

Notably, the ethical use of ICTs and AI must be prioritised in professional development, particularly in lessons on climate change. Misinformation, bias and emotional responses to environmental issues are significant current concerns. Teachers thus require appropriate training to use AI responsibly without breaching data privacy laws, and to be competent in critically evaluating AI-generated content – this, to ensure that technology enhances learning without undermining ethical and pedagogical principles (UNESCO, 2022). Without such support, teachers might either avoid AI altogether or use it in superficial ways that do not support the lesson outcomes meaningfully. Ethical integration is thus not an optional add-on, but a fundamental requirement for the meaningful and equitable use of digital technologies in climate change lessons.

Another important ethical concern that must be taken into account is academic integrity. AI tools can easily and quickly generate essays, summaries and answers to assessment questions. This may tempt some learners to submit AI-generated work without fully engaging with the learning process and acknowledging that this is not their own original work. This creates additional challenges for teachers when assessing learners' understanding and ability to problem solve and think critically. In addition, some AI tools require users to upload or share information, which raises data privacy concerns, particularly if learner data is stored on external platforms. To address this, teachers need to be empowered to design process-based assessments that require learners to explain, reflect and discuss, rather than simply relying only on final written answers. In addition, learners should be guided on the responsible and ethical use of AI tools, so that technology supports learning rather than replacing their own thinking and effort.

IMPLICATIONS FOR PRACTICE, TEACHER EDUCATION, AND POLICY

This final section highlights the importance of practical steps, professional development needs and policy considerations necessary to ensure that ICT- and AI-enhanced climate change education is both effective and equitable.

Implications for Classroom Practice

Teachers require clear guidance on how to integrate ICTs and AI meaningfully into climate change lessons while maintaining curriculum alignment and pedagogical intent. Key implications include:

Purposeful and outcomes-aligned integration: Digital tools should directly support the learning outcomes stated in the CAPS curriculum for climate change lessons. ICT tools and AI must not be used as add-ons.

Strengthening inquiry-based pedagogy: ICTs and AI can be incorporated into investigations, simulations, collaborative tasks and scenario-based activities. Teachers need to include structured examples of how to present such lessons across different schooling contexts.

Equity-oriented strategies for low-resourced schools: Offline simulations, printed datasets, mobile phone–based tasks and school-based local networks can help reduce digital inequality.

Assessment integrity: Teachers should design assessments that require learners to show their thinking, use local examples and explain their reasoning rather than simply recall facts. ICT and AI should be used to support learning, not to produce complete answers for learners.

Implications for Teacher Education and Professional Development

Teacher professional development is central to meaningful ICT and AI integration in climate change lessons. Teacher training institutions and in-service training workshops should consider the following:

Digital pedagogy rather than digital skills: Professional development must move beyond teaching basic use of ICTs and AI. It is essential that they include teachers' ability to select appropriate tools, design inquiry-based lessons and critically evaluate AI outputs.

Ethics, data privacy, and responsible AI: Pre-service and in-service programmes must include training on POPIA compliance, ethical data use, transparency, and the identification of algorithmic bias - especially important when teaching sensitive climate change issues.

Communities of practice: School-based and online teacher networks should share climate change resources, model inquiry-based lessons and exchange problem-solving strategies.

Practice-based professional development models: Lesson study, co-teaching, micro-teaching with feedback and reflective digital portfolios align well with both Social Constructivism and DoI principles.

Implications for Policy Makers and Curriculum Planners

Policy makers play a crucial role in ensuring that ICT and AI integration into climate change lessons is successful, follows correct procedures and is ethical. A number of key considerations for policymakers to take into account are:

Clear guidelines for ICT and AI use in climate change lessons. Policy makers need to ensure that national guidance is implemented regarding the appropriate use of ICT and AI in teaching, learning and assessment, including expectations for academic integrity, data security, and transparency.

Infrastructure investment to ensure digital equality: Policy makers should make sure that clear guidelines are set for equitable access to devices, electricity, connectivity and technical support. Policies should prioritise rural and under-resourced schools.

Alignment of CAPS with emerging digital competencies: Policy makers should continually review the CAPS document to ensure it incorporates explicit digital literacy, data interpretation, and ICT-AI-related competencies linked to climate change topics.

Professional development policy: Policies should stipulate structured, ongoing and curriculum-aligned professional development that equips teachers to integrate ICTs and AI in climate change lessons.

Procurement and quality assurance of digital tools: Policy makers should ensure that standards for selecting ethical, inclusive and curriculum-aligned ICT and AI tools are specified in their climate change policies.

CONCLUSION

This chapter investigated how ICTs and AI can support teaching and learning related to climate change. It highlighted the importance of aligning lessons with CAPS (DBE, 2011), which provides opportunities for interdisciplinary approaches. This chapter also showed that effective implementation depends on teachers' ability to integrate digital tools purposefully into lesson planning, pedagogy and assessment practices.

Using social constructivism and the Diffusion of innovation (DoI) theory as guiding frameworks, the chapter demonstrated that ICTs and AI are most effective when they support inquiry-based learning, collaboration and the application of knowledge to real-world contexts. These tools can enhance learners' problem-solving skills, critical thinking and engagement with locally relevant climate challenges, such as water scarcity, food security and environmental sustainability.

The chapter also emphasised that teachers' adoption of ICTs and AI is affected by contextual constraints, their professional development, personal beliefs and institutional support. Finally, as noted here, teachers remain central to making appropriate pedagogical decisions, interpreting what the curriculum requires, and ensuring that the ethical and emotional dimensions of climate change lessons are dealt with sensitively.

Looking ahead, there is a critical need for continued research into how teachers can successfully integrate ICTs and AI into lessons on climate change across the curriculum. Future research should examine the long-term impact of ICTs and AI-supported teaching pedagogies on learners' climate literacy, problem-solving abilities and critical thinking skills across diverse school contexts. In addition, it is essential to address the alignment of teacher training curricula with relevant, up-to-date ethical considerations related to the use of ICTs and AI in the classroom.

These future directions are essential for ensuring that technology-enhanced climate change education is both practical and sustainable.

Ultimately, when ICTs and AI are carefully integrated into lessons on the causes and impact of climate change, they have the potential to enhance the attainment of learning outcomes and cultivate critically aware, ethically responsible and environmentally engaged citizens.

REFERENCES

Abedi, E. A. (2024). Tensions between technology integration practices of teachers and ICT in education policy expectations: Implications for change in teacher knowledge, beliefs and teaching practices. *Journal of Computers in Education, 11*(4), 1215–1234. DOI: 10.1007/s40692-023-00296-6

Ahiaku, P. K. A., Uleanya, C., & Muyambi, G. C. (2025). Rural schools and tech use for sustainability: The challenge of disconnection. *Education and Information Technologies, 30*(9), 1–15. DOI: 10.1007/s10639-024-13311-9

Chasokela, D., & Hlongwane, J. (2025). Leveraging AI and ICT for greener education in the face of climate change in smart classrooms. *Journal of Research in Mathematics, Science, and Technology Education, 2*(1), 36–45. DOI: 10.70232/jrmste.v2i1.24

Chinta, S. V., Wang, Z., Yin, Z., Hoang, N., Gonzalez, M., & Quy, T. L.. (2024). *FairAIED: Navigating fairness, bias, and ethics in educational AI applications.* arXiv Preprint. https://arxiv.org/abs/2407.18745

Chisango, G., & Marongwe, N. (2025). The digital divide at three disadvantaged secondary schools in Gauteng, South Africa. *Journal of Education, 82*(82), 1–18. DOI: 10.17159/2520-9868/i82a09

de Beer, M., & Lautenbach, G. V. (2024). Using online formative assessment tools in South African schools. *South African Journal of Childhood Education, 14*(1), a1438. DOI: 10.4102/SAJCE.v14i1.1438

Department of Basic Education. (2018). *Professional Development Framework for Digital Learning*. DBE. https://www.education.gov.za/DigitalLearningFramework July18.aspxDepartment of Basic Education

Department of Basic Education. (2024). *National Protocol for Assessment grades R–12*. DBE. https://www.education.gov.za/TeachingLearning/ Assessment.aspx

Department of Basic Education (DBE). (2011). *Curriculum and Assessment Policy Statement (CAPS): Geography grades 10–12*. DBE. https://www.education.gov.za/Curriculum/CurriculumAssessmentPolicy Statements(CAPS).aspx

Dewodo, H., Heeralal, D., & Agbetorwoka, J. (2025). Challenges of constructivist approaches in modern learning environments. *Perspectives in Education, 43*(3), 161–173.

Hebe, H. (2024). Uncovering climate change education-enabling themes in the South African Grade 7 Natural and Social Sciences school curricula. *Jurnal Penelitian dan Pembelajaran IPA, 10*(2). https://jurnal.untirta.ac.id/index.php/JPPI/article/view/29026

Intergovernmental Panel on Climate Change (IPCC). (2023). [*Synthesis report.* IPCC.]. *Climatic Change*, •••, 2023.

Khoza, N. G., & van der Walt, F. (2025). A systematic review on AI-enhanced pedagogies in higher education in the Global South. *Frontiers in Education, 10*, 1667884. Advance online publication. DOI: 10.3389/feduc.2025.1667884

Kundariati, M., Ibrohim, I., Rohman, F., & Nida, S. (2025). Navigating climate change education: A problem-oriented project-based learning in secondary schools. *Science Activities*, *62*(2), 156–170. DOI: 10.1080/00368121.2025.2463653

Lenkauskaitė, J., Colomer, J., & Bubnys, R. (2020). Students' social construction of knowledge through cooperative learning. *Sustainability (Basel)*, *12*(22), 9606. DOI: 10.3390/su12229606

Matlala, D. E., Graham, M. A., & Roux, K. (2025). Utilising ICT for classroom teaching and learning: Practices among South African teachers. *Independent Journal of Teaching and Learning*, *20*(2), 70–92. DOI: 10.17159/ppfn4g11

Matsepe, D., & Maluleke, M. (2024). Infusing climate change education into the Curriculum and Assessment Policy Statement (CAPS) as a resilience strategy in South Africa: Towards a theory of change. *E-Journal of Humanities. Arts and Social Sciences*, *5*(9), 73–82.

Mnisi, B. R., Mtshali, T. I., & Moses, M. (2024). Moving beyond the challenges of learning through technologies: The current status of ICT integration in South African schools. *Journal of Education and e-learning Research*, *11*(1), 128–134. DOI: 10.20448/jeelr.v11i1.5396

Mogoale, P. D., Pretorius, A., Mogase, R. C., & Segooa, M. A. (2025). Ethical considerations in artificial intelligence-driven environments for higher education. *South African Journal of Information Management*, *27*(1), a2007. DOI: 10.4102/sajim.v27i1.2007

Mokoena, O. P., & Seeletse, S. M. (2025). AI in rural classrooms: Challenges and perspectives from South African educators. *International Journal of Current Educational Studies* [Advance online publication]. DOI: 10.46328/ijces.199

Molotsi, A. R. (2022). The use of ICT resources to transform teaching at secondary schools in the Bojanala district, North-West Province. *South African Journal of Education*, *42*(1).

Mpungose, C. B. (2020). Emergent transition from face-to-face to online learning in a South African university in the context of the Coronavirus pandemic. *Education and Information Technologies*, *25*, 993–1004. DOI: 10.1007/s10639-020-10190-3

Msila, V. (2015). Teacher readiness and information and communications technology (ICT) use in classrooms: A South African case study. *South African Journal of Education*, *35*(2), 1–12.

Mwapwele, S. D., Marais, M., Dlamini, S., & van Biljon, J. (2019). Teachers' ICT adoption in South African rural schools: A study of technology readiness and implications for the South Africa Connect Broadband policy. *African Journal of Information and Communication*, *24*(24), 1–21. DOI: 10.23962/10539/28658

Nikitina, I., & Ishchenko, T. (2024). New Ukrainian School: AI implementation in education. *Scientific Journal of Polonia University, 66*(6).

Nonkula, Z. (2025). Student teachers' narratives on artificial intelligence-personalised learning in geography and social sciences teaching at a South African university. [UFS Publications]. *Interdisciplinary Journal of Education Research*, *7*(2), a12. Advance online publication. DOI: 10.38140/ijer-2025.vol7.2.12

Nyathi, T., & Joseph, R. M. (2024). Empowering South African educators: Navigating the challenges of digital teaching and learning competencies. *SA Journal of Human Resource Management, 22*, a2591. https://doi.org/. v22i0.2591DOI: 10.4102/sajhrm

Olaniyi, E., Perekunah, E., & Emeka, O. (2025). ICT-enabled climate change education and adaptation in Africa. In Mbah, M. F., Molthan-Hill, P., & Molua, E. L. (Eds.), *Practices, perceptions and prospects for climate change education in Africa* (pp. 93–106). Springer., DOI: 10.1007/978-3-031-84081-4_5

Oubibi, M., Fute, A., Kangwa, D., Barakabitze, A. A., & Adarkwah, M. A. (2024). Interactive technologies in online teacher education in Africa: A systematic review, 2014–2024. *Education Sciences*, *14*(11), 1188. DOI: 10.3390/educsci14111188

Padayachee, R., & Mbatha, B. (2022). Teachers' readiness for digital transformation in South African schools. *Education and Information Technologies*, *27*, 10201–10220. DOI: 10.1007/s10639-022-11037-1

Phiri, R. G., Nsama, L., Walubita, N., Samantaray, S., Patra, S. S., Mishra, M. R., & Gourisaria, M. K. (2025). Climate literacy: Creating a multilevel interactive platform for climate education. *I.J. Modern Education and Computer Science*, *16*(5), 41–61. DOI: 10.5815/ijmecs.2025.05.03

Rogers, E. M. (1962). *Diffusion of innovations.* Free Press of Glencoe.

Rogers, E. M. (2003). *Diffusion of innovations* (5th ed.). Free Press.

Tisoglu, S., Sönmez, E., Kaya, K. Y., & Et, S. Z. (2025). Bridging pedagogy and technology: A systematic review of immersive virtual reality's potential in climate change education. *Environmental Education Research*, •••, 1–31. DOI: 10.1080/13504622.2025.2480661

United Nations Educational, Scientific and Cultural Organization. (2022). *Recommendation on the ethics of artificial intelligence.* UNESCO. https://www.UNESCO .org/en/artificial-intelligence/recommendation-ethics

United Nations Educational Scientific and Cultural Organization. (2024). *Greening curriculum guidance: Teaching and learning for climate action.* UNESCO. https://www.UNESCO.org/en/articles/greening-curriculum-guidance-teaching-and -learning-climate-actionUNESCO

United Nations Educational Scientific and Cultural Organization. (2025). *Climate change education.* UNESCO. https://www.UNESCO.org/en/climate-change/ educationUNESCO

United Nations Educational, Scientific and Cultural Organization (UNESCO). (2018). *ICT Competency Framework for Teachers (ICT CFT) version 3.* UNESCO. https://www. UNESCO.org/en/digital-competencies-skills/ict-cft

Vygotsky, L. S. (1962). *Language and thought.* MIT Press. DOI: 10.1037/11193-000

Vygotsky, L. S. (1978). *Mind in society: The development of higher psychological processes* (Cole, M., John-Steiner, V., Scribner, S., & Souberman, E., Eds.). Harvard University Press.

Wiggins, G., & McTighe, J. (1998). *Understanding by design.* Association for Supervision and Curriculum Development (ASCD).

Wood, D., Bruner, J. S., & Ross, G. (1976). The role of tutoring in problem solving. *Journal of Child Psychology and Psychiatry, and Allied Disciplines*, *17*(2), 89–100. DOI: 10.1111/j.1469-7610.1976.tb00381.x PMID: 932126

KEY TERMS AND DEFINITIONS

Artificial Intelligence (AI): AI refers to a specific type of technology capable of completing tasks that require human intelligence. Such tasks include analysing patterns, making predictions and providing personalised feedback to learners. AI supports teachers in offering individualised learning experiences.

Climate change lessons: Climate change lessons, or lessons about climate change, refer to lessons in which teachers equip their learners with the knowledge and skills to deal with various climate-change-related topics, such as local and global climate challenges, environmental justice and strategies for climate resilience. These lessons also help to develop learners' critical thinking and problem-solving skills, and their environmental awareness, teaching them about responsible decision-making for sustainability.

Diffusion of innovation (DoI) theory: This theory describes how individuals or groups gradually adopt new technologies by progressing through different stages, from awareness to sustained use. In climate change lessons, it helps to interpret variations in ICT and AI adoption across schools. According to DoI, progression through these stages is influenced by how teachers perceive the relative advantage, usefulness and difficulty of using ICTs and AI. The theory can assist in explaining why some teachers, when teaching the topic of climate change, progress faster toward successful ICT uptake than others teaching in comparable educational contexts.

Digital literacy: Digital literacy refers to the knowledge, skills and attitudes required to use digital technologies effectively, critically and responsibly. It includes the ability to access, evaluate, create and communicate information using digital tools. In climate change lessons, digital literacy enables learners to interpret climate data, evaluate online sources and engage critically with digital and AI-generated information.

Information and Communication Technologies (ICTs): ICTs include digital tools such as computers, mobile devices and software platforms. These tools are used to support the teaching and learning processes. In climate change lessons, ICTs can provide interactive simulations, access to live data and innovative learning experiences to enrich the learning journey.

Inquiry-Based Learning: Inquiry-based learning refers to a teaching and learning approach that is learner-centred. Learners actively investigate questions, problems, or scenarios rather than passively receiving information. It emphasises questioning, exploration, data analysis, reflection and evidence-based reasoning. In climate change lessons, inquiry-based learning enables learners to investigate real-world environmental issues, analyse climate data and develop informed responses to sustainability challenges.

Pedagogical integration: Pedagogical integration refers to the deliberate and specific use of ICTs in a particular lesson. Such technologies are carefully aligned with the lesson outcomes to enhance teaching and learning. In climate change lessons, pedagogical integration ensures that the digital tools support the successful achievement of the lesson outcomes and enhance learner engagement.

Social constructivism: Social constructivism is a specific theoretical framework that describes learners' ability to co-create knowledge within a social and cultural context. It highlights collaboration, critical reflection and the use of mediating tools, such as language and technology. In climate change lessons, learners examine both local and global environmental issues critically, engage in dialogue about sustainable practices and co-develop strategies to deal with climate-related challenges.

Teachers' professional development: This term refers to structured, targeted and ongoing training opportunities aimed at improving teachers' pedagogical knowledge, content expertise and technological skills. Effective professional development can support teachers in integrating ICTs and AI meaningfully into lessons on climate change.

Chapter 13
Teachers and Teacher Educators' Readiness for Integrating AI in the Teaching of Climate Change Education

Manoah Muchanga
http://orcid.org/0000-0002-1156-4245
University of Zambia, Zambia

Garikai Membele
University of Zambia, Zambia

Matilda K. Nakazwe
University of Zambia, Zambia

Everisto Simweene
University of Zambia, Zambia

Chipo Moonga
University of Zambia, Zambia

Pauline N. Monde
University of Zambia, Zambia

Liberty Mweemba
University of Zambia, Zambia

Steria M. Simooya
University of Zambia, Zambia

Kabwe H. Mubanga
University of Zambia, Zambia

Inonge K. Milupi
University of Zambia, Zambia

Tamara Chansa-Kabali
University of Zambia, Zambia

Kaiko L. Mubita
University of Zambia, Zambia

Progress H. Nyanga
University of Zambia, Zambia

Marvin Kapenda
FAO, Zambia

DOI: 10.4018/979-8-3373-6975-4.ch013

Vincent Ziba
FAO, Zambia

Tiwine Z. Muchipa
http://orcid.org/0009-0009-0774-4209
FAO, Zambia

Amos Sikayile
UNESCO, Zambia

Ecloss Munsaka
University of Zambia, Zambia

ABSTRACT

The rapid advancement of artificial intelligence (AI) is reshaping teaching and learning across education systems worldwide. This study examined teachers, teacher educators, and student teachers readiness to integrate AI into climate change and sustainability education in Zambia. Drawing on a descriptive survey of 2,634 participants across ten provinces, the research reveals that while awareness and positive attitudes toward AI are widespread, pedagogical integration remains limited. AI use is largely exploratory and concentrated on tasks such as lesson preparation rather than transformative teaching practices. A significant gap exists in climate change training, with the majority of educators lacking recent professional development. The findings highlight that educator readiness is shaped by knowledge, pedagogical competence, ethical awareness, and institutional support. Structural challenges, including uneven infrastructure and limited policy guidance, further constrain progress.

BACKGROUND

The unprecedented global advancement of artificial intelligence (AI) is reshaping pedagogies and andragogies across education systems worldwide, creating new possibilities for personalised instruction, automated feedback, and data-informed decision-making in classrooms. Studies have shown that AI tools can support differentiated learning, enhance assessment practices, and reduce teachers' administrative workload, thereby allowing educators to focus more on pedagogy and learner engagement (Tan et al., 2025; Organisation for Economic Co-operation and Development (OECD), 2021). At the same time, international policy frameworks emphasise that effective integration of AI in education requires not only access to technology but also the development of appropriate competencies, ethical awareness, and pedagogical readiness among educators (UNESCO, 2026). This chapter argues that while AI holds significant potential to enhance climate change and sustainability education, its impact in low-resource contexts is constrained by uneven educator readiness, limited institutional support, and gaps in climate pedagogy. Addressing these challenges is essential for ensuring that AI contributes meaningfully to both educational transformation and climate action globally. Specifically, this study is guided by two key research questions, to what extent are teachers, teacher educators, and student teachers prepared to integrate AI into climate change and sustainability education?, and what institutional, pedagogical, and capacity-related factors shape this readiness in low-resource contexts? This chapter contributes to global debates by providing empirical evidence and a multidimensional framework for understanding educator readiness at the intersection of AI and climate education, particularly from an underrepresented Sub-Saharan African context.

In parallel, climate change and sustainability education has emerged as a global priority, with education systems expected to equip learners with the knowledge, skills, values, and behaviours necessary to respond to environmental challenges. Research in sustainability education highlights that behavioural change, critical thinking, and interdisciplinary learning are essential outcomes of climate education, yet achieving these outcomes depends heavily on teachers' capacity to design meaningful and contextually relevant learning experiences (OECD, 2021). Artificial intelligence has been identified as a promising tool for strengthening climate education by supporting inquiry-based learning, modelling environmental systems, and enabling the analysis of complex data (UNESCO, 2026). Despite these opportunities, the literature consistently reports that teachers' readiness to integrate AI remains uneven (Ayanwale et al., 2022). Readiness is widely conceptualised as a multidimensional construct encompassing knowledge, skills, attitudes, ethical understanding, and institutional support (Tan et al., 2025). However, in many contexts particularly low-resource settings this readiness is further shaped by structural constraints, including limited access to professional development, inadequate infrastructure, and the absence of clear institutional or national policies guiding AI integration. Many educators demonstrate positive attitudes toward AI but lack structured training and pedagogical guidance, which limits meaningful classroom integration (Ayanwale et al., 2022). Teacher educators face additional complexity, as they must simultaneously develop their own competencies while preparing future teachers for AI-enhanced learning environments (Tondeur et al., 2017; Trust et al., 2023). Similarly, pre-service teachers often exhibit familiarity with digital tools but require systematic preparation to translate this familiarity into professional and pedagogical readiness (Ghimire et al., 2024). These challenges are further compounded in the domain of climate change education. Evidence from developing-country contexts shows that teachers often lack adequate pedagogical content knowledge in climate education (Muchanga & Nakazwe, 2015; Muchanga & Nkhata, 2023). Consequently, the introduction of AI risks becoming an additional layer of complexity rather than a transformative solution. The integration of AI into climate education therefore requires not only technological competence but also strong grounding in sustainability concepts and pedagogical strategies that promote behavioural change.

Findings from the present study reinforce this concern, indicating that a large proportion of educators have not received recent training in climate change education, while AI adoption remains largely exploratory rather than institutionalised. This suggests that without targeted interventions, the potential of AI to enhance climate education may remain unrealised, particularly in contexts where foundational capacities are still developing. The study therefore examines the level of readiness among teachers, teacher educators, and student teachers to integrate AI into climate change and sustainability education; assesses educators' familiarity with AI tools and their patterns of use; and identifies training needs, capacity gaps, and institutional factors influencing integration. In addition, it explores how AI can support teaching, learning, and behavioural change in sustainability education, with the aim of generating evidence-based recommendations for strengthening educator readiness and promoting responsible AI integration. By focusing on the intersection of artificial intelligence and climate change education, this study addresses a critical but underexplored area of research. While existing literature often treats these domains separately, this study brings them together, offering a more integrated understanding of how technological innovation can support broader social and environmental goals. Furthermore, by conceptualising readiness as a multidimensional construct including pedagogical competence, ethical awareness, institutional support, and professional confidence the study provides a more comprehensive framework for analysing educational transformation. Finally, the study contributes large-scale empirical evidence from a Sub-Saharan African context, a region that remains underrepresented in research on AI in education. The findings provide contextually grounded insights into patterns of awareness, usage, and training needs across different educator groups, thereby enriching global discussions on equitable and sustainable pedagogical innovation.

LITERATURE REVIEW

As AI tools become more accessible, attention has increasingly shifted toward understanding teachers', teacher educators', and student teachers' prior familiarity with AI and their readiness to integrate such technologies into pedagogical practice. In educational research, readiness is widely conceptualized as a multidimensional construct that includes knowledge, skills, attitudes, ethical awareness, and the institutional conditions necessary for effective implementation (Tan et al., 2025; UNESCO, 2026). While familiarity with AI tools has grown substantially, research consistently indicates that professional readiness remains uneven and context-dependent. Prior familiarity extends beyond awareness or occasional use to include conceptual understanding of how AI systems function, their limitations, and their implications for teaching, learning, and assessment, whereas readiness encompasses confidence, pedagogical integration skills, and enabling institutional conditions. Although closely related, these constructs remain analytically distinct. Recent studies on generative AI show that high levels of surface familiarity particularly with tools such as ChatGPT often coexist with limited conceptual understanding, widening the gap between perceived and actual readiness (Kasneci et al., 2023; Ali et al., 2024). Comparative analyses further indicate that readiness is stratified across contexts, with educators in low- and middle-income settings reporting lower preparedness due to infrastructural, policy, and professional learning constraints (OECD, 2021; Holmes et al., 2019).

Several theoretical models provide important analytical lenses for understanding educator readiness. The Technology Acceptance Model (TAM) explains how perceived usefulness and ease of use shape adoption decisions, helping to account for the gap between positive attitudes and limited classroom integration (Davis, 1989), while Diffusion of Innovations theory highlights the role of institutional support, compatibility, and relative advantage in shaping uneven uptake across systems (Rogers, 2003). Complementing these perspectives, the Technological Pedagogical Content Knowledge (TPACK) framework emphasizes that meaningful integration depends on the intersection of technological knowledge, pedagogy, and subject content (Mishra & Koehler, 2006). Synthesizing these frameworks suggests that educator readiness emerges from the interaction between individual beliefs, institutional and social dynamics, and pedagogical competence, thereby addressing a key limitation in existing research, which often treats readiness as a single-dimensional construct rather than a systemic and relational phenomenon.

An important gap in the literature concerns how African countries are integrating AI within education systems, with emerging evidence showing that countries are pursuing diverse pathways shaped by policy priorities, infrastructure, and skills agendas (African Union, 2024). At the continental level, the African Union's Continental AI Strategy positions AI as central to development and emphasizes ethical and inclusive adoption, while national approaches vary considerably. South Africa illustrates a curriculum-linked pathway through the integration of AI into coding, robotics, and digital skills reforms (Department of Communications and Digital Technologies, 2020; 2022), Nigeria reflects a policy-driven approach embedding AI in digital learning systems and teacher preparation (Federal Ministry of Education, 2023; Adejumo et al., 2026), Rwanda demonstrates a competency-based model incorporating AI into curriculum frameworks, higher education standards, and institutional policies (Higher Education Council, 2024a; 2024b; Sebihi et al., 2025), and Egypt represents a phased approach introducing AI incrementally through secondary education and teacher training initiatives (Sharawy, 2023). Collectively, these cases show that AI integration in Africa is localized rather than uniform, but also reveal that many initiatives remain concentrated in policy, pilot programmes, or higher education, with limited evidence of systematic integration across teacher education and classroom practice, thereby highlighting a critical gap between policy ambition and implementation.

A substantial body of research shows that teachers recognize the potential of AI to enhance personalized learning, automate administrative tasks, and improve feedback, yet lack formal training to integrate these technologies effectively (Ayanwale et al., 2022). Readiness is influenced by digital competence, self-efficacy, and perceived usefulness, with stronger skills and confidence associated with higher readiness levels; however, evidence also points to a pattern of partial adoption, where teachers use AI primarily for operational tasks such as lesson planning while remaining hesitant to apply it in assessment due to concerns about accuracy, bias, and ethical implications (Kasneci et al., 2023; Cotton et al., 2023). Research further indicates that sustained pedagogical integration depends less on perceived usefulness and more on pedagogical knowledge and opportunities for practice (Mishra & Koehler, 2006; Ertmer & Ottenbreit-Leftwich, 2013). Institutional factors including access to professional development, supportive leadership, and clear policies play a decisive role, as their absence often results in fragmented and exploratory use, while cross-system comparisons show that coherent policy frameworks are associated with higher levels of teacher confidence and more consistent classroom integration (UNESCO, 2021; OECD, 2021).

Ethical considerations are increasingly central to discussions of AI in education, with key issues including data privacy, algorithmic bias, transparency, and academic integrity. AI systems often rely on large datasets containing sensitive information, raising concerns about surveillance and data protection (Williamson & Eynon, 2020; UNESCO, 2021), while biased training data can produce inequitable outcomes, particularly for underrepresented contexts (Holmes et al., 2019). Generative AI tools introduce additional challenges related to authorship, originality, and the detection of AI-generated work (Cotton et al., 2023). These concerns significantly influence teachers' willingness to adopt AI and often act as stronger barriers than technical limitations, yet they remain insufficiently integrated into models of educator readiness. As a result, embedding ethical awareness within teacher education and professional development is increasingly recognized as essential for responsible and sustainable AI integration (Zhou et al., 2025).

Teacher educators play a pivotal role in shaping future teachers' competencies, yet remain comparatively underrepresented in the literature. Existing studies suggest that while they recognize the importance of AI, many lack sufficient training and institutional support to integrate it effectively into teacher education programmes (Tondeur et al., 2017). Their role extends beyond personal technological competence to include curriculum design, modelling of practice, and ethical supervision, and research shows that they act as "gatekeepers" of innovation, with their beliefs and competencies significantly influencing whether AI is meaningfully embedded across programmes (Uerz et al., 2018). Where institutional support is limited, integration tends to remain superficial and fragmented. Evidence grounded in social learning theory further demonstrates that modelling plays a critical role in professional learning, as student teachers are more likely to develop competence and confidence when teacher educators demonstrate effective and ethical AI use (Bandura, 1986; Instefjord & Munthe, 2017), underscoring the importance of targeted professional development at this level.

Research on student teachers indicates that while they often display positive attitudes and frequent use of AI tools, their professional readiness remains limited. Pre-service teachers commonly use AI for academic tasks but lack structured preparation in pedagogy, ethics, and assessment (Ghimire et al., 2024), and studies on generative AI show that they often lack critical AI literacy, including the ability to evaluate outputs, detect bias, and align AI use with pedagogical goals (Ali et al., 2024; Kasneci et al., 2023). This suggests that exposure to AI does not automatically translate into professional competence, highlighting a gap in teacher education programmes where technological familiarity is not sufficiently developed into pedagogically meaningful practice.

Conceptual and policy-oriented literature emphasizes that readiness requires integrated competencies across technical, pedagogical, and ethical domains. The UNESCO AI Competency Framework highlights a human-centered approach that includes critical thinking, inclusivity, and responsible use of AI (UNESCO, 2026), while comparative analyses show convergence around these domains across frameworks (Holmes et al., 2019; OECD, 2021). Research further demonstrates that sustained, practice-based professional development is more effective than short-term training in building readiness (Desimone & Garet, 2015; Tan et al., 2025), reinforcing the view that readiness is a developmental process requiring ongoing institutional support, collaboration, and opportunities for applied learning.

Despite the growing body of research on AI in education, significant gaps remain. Much of the literature focuses on high-income contexts, with limited empirical attention to African education systems, while existing studies tend to emphasize individual factors such as attitudes and digital skills rather than the institutional and policy conditions shaping readiness. The role of teacher educators also remains insufficiently explored, despite their central influence on future teachers' competencies. Furthermore, although familiarity with AI is increasing, it often reflects surface-level use rather than deep pedagogical readiness, indicating a persistent gap between exposure and effective integration. There is therefore a need for context-specific, multi-level research that examines educator readiness holistically across teachers, teacher educators, and student teachers, particularly within underrepresented regions.

METHODOLOGY

This study adopted a descriptive survey design to generate a comprehensive national baseline of educators' climate change awareness, attitudes, and practices, including the use of artificial intelligence (AI) in teaching and teacher education. The descriptive design was appropriate for systematically capturing current conditions, trends, and patterns across a large and geographically dispersed population. By combining quantitative and qualitative data, the study provides both numerical summaries and contextual insights, thereby enhancing the robustness of the findings (Creswell & Creswell, 2018).

The study involved a total of 2,634 participants, comprising in-service teachers, student teachers, and teacher educators drawn from all ten provinces of Zambia. Particular emphasis was placed on districts identified as priorities under the FACE-NDC Project framework, including Lukulu, Kalomo, Choma, Lusaka, Mumbwa, Chibombo, Kapiri Mposhi, and Nyimba. A cluster-based voluntary sampling strategy was employed, whereby educational institutions such as schools, colleges, and universities served as primary sampling units, and participants within these clusters opted into the study. As a non-probability sampling approach, the sample is not statistically representative, and the findings are therefore not generalizable to the entire population of educators in Zambia, but can be transferable. However, efforts were made to ensure diversity across gender, geographic regions, and professional categories in order to provide a broad and indicative overview of patterns and trends.

Data were collected using structured questionnaires and document review techniques. The structured questionnaire served as the primary instrument and was designed to capture demographic characteristics, educators' use of digital technologies, and their readiness to integrate AI into teaching and learning. Educator readiness was operationalised as a multidimensional construct comprising four key dimensions: technical familiarity (awareness and frequency of AI tool use), pedagogical readiness (ability to integrate AI into teaching, assessment, and learning design), ethical awareness (understanding of issues such as bias, privacy, and academic integrity), and institutional support (access to training, policies, and infrastructural resources). These dimensions were measured through a series of Likert-scale items, categorical questions, and selected open-ended responses that captured both self-reported competence and contextual conditions influencing AI use.

The instrument was developed through a review of existing literature on AI in education and educator readiness, including competency frameworks and prior empirical studies. Items were adapted and contextualised to the Zambian education system and the focus on climate change education. The questionnaire was piloted with a small group of educators and students in Chongwe, Lusaka and Chilanga Districts to assess clarity, relevance, and contextual appropriateness, leading to minor revisions in wording and structure. While formal statistical validation (e.g., factor analysis) was not conducted, steps were taken to enhance content validity through alignment with established constructs in the literature and expert-informed review during the piloting phase.

To enhance efficiency and inclusivity, the questionnaire was digitized and administered using KoboToolbox (KoboTool Collect), an open-source platform that supports both online and offline data collection. This enabled a hybrid approach, allowing data collection in areas with and without internet connectivity, with subsequent synchronization to a central database. The use of KoboToolbox was instrumental in ensuring participation across all ten provinces, particularly in resource-constrained settings, while also improving data accuracy and enabling real-time monitoring of fieldwork progress (Dillman et al., 2014; Groves et al., 2009).

In addition to primary data collection, a document review was conducted to provide contextual depth and align findings with national policy frameworks. Key documents reviewed included the National Curriculum Framework, Zambia's Nationally Determined Contributions (NDC), the National Climate Change Learning Strategy (NCCLS), and relevant scholarly literature. Data collection was conducted over a four-month period from June to September 2025 and followed a structured, phased process, including instrument development and piloting, field data collection supported by 17 trained research assistants, and data cleaning and verification.

Data was analysed using descriptive statistical techniques, including frequencies and percentages, to summarise patterns and establish baseline indicators. Qualitative data derived from open-ended responses and document review were analysed using a thematic analysis approach. This involved an initial phase of open coding to identify recurring concepts, followed by the grouping of codes into broader categories aligned with the study's key dimensions of readiness, and finally the development of overarching themes that captured patterns across participant responses. This iterative coding process allowed for the identification of both expected and emergent insights related to AI use and educator readiness.

Ethical approval for the study was obtained from the University of Zambia Humanities and Social Sciences Research Ethics Committee (UNZA-HSSREC). Participation was voluntary, and informed consent was obtained from all participants. Measures were taken to ensure confidentiality and anonymity, including the exclusion of personally identifiable information from the dataset. All data were securely stored and used solely for research purposes, and research assistants were trained to adhere to ethical standards throughout the data collection process.

RESULTS AND DISCUSSION

Professional Profile, Qualifications, and Baseline Capacity Conditions

The baseline study engaged a total of 2,634 participants drawn from all ten provinces of Zambia, comprising in-service teachers, student teachers, and teacher educators (Table 1). Participation was uneven across provinces, with Lusaka, Southern, and Copperbelt recording the highest numbers, while Muchinga, Luapula, and North-Western were less represented, see Table 2. Distribution of Participants in the FACE-NDC Priority Districts, participants accounted for 57% of the total sample, with Lusaka emerging as the main hub of participation (Table 2). Teachers formed the largest participant group, and the overall gender distribution was relatively balanced, although women were more visible in classroom teaching than in teacher education positions. Because the study relied on a non-probability voluntary sample, these patterns should be interpreted as indicative rather than statistically generalizable to all educators in Zambia. Even so, the demographic distribution is analytically important because it provides the institutional and geographic context within which readiness for AI and climate pedagogy should be understood. Similar survey-based educational studies have emphasized that participant composition matters for interpreting system readiness, especially where infrastructure, access, and institutional support vary across regions (Groves et al., 2009).

Table 1. Distribution of Participants in the Baseline Study

Province	Teachers		Student Teachers		Teacher Educators		Provincial Total
	F	M	F	M	F	M	
Central	66	74	21	16	5	12	**194**
Copperbelt	55	89	93	69	16	10	**332**
Eastern	117	94	37	39	15	17	**319**
Luapula	5	3	8	41	12	14	**83**
Lusaka	250	210	85	93	101	47	**786**
Muchinga	5	5	4	8	0	6	**28**
Northern	10	7	30	23	7	14	**91**
North-Western	1	4	5	13	6	7	**36**
Southern	231	200	10	28	14	19	**502**
Western	122	110	11	17	3	0	**263**

continued on following page

Table 1. Continued

Province	Teachers		Student Teachers		Teacher Educators		Provincial Total
	F	M	F	M	F	M	
Zambia	**862**	**796**	**304**	**347**	**179**	**146**	**2634**
TOTAL PER CATEGORY OF PARTICIPANTS	**1658**		**651**		**325**		

Table 2. Distribution of Participants in the FACE-NDC Priority Districts

Province	District	Teachers		Student Teachers		Teacher Educators		District Total
		Female	Male	Female	Male	Female	Male	
Central	Chibombo	19	27	0	0	0	0	46
	Kapiri Mposhi	6	6	5	2	1	0	20
	Mumbwa	17	10	3	3	0	0	33
	Subtotal	**42**	**43**	**8**	**5**	**1**	**0**	**99**
Eastern	Nyimba	105	75	0	0	0	0	180
	Subtotal	**105**	**75**	**0**	**0**	**0**	**0**	**180**
Lusaka	Lusaka	235	208	59	73	95	45	715
	Subtotal	**235**	**208**	**59**	**73**	**95**	**45**	**715**
Southern	Choma	85	72	2	1	0	1	161
	Kalomo	94	81	0	4	1	1	181
	Subtotal	**179**	**153**	**2**	**5**	**1**	**2**	**342**
Western	Lukulu	82	88	0	0	0	0	170
	Subtotal	**82**	**88**	**0**	**0**	**0**	**0**	**170**
% Gender/category of respondents		**53% (643)**	**47% (567)**	**45% (69)**	**55% (83)**	**67% (97)**	**33% (47)**	
TOTAL PER RESPONDENTS' CATEGORY		**1210**		**152**		**144**		**1506**

The subject specialisation profile further clarifies the institutional context for interpreting AI readiness. Excluding student teachers, the largest proportion of respondents came from ECE and Primary Education (23.60%), followed by English Language and Literature (20.27%) and Social Sciences (18.31%), while Mathematics (14.67%), Science (9.93%), Expressive Arts (5.24%), Computer Science (4.29%), Home Economics (2.62%), and Local Languages (1.06%) were less represented. This matters because climate change education in Zambia is framed as a cross-cutting issue in the 2023 curriculum, meaning that readiness for climate pedagogy and AI integration is not confined to one disciplinary area (Ministry of Education, 2023). A strong representation of language, and social science educators suggests that the study may capture more strongly the forms of AI use associated with content generation, communication, and classroom support than those associated with modelling, simulation, or data analysis. The implication is not that one subject area is inherently more prepared than another, but that readiness is likely to take different forms across disciplines, consistent with broader work on Technological Pedagogical Content Knowledge (Mishra & Koehler, 2006).

The qualification and professional profile of participants reveal a workforce with moderate formal qualifications and limited progression to advanced academic levels. Bachelor's degrees were the dominant qualification for both women and men, while diploma qualifications were also common. Doctoral qualifications were very low for both genders (Fig 1).

Figure 1.

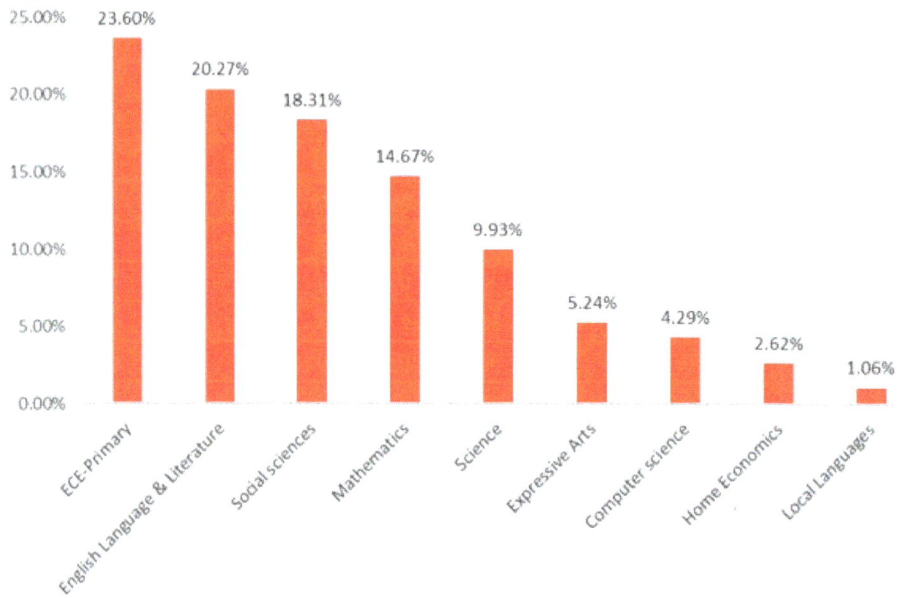

The evidence shows that bachelor's degree qualifications dominated across both genders, with males showing a slightly higher proportion at bachelor's level and females a somewhat stronger concentration at diploma level. However, the more important pattern is the very limited progression to Master's and doctoral qualifications. This finding is broadly consistent with wider Sub-Saharan African patterns, where participation in postgraduate education remains constrained by structural and economic barriers (UNESCO, 2021). It may also reflect the limited professional incentives associated with advanced qualifications in some education systems. Research has shown that when salary progression and promotion are weakly linked to postgraduate attainment, teachers often have reduced motivation to pursue higher degrees (Bennell & Akyeampong, 2007). In the Zambian case, this concern is also reflected in the observation that, advanced qualifications do not necessarily result in differentiated salary or promotion outcomes within the Ministry of Education (Muchanga, 2026). The implication for AI integration should, however, be stated cautiously. The present descriptive data do not demonstrate that lower qualification levels cause lower AI readiness. Rather, they suggest that limited progression to advanced professional learning may constrain the wider knowledge environment needed for sustained innovation. Existing research indicates that effective integration of AI in education often requires more than basic digital familiarity; it also benefits from stronger capacities in curriculum design, critical evaluation, research literacy, and reflective pedagogy (Holmes et al., 2019; UNESCO, 2023). Thus, the qualification profile may be understood as part of the broader institutional conditions shaping readiness rather than as a direct determinant. Teacher union affiliation provides an additional window into the professional landscape. Representation was heavily concentrated in a small number of unions, with BETUZ and SESTUZ together accounting for the largest share (77%) and ZNUT contributing an additional portion (Fig 2).

Figure 2.

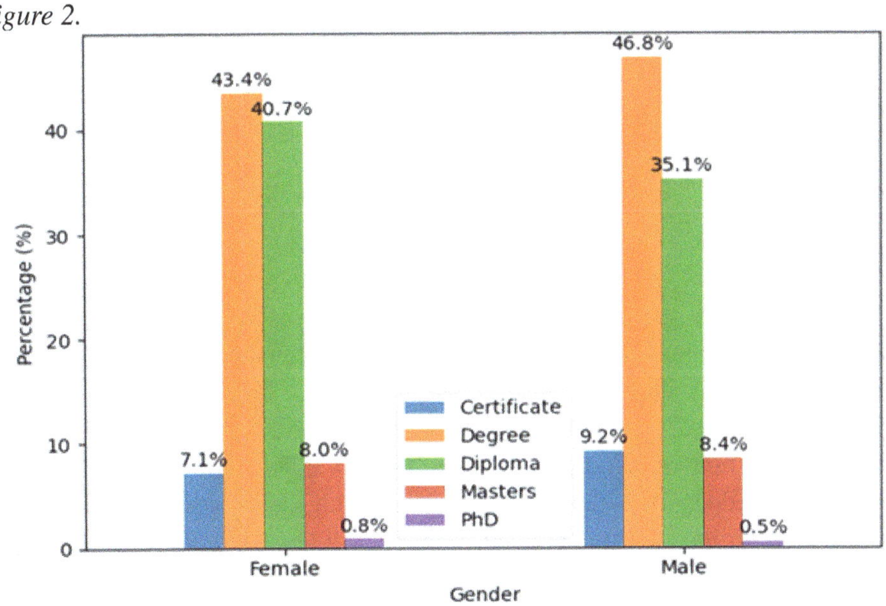

This concentration is noteworthy because comparative evidence suggests that large unions often have stronger institutional legitimacy, broader membership reach, and greater capacity to support professional development and policy engagement (Mccollow, 2017; OECD, 2021). Similar patterns of union concentration have been documented in South Africa, the United States, and the United Kingdom, where a few major organizations shape teacher representation and professional learning agendas. In this study, the data show concentration, but they do not establish that unions are currently driving AI-related training (Stromquist, 2018). The more cautious implication is that the dominant unions may represent useful channels for scaling Continuous Professional Development (CPD) on AI and climate pedagogy, provided such efforts are aligned with ministry priorities and supported by relevant expertise (UNESCO, 2021; OECD, 2021, Habeenzu, 2016). The workforce profile by teaching level and years of experience also reveals conditions relevant to system readiness. Most respondents were concentrated at the Upper Secondary level (37.7%), with lower participation from ECE to Lower Primary and the tertiary level (Fig 3).

Figure 3.

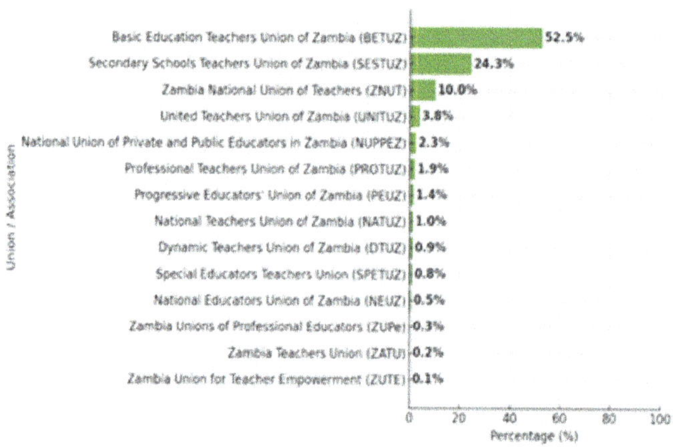

The distribution in Fig 4 suggests that the dataset is weighted more strongly toward secondary education. This has implications because the literature indicates that AI integration is often more visible at secondary level, where structured subject specialization, formal assessment systems, and digital experimentation are more developed than in early childhood settings (OECD, 2021; Zawacki-Richter et al., 2019). Primary education is increasingly engaging with AI literacy and STEM-related applications, but the field remains less mature, while early childhood education continues to show more limited and exploratory uses of AI (Long & Magerko, 2020; Chen et al., 2020; UNESCO, 2021). The implication is that the study may capture more clearly those forms of readiness associated with secondary schooling than those relevant to foundational learning.

Figure 4.

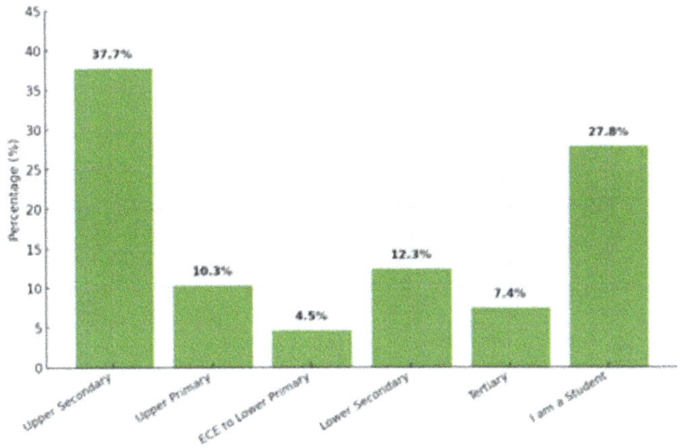

The teaching experience profile shows that approximately 80% of both female and male teachers had fewer than ten years of experience. This suggests that the workforce is heavily concentrated in early- and mid-career stages. International evidence often associates early-career teachers with greater openness to digital innovation and emerging technologies, including AI, partly because of their higher exposure to digital environments (OECD, 2021; Chen et al., 2020). However, professional development research also cautions that openness to technology does not automatically translate into the pedagogical capacity required for meaningful instructional integration, especially in complex domains such as climate change education (UNESCO, 2021; Long & Magerko, 2020). From a TPaCK perspective, this means that the current workforce profile may offer potential for scaling innovation, but that potential depends on strengthening pedagogical content knowledge and context-sensitive application (Mishra & Koehler, 2006). The implication is that the workforce appears promising for AI adoption, but not yet necessarily prepared for deep pedagogical use. One of the clearest findings from the baseline concerns the limited recent exposure to climate change training. Across teachers, student teachers, and teacher educators, 87.5% reported that they had not received climate-change training in the two years prior to 2025 (Fig 5).

Figure 5.

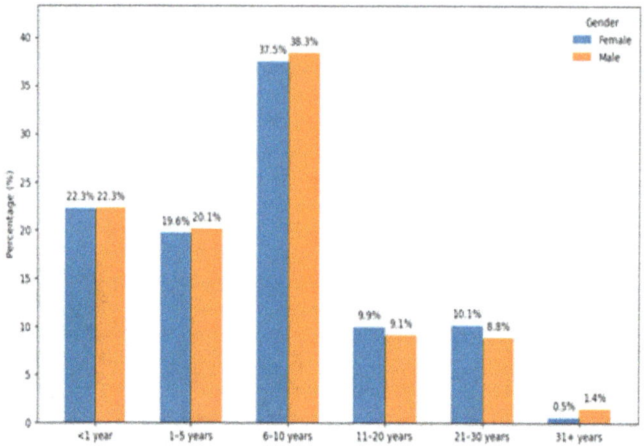

What the data show is a substantial training gap in climate change awareness and pedagogy. This finding is particularly important because the study focuses not merely on AI use in general, but on the integration of AI into climate change and sustainability education. The result suggests that many educators may be trying to engage with AI in a context where the foundational content and pedagogical preparation for climate education remains limited. This interpretation is consistent with prior work showing that teachers in the region have often been inadequately equipped with pedagogical content knowledge in climate change education (Muchanga & Nakazwe, 2015; Muchanga & Nkhata, 2023). It is also consistent with broader literature indicating that AI tools are most educationally useful when teachers already possess strong content knowledge and pedagogical strategies for guiding learning (UNESCO, 2023). The implication should nevertheless be framed carefully. The descriptive data do not prove that lack of climate training causes weak AI integration. They do, however, point to an important coexistence of gaps namely, limited recent climate pedagogy training alongside emerging but still shallow AI use. In practice, this means that AI is unlikely to have transformative value for climate education where foundational teacher preparation in climate content is still weak. Under such conditions, AI may be used mainly for general support tasks rather than to deepen inquiry, systems thinking, or behaviour-oriented climate learning. AI-use patterns reinforce this interpretation. Across all three participant groups, the most common pattern was occasional use rather than regular or institutionalized use (Fig 6).

Figure 6.

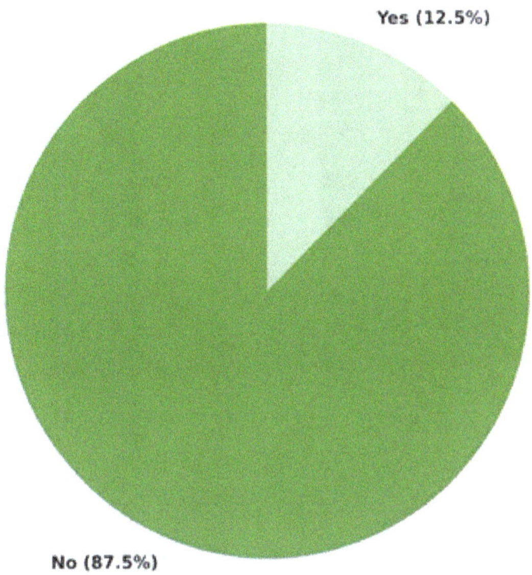

The predominance of occasional use as shown in Fig 7 indicates that educators are experimenting with AI, but that such use has not yet become routine or deeply embedded in curriculum planning, assessment, or instructional design. This pattern closely resembles broader international findings showing that educators often move first through exploratory and low-risk engagement before integrating AI more systematically (Tan et al., 2025; Zhai et al., 2023). In this sense, the present findings point more to emerging familiarity than to consolidated readiness.

Figure 7.

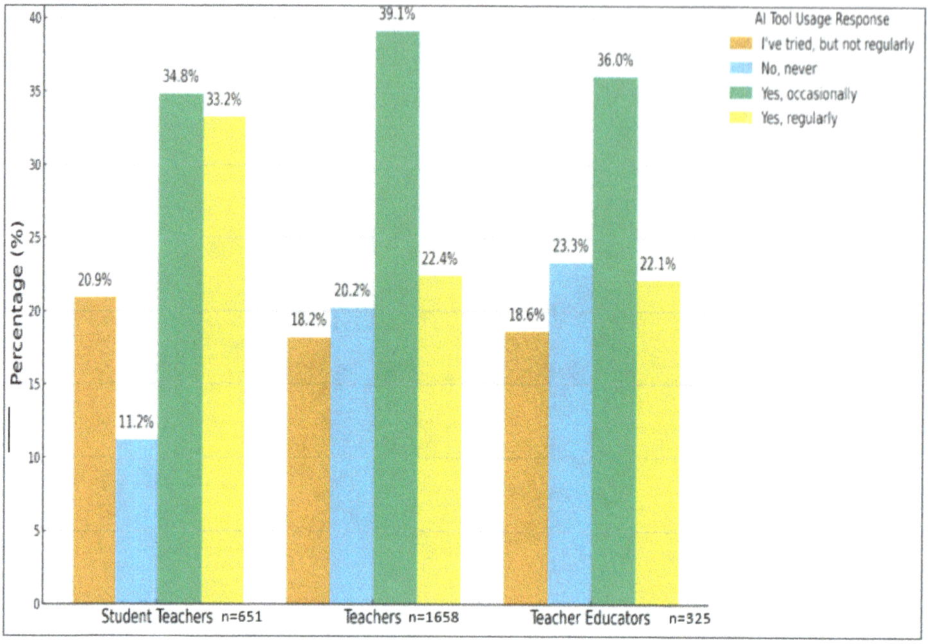

The variation across groups is also instructive. Student teachers reported somewhat higher levels of regular use than teachers and teacher educators. This may plausibly reflect generational and experiential differences, as younger or pre-service educators often demonstrate greater comfort with digital tools and technology-mediated learning environments (Ghimire et al., 2024). However, higher frequency of use should not be taken as evidence of stronger professional readiness. Prior studies show that pre-service teachers often require more structured preparation in ethics, pedagogy, and assessment before AI use becomes educationally meaningful (Ali et al., 2024; UNESCO, 2026). By contrast, the relatively higher levels of non-use among teachers and teacher educators may reflect time constraints, uncertainty about curriculum alignment, concern about reliability, or caution regarding academic integrity and professional responsibility (Ayanwale et al., 2022; OECD, 2021; Cukurova et al., 2023). The implication is that the study identifies a readiness gap rather than a simple adoption problem. Awareness of AI tools appears to be spreading across all three groups, but readiness lags because it requires a deeper combination of pedagogical competence, ethical judgment, institutional support, and subject grounding. This distinction between familiarity and readiness is well established in the literature (Tan et al., 2025; UNESCO, 2026), and the current findings support that distinction in the Zambian context.

AI Tools Used, Preferred Features, and Differentiated Training Needs

All participants reported using at least one artificial intelligence tool regardless of professional category, suggesting a general openness toward the use of AI in teaching and learning processes (see Fig 8). The similarity in usage scores across teachers, teacher educators, and student teachers suggests that familiarity with widely accessible AI tools is becoming broadly distributed rather than confined to a particular group. This pattern is consistent with earlier research showing that awareness and exploratory use of generative AI tools have expanded rapidly across educational levels, with educators and students increasingly engaging in tasks such as drafting instructional materials, summarizing content, and supporting learning activities (Ghimire et al., 2024; Tan et al., 2025). ChatGPT and Google Classroom recorded the highest uptake scores, while tools such as Grammarly and Gemini were used at moderate levels (Fig 8). The prominence of generative AI chatbots aligns with broader evidence indicating that such tools are currently among the most widely adopted educational AI applications because they are accessible, easy to use, and applicable to common instructional tasks (Zhai et al., 2023). Writing-support tools such as Grammarly are similarly adopted because they enhance activities already embedded in academic practice, particularly writing and feedback processes (Phokoye et al., 2025).

The diversity of AI tools used is further illustrated in Figure 8.1, which indicates that, rather than relying on a single system, participants appear to be experimenting with multiple AI applications that serve different functions, including content generation, writing support, information retrieval, and classroom management. Systematic reviews of artificial intelligence in education indicate that such diversity is characteristic of the early stages of adoption as exemplified in the Technological Adoption Model (TAM) (Davis, 1989), when educators explore various tools before identifying those most suitable for sustained pedagogical use (Tan et al., 2025). This pattern suggests that participants are developing functional familiarity with different AI applications, which may facilitate deeper integration in the future. This diversity also indicates a favourable starting point for integrating artificial intelligence into interdisciplinary areas such as climate change education, behaviour change promotion, and climate action initiatives. Earlier studies on digital and AI-supported learning environments suggests that such tools can support inquiry-based learning, data interpretation, and problem-solving related to environmental and sustainability issues (OECD, 2021). When used within structured pedagogical frameworks, AI applications can enhance engagement and enable learners to explore complex global challenges through simulation, analysis, and collaborative learning (UNESCO, 2026). Despite this encouraging pattern, the other studies caution that widespread access to multiple AI tools does not automatically lead to meaningful pedagogical integration. Studies show that educators frequently use AI primarily for supportive or peripheral tasks, such as editing, brainstorming, or generating outlines, rather than for deeper curriculum transformation (Zhai et al., 2023; Tan et al., 2025). So, while the usage patterns demonstrate growing familiarity and openness, sustained and systematic integration will depend on targeted professional development, institutional guidance, and the development of pedagogical competencies for responsible AI use. Meaningful pedagogical transformation requires moving beyond functional use toward designing interactive and student-centred learning experiences. AI can support project-based learning, climate simulations, collaborative inquiry, and real-world problem-solving tasks, thereby enhancing user engagement.

Figure 8.

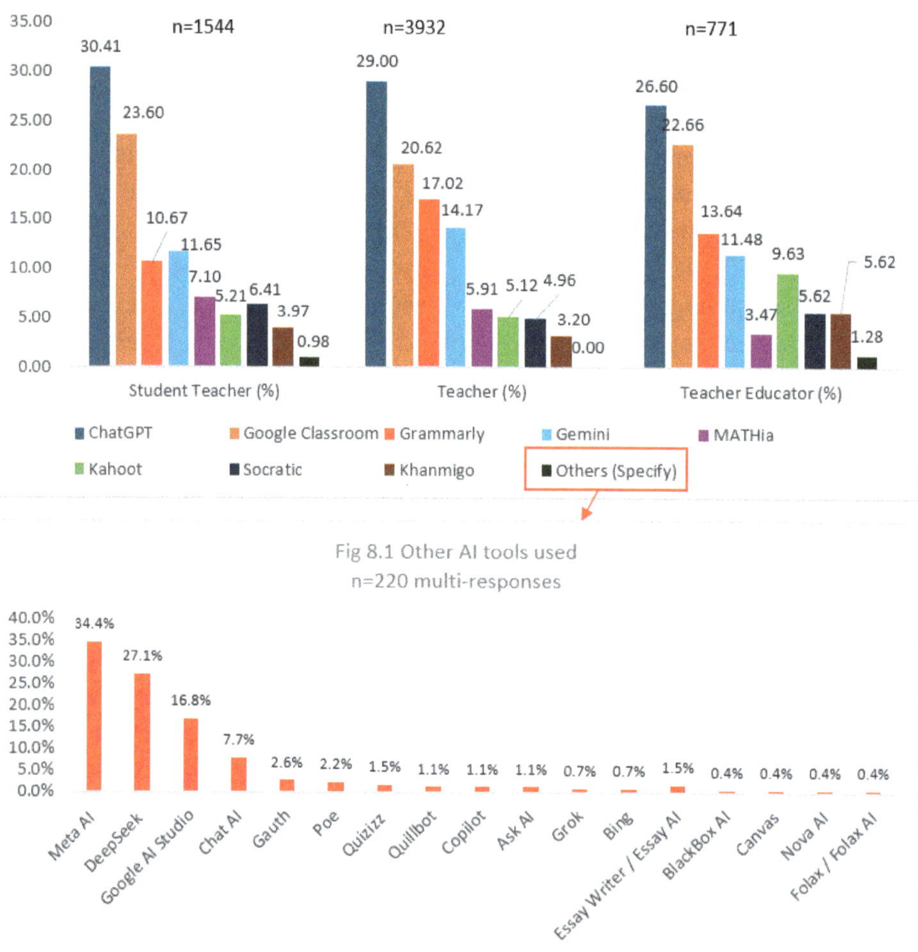

In terms of preferred features that future AI tools for climate change education should contain, the choices were diverse. The highest preferences were AI-generated lesson plans and personalized learning recommendation systems, followed by moderate support for assessment question generation, progress tracking tools, and automated grading systems (Fig 9-10).

Figure 9.

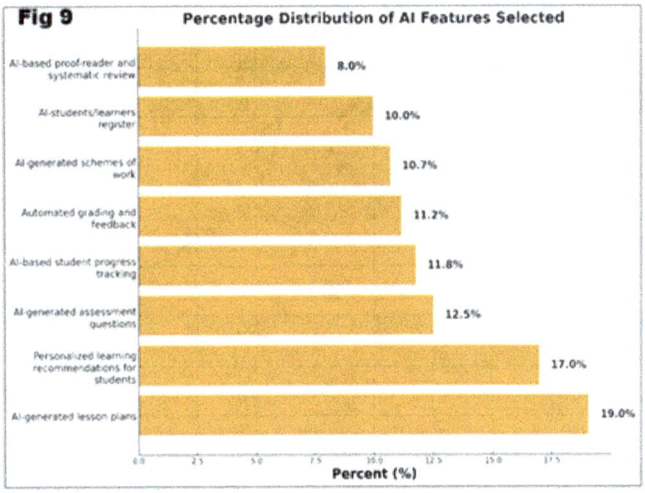

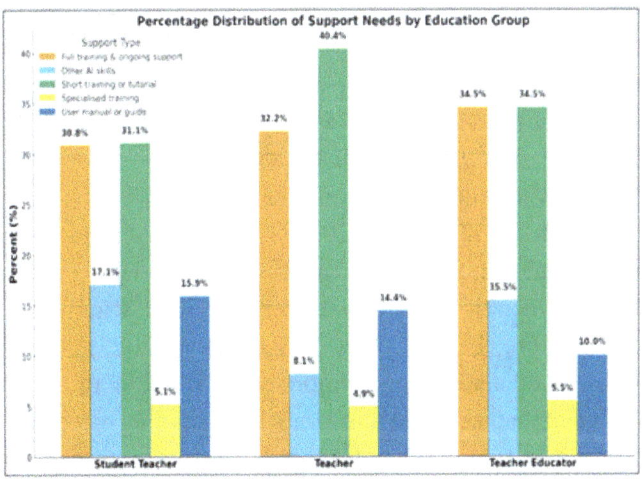

Figure 10.

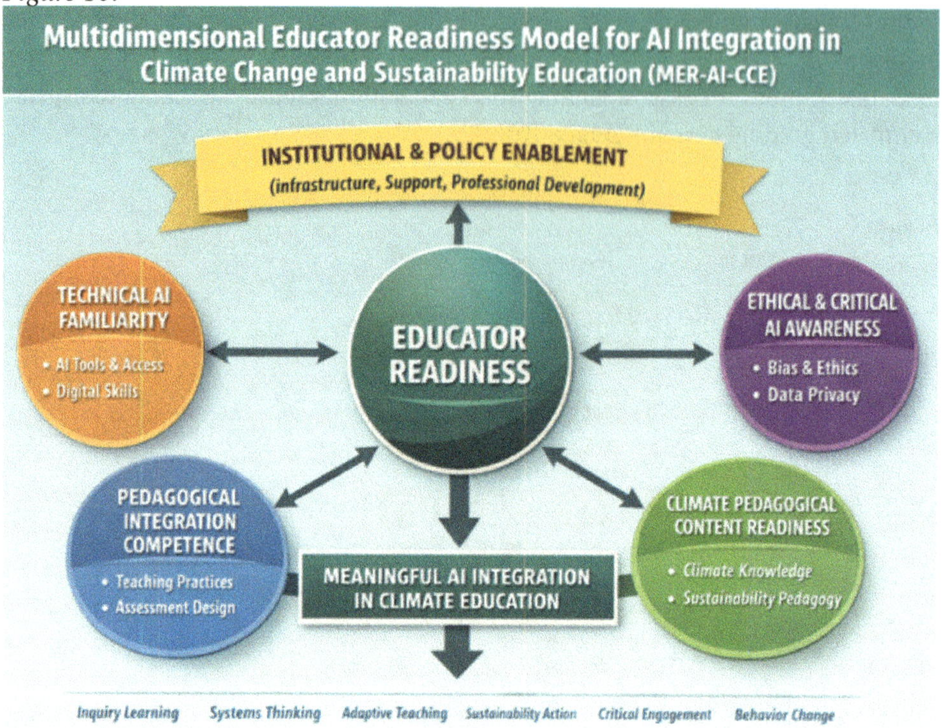

What this shows is that educators prioritize AI functions that directly support teaching preparation (Fig 10). Existing research shows that teachers are especially likely to adopt AI where the immediate pedagogical or workload benefits are visible, particularly in lesson preparation and adaptive support for learners (Tan et al., 2025; OECD, 2021; Zhai et al., 2023). The somewhat lower preference for automated grading and administrative functions may reflect caution about reliability, fairness, or academic integrity, concerns that are widely discussed in the AI-in-education literature (UNESCO, 2026; Zhai et al., 2023). A careful interpretation is therefore that educators appear more comfortable with AI as a support for professional judgment than as a substitute for it. The training-needs findings further strengthen this interpretation. Student teachers showed the strongest preference for short training or tutorials, teachers indicated the strongest need for structured and ongoing training, whereas teacher educators were split between short-full training and or ongoing capacity building. These patterns suggest that professional roles shape the form of support required. Pre-service teachers appear to need orientation and confidence-building, consistent with studies showing that novice educators benefit most from guided, practice-based exposure embedded in coursework (Trust et al., 2023; Zawacki-Richter et al., 2019). Practicing teachers, by contrast, appear to need sustained support because they must align AI with curriculum, classroom management, assessment, and workload demands (OECD, 2021; UNESCO, 2023; Nabi et al., 2025). Teacher educators require a more complex mix of foundational and advanced support because they are expected not only to use AI but also to model it, critique it, and prepare others to use it responsibly (European Commission, 2020; Trust et al., 2023).

The relatively low preference for manuals or guides across groups is also revealing. It suggests that participants view AI competence as something best developed through interactive, practice-oriented learning rather than through passive reference materials. This is consistent with established research on teacher professional development, which shows that active, collaborative, and sustained learning opportunities are more effective than one-off or purely informational interventions (Desimone & Garet, 2015). The implication, therefore, is that capacity-building strategies for AI in education should be differentiated by role, practice-based, and ongoing, rather than generic or documentation-heavy.

Synthesised Implications of findings on Pedagogy, Curriculum Design, and Climate Education

The findings indicate that educators across Zambia could be becoming increasingly aware of AI tools and are beginning to use them, but that use remains mostly exploratory. The strongest current interest lies in applications that support lesson planning, content preparation, personalization, and selected assessment tasks. This suggests that AI is presently being positioned more as a pedagogical support tool than as a transformative force in teaching practice. Such a pattern is consistent with broader international evidence showing that teachers typically adopt emerging technologies first in areas that provide clear efficiency gains and fit existing routines, before moving toward more complex pedagogical redesign (OECD, 2021; Trust et al., 2023; Zhai et al., 2023). The implications for curriculum design are promising but should not be overstated. AI may potentially support Zambia's shift toward competency-based education by enabling more adaptive instruction, interdisciplinary learning, and data-informed feedback. In climate change education specifically, AI tools may help support inquiry-based learning, simulation, problem-solving, and the interpretation of complex environmental information, all of which are relevant to behavioural change and climate action (UNESCO, 2023; Zhai et al., 2023; OECD, 2021). However, the present findings do not demonstrate that such transformation is already taking place. Rather, they indicate that the conditions for it are uneven and still developing. The more evidence-based conclusion is that AI has potential to strengthen climate pedagogy if teacher capacity, ethical awareness, institutional guidance, and curriculum alignment are simultaneously strengthened.

The discussion of ethics also requires careful phrasing. Concerns about data privacy, reliability, bias, and academic integrity remain relevant to AI integration, but such concerns should not be generalized beyond the evidence. For example, it would be too broad to claim that large language models are universally or inherently Eurocentric in all outputs. A more defensible statement is that AI systems may reflect biases present in their training data and design, which can produce uneven or contextually inappropriate outputs, especially where local languages, contexts, and knowledge traditions are underrepresented (Holmes et al., 2019; UNESCO, 2021; Williamson & Eynon, 2020). This more cautious framing better reflects the current state of evidence and avoids overgeneralization. The main implication of the findings is that AI integration in this context should be approached as a capacity-building and system-development issue, not merely as a technology-access issue. Training in AI should be linked to climate pedagogy, curriculum reform, and differentiated professional roles. Without such alignment, AI is likely to remain at the level of occasional experimentation rather than developing into a meaningful support for teaching, learning, and climate action.

What This Study Means Globally

Although this study is situated in Zambia, several findings are likely to have relevance for other low-resource education contexts. First, the gap between high awareness of AI tools and lower pedagogical readiness is unlikely to be unique to Zambia. Similar patterns have been reported in settings where educators have growing access to public AI tools but limited institutional support, policy guidance, or sustained professional development (OECD, 2021; UNESCO, 2026; Tan et al., 2025). The transferable insight here is that access to AI tools may spread more rapidly than the professional conditions required for meaningful educational use. Second, the findings suggest that AI readiness in low-resource contexts is shaped not only by technology availability, but also by the interaction of training systems, curriculum structures, subject specialization, policy clarity, and teacher educator capacity. This is an important transferable lesson because it shifts attention away from narrow technology-centered explanations toward broader institutional and pedagogical conditions. Research from other developing and transitional education systems similarly shows that educational innovation depends on alignment between tools, teacher preparation, and organizational support rather than on access alone (OECD, 2021; European Commission, 2020). Third, the study suggests that the integration of AI into climate change and sustainability education may be especially constrained where educators have had limited recent preparation in climate pedagogy itself. This insight is likely to matter beyond Zambia. In other low-resource contexts, AI may have limited value for climate learning unless teachers are already equipped with sufficient subject knowledge, pedagogical strategies, and ethical awareness to use these tools critically and purposefully (UNESCO, 2023; Muchanga & Nkhata, 2023). The global relevance of the study therefore lies less in making broad universal claims and more in identifying transferable enabling conditions; role-specific professional development, support for teacher educators, curriculum alignment, and integration of AI capacity-building with subject-specific pedagogy.

Limitations and Contribution

Despite the breadth of the dataset and its national reach, several limitations should be acknowledged. First, the study used a non-probability, voluntary, cluster-based sample, meaning that the findings are not statistically generalizable, but quite transferable to all educators in Zambia, . Second, participation was uneven across provinces, with some areas underrepresented because funds could not permit fully balanced coverage. Third, the study relied heavily on self-reported data, which may introduce response bias, especially on questions of prior training and AI use. Fourth, as a baseline descriptive study, it does not establish causal relationships or show how practices change over time. These limitations are important because they define the boundaries of interpretation; the study is most useful for identifying patterns, gaps, and priorities rather than for making causal or nationally generalisable claims.

Notwithstanding these limitations, the novelty of this study lies in its integrated examination of artificial intelligence (AI) and climate change education within a single analytical framework, an area that has been largely treated separately in existing research. Unlike prior studies that focus on isolated aspects of technology adoption, this research conceptualises educator readiness as a multidimensional construct incorporating technical, pedagogical, ethical, and institutional dimensions, thereby offering a more holistic understanding of integration. It further contributes large-scale empirical evidence (n=2,634) from a Sub-Saharan African context, addressing a significant geographical gap in the literature . The study is also novel in its comparative inclusion of teachers, teacher educators, and student teachers, highlighting differentiated readiness and training needs across professional groups. Additionally, it reveals the coexistence of AI adoption with a critical gap in climate pedagogy, an underexplored intersection that shapes meaningful integration. Finally, the study advances the field by positioning AI integration not merely as a technological issue, but as a systemic capacity-building and curriculum alignment challenge, particularly relevant for low-resource education systems. Although this may be contested, it is what we considered to be a contribution to ongoing general debates around AI integration in formal education.

This study further proposes a Multidimensional Educator Readiness Model for AI Integration in Climate Change and Sustainability Education (MERAICCE). The model conceptualises readiness not as a single attribute or technical skill, but as a relational construct emerging from the interaction of five dimensions namely, technical AI familiarity, pedagogical integration competence, ethical and critical AI awareness, climate pedagogical content readiness, and institutional-policy enablement. The model further argues that meaningful integration of AI in climate education depends not only on educators' access to and awareness of AI tools, but also on their ability to use such tools in pedagogically purposeful, ethically responsible, and climate-relevant ways under supportive institutional conditions. In this view, readiness is both developmental and systemic, shaped simultaneously by individual competencies and by the organisational and policy environments in which educators work

CONCLUSION AND RECOMMENDATIONS

This study provides baseline evidence that, although awareness of artificial intelligence (AI) among educators in Zambia is widespread and attitudes toward its potential are generally positive, meaningful pedagogical integration remains limited and largely exploratory. Current uses of AI are predominantly efficiency-oriented, supporting tasks such as lesson preparation and content generation rather than enabling deeper instructional transformation. This pattern reflects a broader readiness gap shaped by the interaction of pedagogical competence, ethical awareness, institutional support, and subject-specific knowledge. From a climate education perspective, the study positions AI not merely as a technological tool, but as a cognitive and pedagogical enhancer with the potential to support systems thinking, inquiry-based learning, and behaviour change. However, this potential remains constrained by a parallel deficit in climate change education itself. Limited recent training in climate pedagogy restricts educators' ability to engage meaningfully with sustainability issues, suggesting that without strengthening pedagogical content knowledge (PCK), particularly in interdisciplinary domains, AI risks reinforcing surface-level learning rather than enabling transformative educational outcomes.

The findings further show that limited AI integration is not incidental, but rooted in broader structural and pedagogical constraints, including gaps in teacher preparation, insufficient opportunities for sustained professional development, and the absence of coherent institutional and policy frameworks. Readiness is uneven across the system, varying by professional role, experience, and access to learning opportunities. Student teachers require structured foundational exposure to AI and its pedagogical applications, practicing teachers need sustained and practice-oriented professional development to support curriculum integration, and teacher educators require advanced competencies to model, guide, and critically engage with AI in teacher education programmes. As such, teacher educators emerge as a critical leverage point for systemic change, given their multiplier effect across pre-service training. These insights indicate that AI integration should not be approached as a standalone technological reform, but as part of a broader process of systemic transformation. Effective integration particularly within climate change and sustainability education requires alignment across curriculum design, teacher education, professional development, and policy frameworks. This includes embedding AI literacy and climate competencies within pre-service education, providing continuous and practice-based professional development for in-service teachers, and strengthening institutional support through clear policies, infrastructure, and leadership. Addressing regional and institutional inequalities is equally essential to ensure that innovation does not reproduce existing disparities.

This study offers a critical global lesson, which show that, access to AI tools is expanding far more rapidly than the pedagogical, institutional, and disciplinary conditions required for meaningful educational use. The Zambian case demonstrates that awareness does not equate to readiness, and that technological adoption without corresponding investment in teacher capacity, curriculum alignment, and subject-specific pedagogy is unlikely to produce transformative learning outcomes. It further shows that in domains such as climate change education, AI integration is doubly dependent, not only on technological competence, but also on the strength of underlying pedagogical content knowledge. In this sense, Zambia reflects a broader pattern observable across many low- and middle-income education systems, where innovation is constrained less by access than by systemic capacity. The transferable insight is that AI in education should be understood primarily as a pedagogical and institutional reform challenge rather than a technological one.

Given the study's descriptive and non-probability design, the findings are not statistically generalisable but offer analytically transferable insights for other low-resource and transitional contexts facing similar challenges. The contribution of the study lies in documenting a critical transitional moment in global education, where AI awareness is accelerating while pedagogical readiness remains uneven and underdeveloped. Ultimately, the integration of AI in education requires deliberate alignment between technological innovation, teacher professional development, curriculum reform, and policy support. When grounded in strong pedagogical foundations and supported by coherent institutional frameworks, AI holds significant potential to enhance interdisciplinary learning, strengthen critical and systems thinking, and support the behavioural changes necessary for sustainable development. Realising this potential, however, will require sustained, context-sensitive investment in educator capacity, positioning teachers and teacher educators not as passive adopters of technology, but as central agents in shaping the educational purposes and societal impact of artificial intelligence.

FUNDING

IKI (International Climate Initiative) under FACENDC WP4 intervention. Project No. FACE-NDC 22_I_492_ZMB_ ACE Climate Education.

CONFLICT OF INTEREST

Authors declare no conflict of interest.

ACKNOWLEDGEMENT

All authors contributed to the chapter at varying levels of effort.

REFERENCES

Adejumo, A. A., Oyelere, S. S., Sanusi, I. T., & Suhonen, J. (2026). A systematic review of the impact of generative AI on learning performance, AI hallucinations, and problem-solving in computer science education. *Computers and Education: Artificial Intelligence, 10*, 100570. DOI: 10.1016/j.caeai.2026.100570

African Union. (2024). *Continental artificial intelligence strategy*. https://au.int/en/documents/20240809/continental-artificial-intelligence-strategy

Ali, D., Fatemi, Y., Boskabadi, E., Nikfar, M., Ugwuoke, J., & Ali, H. (2024). ChatGPT in teaching and learning: A systematic review. *Education Sciences, 14*(6), 643. DOI: 10.3390/educsci14060643

Ayanwale, M. A., Sanusi, I. T., Adelana, O. P., Aruleba, K. D., & Oyelere, S. S. (2022). Teachers' readiness and intention to teach artificial intelligence in schools. *Computers and Education: Artificial Intelligence, 3*, 100099. DOI: 10.1016/j.caeai.2022.100099

Bandura, A. (1986). *Social foundations of thought and action: A social cognitive theory*. Prentice-Hall.

Bennell, P., & Akyeampong, K. (2007). *Teacher motivation in Sub-Saharan Africa and South Asia (Researching the issues No. 71)*. Department for International Development (DFID). https://www.gov.uk/research-for-development-outputs/teacher-motivation-in-sub-saharan-africa-and-south-asia-researching-the-issues-71

Chen, L., Chen, P., & Lin, Z. (2020). Artificial intelligence in education: A review. *IEEE Access : Practical Innovations, Open Solutions, 8*, 75264–75278. DOI: 10.1109/ACCESS.2020.2988510

Cotton, D. R. E., Cotton, P. A., & Shipway, J. R. (2023). Chatting and cheating? Ensuring academic integrity in the era of ChatGPT. *Innovations in Education and Teaching International*. Advance online publication. DOI: 10.1080/14703297.2023.2190148

Creswell, J. W., & Creswell, J. D. (2018). *Research design: Qualitative, quantitative, and mixed methods approaches* (5th ed.). Sage.

Cukurova, M., Miao, F., & Brooker, R. (2023). Adoption of artificial intelligence in schools: Unveiling factors influencing teachers' engagement. *Lecture Notes in Computer Science, 4*, 100131. DOI: 10.1007/978-3-031-36272-9_13

Davis, F. D. (1989). Perceived usefulness, perceived ease of use, and user acceptance of information technology. *Management Information Systems Quarterly, 13*(3), 319–340. DOI: 10.2307/249008

Department of Communications and Digital Technologies. (2020). *National digital and future skills strategy*. Government of South Africa.

Department of Communications and Digital Technologies. (2022). *Implementation programme for the national digital and future skills strategy of South Africa*. Government of South Africa.

Desimone, L. M., & Garet, M. S. (2015). Best practices in teachers' professional development in the United States. *Psychology, Society, &. Education, 7*(3), 252–263.

Dillman, D. A., Smyth, J. D., & Christian, L. M. (2014). *Internet, phone, mail, and mixed-mode surveys: The tailored design method* (4th ed.). Wiley. DOI: 10.1002/9781394260645

Ertmer, P. A., & Ottenbreit-Leftwich, A. T. (2013). Removing obstacles to the pedagogical changes required by authentic technology-enabled learning. *Computers & Education, 64*, 175–182. DOI: 10.1016/j.compedu.2012.10.008

European Commission. (2020). *Digital education action plan (2021–2027): Resetting education and training for the digital age*. https://eur-lex.europa.eu/legal-content/EN/TXT/?uri=celex:52020DC0624

Federal Ministry of Education. (2023). *National digital learning policy*. Government of Nigeria.

Ghimire, A., Prather, J., & Edwards, J. (2024). Generative artificial intelligence in education: Awareness, sentiments, and influencing factors among educators. *arXiv*. https://doi.org//arXiv.2403.15586DOI: 10.48550

Groves, R. M., Fowler, F. J., Couper, M. P., Lepkowski, J. M., Singer, E., & Tourangeau, R. (2009). *Survey methodology* (2nd ed.). Wiley.

Habeenzu, M. (2016). *Factors influencing the formation of splinter unions from the Zambia National Union of Teachers (ZNUT)* (Master's thesis, University of Zambia). https://dspace.unza.zm/bitstreams/f416a018-f158-493b-825d-450a4673a16a/download

Higher Education Council. (2024a). *Guidelines and assessment tools for virtual learning in higher learning institutions*. Government of Rwanda.

Higher Education Council. (2024b). *Benchmarks and guidelines for education-related programmes*. Government of Rwanda.

Holmes, W., Bialik, M., & Fadel, C. (2019). *Artificial intelligence in education: Promises and implications for teaching and learning*. Center for Curriculum Redesign.

Instefjord, E. J., & Munthe, E. (2017). Educating digitally competent teachers: A study of integration of professional digital competence in teacher education. *Teaching and Teacher Education, 67*, 37–45. DOI: 10.1016/j.tate.2017.05.016

Kasneci, E., Sessler, K., Küchemann, S., Bannert, M., Dementieva, D., Fischer, F., Gasser, U., Groh, G., Günnemann, S., Hüllermeier, E., Krusche, S., Kutyniok, G., Michaeli, T., Nerdel, C., Pfeffer, J., Poquet, O., Sailer, M., Schmidt, A., Seidel, T., & Kasneci, G. (2023). ChatGPT for good? On opportunities and challenges of large language models for education. *Learning and Individual Differences, 103*, 102274. DOI: 10.1016/j.lindif.2023.102274

Long, D., & Magerko, B. (2020). What is AI literacy? Competencies and design considerations. In *Proceedings of the 2020 CHI Conference on Human Factors in Computing Systems*. https://doi.org/DOI: 10.1145/3313831.3376727

McCollow, J. (2017). Teacher unions. *Oxford Research Encyclopedia of Education*. https://www.researchgate.net/publication/320046867_Teacher_Unions

Ministry of Education. Zambia. (2023). *Zambia education curriculum framework (ZECF)*. Government of the Republic of Zambia.

Mishra, P., & Koehler, M. J. (2006). Technological pedagogical content knowledge: A framework for teacher knowledge. *Teachers College Record, 108*(6), 1017–1054. DOI: 10.1111/j.1467-9620.2006.00684.x

Muchanga, M. (2026). *Baseline survey of learners' climate behaviours among selected schools to inform the FACENDC intervention in Zambia*. Food and Agriculture Organization (FAO).

Muchanga, M., & Nakazwe, K. M. (2015). Climate change education in Zambia. In Nhamo, G., & Shava, S. (Eds.), *Climate change education in the SADC school curriculum*. Africa Institute of South Africa.

Muchanga, M., & Nkhata, B. (2023). *Preliminary assessment of the status of climate change education and action in the Zambian educational system for the ZEPCA project*. Food and Agriculture Organization (FAO).

Nabi, F., Vortia, W., & Shardey, E. (2025). Teacher readiness for AI and digital tools in K–12 classrooms: A review of professional development trends and gaps. *International Journal for Multidisciplinary Research*.

Organisation for Economic Co-operation and Development (OECD). (2021). *Teachers and school leaders as lifelong learners (TALIS).* OECD Publishing.

Phokoye, S. P., Dlamini, S., Mthalane, P. P., Luthuli, M., & Moyane, S. P. (2025). A comprehensive review of ChatGPT in teaching and learning within higher education. *Informatics (MDPI), 12*(3), 74. DOI: 10.3390/informatics12030074

Rogers, E. M. (2003). *Diffusion of innovations* (5th ed.). Free Press.

Sebihi, A., Schoelen, L., & Uwamwezi, B. (2025). Unveiling AI's impact in Rwandan higher education. *Discover Education, 4*(1), 166. DOI: 10.1007/s44217-025-00576-6

Sharawy, F. S. (2023). *The use of artificial intelligence in higher education: A study on faculty perspectives in universities in Egypt* (Master's thesis, The American University in Cairo).

Stromquist, N. P. (2018). *The global status of teachers and the teaching profession.* Education International. https://www.ei-ie-al.org/sites/default/files/docs/2018_ei_research_statusofteachers_eng_final.pdf

Tan, X., Cheng, G., & Ling, M. H. (2025). Artificial intelligence in teaching and teacher professional development: A systematic review. *Computers and Education: Artificial Intelligence.*

Tondeur, J., van Braak, J., Ertmer, P. A., & Ottenbreit-Leftwich, A. (2017). Understanding the relationship between teachers' pedagogical beliefs and technology use in education. *Educational Technology Research and Development, 65*(3), 555–575. DOI: 10.1007/s11423-016-9481-2

Trust, T., Whalen, J., & Mouza, C. (2023). ChatGPT: Challenges, opportunities, and implications for teacher education. *Contemporary Issues in Technology & Teacher Education, 23*(1), 1–23.

Uerz, D., Volman, M., & Kral, M. (2018). Teacher educators' competences in fostering student teachers' proficiency in teaching and learning with technology. *Teaching and Teacher Education, 70*, 12–23. DOI: 10.1016/j.tate.2017.11.005

United Nations Educational, Scientific and Cultural Organization (UNESCO). (2021). *AI and education: Guidance for policy-makers.* https://unesdoc.unesco.org/ark:/48223/pf0000376709

United Nations Educational, Scientific and Cultural Organization (UNESCO). (2023). *Guidance for generative AI in education and research.* https://unesdoc.unesco.org/ark:/48223/pf0000386693

United Nations Educational, Scientific and Cultural Organization (UNESCO). (2026). *Artificial intelligence competency framework for teachers*. https://www.unesco.org/en/articles/ai-competency-framework-teachers

Williamson, B., & Eynon, R. (2020). Historical threads, missing links, and future directions in AI in education. *Learning, Media and Technology*, *45*(3), 223–235. DOI: 10.1080/17439884.2020.1798995

Zawacki-Richter, O., Marín, V. I., Bond, M., & Gouverneur, F. (2019). Systematic review of research on artificial intelligence applications in higher education. *International Journal of Educational Technology in Higher Education*, *16*(1), 39. DOI: 10.1186/s41239-019-0171-0

Zhai, X., Neumann, K., & Krajcik, J. (2023). AI for tackling STEM education challenges. *Frontiers in Education*, *8*, 1183030. DOI: 10.3389/feduc.2023.1183030

Zhou, T., Tondeur, J., & Howard, S. (2025). AI competency frameworks for teachers: A systematic review. In *Proceedings of EdMedia + Innovate Learning 2025* (pp. 636–642).

Chapter 14
Decoding Climate Complexity: The Big Data Revolution in Climatology

Arun Sampaul Thomas
http://orcid.org/0000-0002-5438-5781
Anurag University, India

S. Muthukaruppasamy
http://orcid.org/0000-0002-7978-4236
Department of EEE, Velammal Institute of Technology, Panchetti, India

S. Sathish Kumar
http://orcid.org/0000-0002-1272-8440
Department of Artificial Intelligence and Machine Learning, J.B. Institute of Engineering and Technology, Hyderabad, India

J. Nandha Gopal
http://orcid.org/0000-0003-3301-484X
Department of EEE, Velammal Institute of Technology, Panchetti, India

K. Saravanan
http://orcid.org/0000-0001-9517-9792
Department of CSE, College of Engineering Guindy, Anna University, Chennai, India

ABSTRACT

Big data analytics allows researchers study satellite images, climate models, and weather station data. Climate division. Big data analytics and climateology need good

DOI: 10.4018/979-8-3373-6975-4.ch014

data. Weather stations, satellites, and climate models collect data. Errors, biases, and arguments may occur. Researchers streamlined data merging security. Researchers use ML to modify climate data. Updated protocols facilitate data sharing and use. Climate modeling is hard with nonlinear dynamics. Analysing and modeling data is tricky. Climate data rarely change linearly. To solve these difficulties, scientists are creating new climate data management and interpretation technologies. Climate data is simplified by T-SNE. Nonlinear, non-stationary machine learning is studied. Climate scientists use big data analysis to identify complicated patterns and trends in vast databases. Big data analytics may benefit climate researchers despite its downsides. Big data analytics and new innovation can help scientists fight climate change. Future sustainability would improve.

INTRODUCTION

Ways to Use Big Data Analytics Methods to Analyze Climate Data

Climatologists may now use big data analytics to sort through huge amounts of data from weather stations, satellite photos, and climate models. Technology is now an important instrument for solving environmental concerns because climate change is still continuing strong. Data science is one of the best tools we have for fighting environmental problems. Data science uses big data, AI, and predictive analytics to figure out how to stop climate-related calamities from happening and lessen their effects when they do. Climate tech is based on data science and focuses on developing early warning systems and ways to make the most of renewable energy.

Here are some of the most essential big data analytics methods that climatologists are adopting.

Outline of Machine Learning and Deep Learning Techniques

Researchers studying the environment are increasingly using machine learning and deep learning to analyze large databases for patterns and trends. Using these methodologies could lead to better climate models, more accurate predictions of extreme weather, and the ability to find areas where climate change is happening the most.

We utilized deep learning and machine learning. Statistical modeling, big data analytics, and machine learning are examples of data science methods that can help us better understand complicated weather patterns. Researchers, businesses, and governments may use the beneficial information that comes from data analysis to make smart choices about how to lower the risks of climate change.

a) Neural networks: Neural networks are a variety of machine learning algorithm that can predict things that constantly happen, like the weather or the temperature. With their help, we may now be able to better predict hurricanes, droughts, and other extreme weather events.
b) Decision trees: Decision trees are a category of machine learning that may be used to sort data into the right categories. These technologies have been used to assess climate change's impact on ecosystems and at-risk areas.

Convolutional neural networks (CNNs) are deep learning algorithms that can deal with satellite photos and other kinds of visual data. They have been used to study deforestation, ocean currents, and predictions of future sea levels. (Soni et al., 2025)

There are several ways to study climate change, such as

a) Being aware of the situation:

Data science predictive analytics makes it feasible to predict hurricanes, floods, and wildfires. Machine learning models find outliers in data and utilize it to produce accurate predictions about future disasters. They achieve these goals by utilizing real-time data from sensors and climate records. The models are useful tools for governments and businesses to get ready for and deal with problems caused by climate change. (Ikegwu et al., 2024)

b) Getting weather data and showing it

Data science tools for measuring temperature help make climate models more accurate and reliable. Scientists can better predict how climate change will affect the world in the long term by using simulated climate scenarios created by very complicated algorithms that look at huge databases. Researchers can use simulation technologies to make better decisions about policies, how to manage resources, and how to prepare for disasters.

c) Getting the most out of eco-friendly power generation

As part of the global effort to fight climate change, people are moving toward more sustainable ways of living, such as using solar and wind power. Data scientists' top goal is to make it easier to generate and send energy. Predictive maintenance models may notify operators when equipment is likely to fail, which lets them keep the power on all the time. AI analytics makes the electrical system more stable, which means that outages caused by weather-related problems happen less often and are less severe. (Badawy et al., 2024)

d) Smart disaster response and smart use of resources

When responding to disasters caused by climate change, emergency responders can better use their resources by merging AI with data science models. First responders can utilize predictive analytics to find weak spots and focus their efforts on evacuating people and distributing resources there. Police can use geographical data to find places that need help the most and plan their response better.

e) Monitoring and reducing carbon emissions is crucial.

Using data science, teams can keep track of carbon footprints and find patterns that show how climate change can affect things in the future. AI systems process data on things like industrial pollution, car pollution, and forest loss rates. This knowledge is then used for making decisions and running businesses.

APPLICATIONS OF NATURAL LANGUAGE PROCESSING IN CLIMATE RESEARCH

Natural language processing (NLP) is an AI field that explores how humans and computers communicate. Natural language processing is sorting through massive amounts of news stories, reports, and social media posts for climate experts. Natural Language Processing (NLP) helps climate change researchers uncover valuable information in massive material. Since there are so many climate-related scientific papers, reports, and articles online, NLP is becoming more popular.

Natural language processing is used a lot in climate change research for sentiment analysis. By looking at writings that talk about climate change, natural language processing (NLP) can help us understand how people feel, think, and act about it. This information can help politicians understand what people think and come up with better ways to communicate with them. Natural language processing is also useful for the process of extracting information. Natural language processing technologies can automatically get important climate-related information from texts, such as temperature tendencies, greenhouse gas radiations, and outcomes on biodiversity. Researchers can save a lot of time by automating this process with natural language processing methods. They can then use that time to look at the data they got back and draw conclusions. (Do et al., 2024)

NLP is especially helpful for looking at the effects of climate change. By looking at climate-related information, natural language processing (NLP) can sort the different effects of climate change, such as accumulative sea levels, habitat forfeiture, and intense weather. This information lets us figure out which areas or sectors are most at risk from climate change and how to best deal with it. Using NLP approaches can also help with climate change communication and education projects. Natural language processing tools can find relevant concepts, words, and links in writing about climate change. This information can help raise awareness about climate change, improve instructional materials, and help people understand how complicated climate science is.

Natural language processing (NLP) has been used in climate research in several different ways:

a) Sentiment analysis: This NLP method can be used to figure out the tone of text data, like news articles or postings on social media. It has been used to keep an eye on public opinion about climate change and areas of concern.
b) Topic modeling: This NLP method can be used on huge amounts of text data to find common themes or topics that come up again and again. It has been used to find the most important areas of research in climate science and to track how climate-related research has changed over time.

Using NLP to Study Climate Change Methods and Tools for Analyzing Data

Natural language processing also gives us sentiment analysis, which is a way to look at data about climate change. Sentiment analysis is the process of figuring out the emotional tone or sentiment behind written documents. Researchers can use sentiment analysis to learn what people think, feel, and believe about climate change by looking at texts that talk about it. This process can help groups and policymakers figure out what people think and change their messages accordingly. Named entity recognition (NER) and other natural language processing techniques or methodologies may also be used to find and remove particular references from climate change data. Anything can be considered an entity, including places, organizations, dates, and more. By automatically finding and classifying these kinds of things, researchers may learn more about how they are connected to climate change, what patterns are happening, and what events are happening.

Many natural language processing frameworks and packages can be used to analyze climate change data. The popular Natural Language Toolkit (NLTK) offers several NLP algorithms and tools. The Python-based NLTK prepares and analyzes climate change data by tokenizing, stemming, and tagging sections of speech. Gensim, spaCy, and Stanford CoreNLP are among popular climate change NLP tools. Dependency parsing, named entity recognition, and subject modeling make these tools useful for analyzing climate change texts.

The Influence of Sentiment Analysis in Climate Change Analogy

Sentiment analysis helps us figure out how people feel and what they think about climate change when we talk about it. Sentiment analysis looks at social media posts, online articles, and other written material to figure out how people feel about climate change, concern, anticipation, skepticism, or triviality.

We can see how people feeling on climate change vary over time with sentiment analysis. This can help lawmakers, scientists, and communicators keep track of how much people know about and care about climate change. Businesses can better customize their messages and campaigns if they understand how people feel. Sentiment analysis can help find people who are affecting climate change. By looking at how people feel about things on social media, we can find out who has a lot of power. We can use this information to work with these influencers in a smart way and build climate action alliances. (Amato et al., 2020)

Sentiment analysis can also be used to evaluate efforts in climate change communication. By looking into sentiment, communication specialists can tell if their campaigns or messages are getting to the right people. They might adjust their plan based on facts to have the biggest effect. Sentiment analysis can do more than just look at how people feel; it can also find false information and dispel myths about climate change. We can swiftly fix false stories and misunderstandings in online chats by giving them precise information through our sentiment analysis. This helps people trust each other and talk about climate change. (Andersson et al., 2020)

Climate Change Discourse: Analyzing Emotions That Change the Game

Emotion analysis has changed the way people talk about and deal with climate change. This unique research examines the emotions of individuals engaged in discussions, arguments, and debates about climate change. Emotion analysis helps us understand the psychological and social sides of the climate change issue by looking at how people, policymakers, and other important people feel.

NLP algorithms are used in emotion analysis to sort emotions in text into groups like anger, fear, hope, and sadness. These algorithms look at attitude, terminology, and stance to figure out just how emotionally charged climate change discussions are. This study gives us more information about how people feel about things, how their moods fluctuate, and how climate change is talked about. There are a number of effects that using emotion analysis in climate change discourse has. First, it helps politicians and climate change activists adapt the way they talk and what they say to better connect with people's feelings and worries. By focusing on common feelings like anxiety and dissatisfaction, stakeholders may get more people involved and take good action.

In discussions on climate change, emotion analysis may also assist find and understand the emotional biases and divisions that people show. Policymakers and researchers may find it advantageous to analyze emotional trends and patterns when addressing socioeconomic disparities and emotional barriers that hinder consensus on climate policy. Analyzing emotions in climate change talks can assist in judging climate-related campaigns, initiatives, and policies, as well as modify how people talk to each other. Stakeholders can adapt their plans based on how people feel about climate change initiatives by keeping track of how people's moods and emotions evolve over time. It is important to know what emotion analysis can't do in this situation. It's hard to put words to complex and subjective feelings. Differences in language and culture can change how emotions are expressed and understood; thus, analysis and adaptation must take these into account. (Avand & Moradi., 2021)

Climate Change Topic Modeling: Main Ideas and Patterns

The fundamental goal of climate change topic modeling is to uncover big themes and patterns in the huge amount of data. This approach employs machine learning to look at large text files about climate change, such as scientific articles, research reports, and social media posts. Many people use Latent Dirichlet Allocation to model subjects. LDA thinks that every text has more than one subject and that each subject has its word distribution. LDA can help us find hidden topics and word distributions automatically.

Before analysis, preprocessing the text data gets rid of stop words, punctuation, and information that isn't needed. Next, we change the text into a BoW (Bag-of-Words) or TF-IDF matrix to find out how often each word is used or how important it is. You use LDA on the prepared data and tell it how many subjects you want. The system allocates words to subjects and documents to topic mixes over and over again until it finds the best match. We can look at the most popular terms and probabilities for each topic when the algorithm converges. (Avand et al., 2021)

The highlighted topics can help us comprehend the themes and patterns in climate change discourse. We might learn about how many greenhouse gases are released into the air, renewable energy, how to adapt, or international climate agreements. You may also see trends in how these themes are spread out throughout time.

NLP and Climate Change Policy: Perceptions and Effects

NLP and climate change policy discoveries and impacts abound. You can analyze enormous amounts of climate change and policy text data with NLP. Policymakers can learn how climate change worries people using NLP. The effectiveness of climate change policies can be assessed using NLP. NLP algorithms can identify patterns, gaps, and trends in climate change policy across social media, news, and scientific papers. We may use this data to evaluate present policies and inform future ones.

NLP is also useful for keeping an eye on climate change policy. By looking at text data from government articles, official intelligences, and other sources. NLP algorithms can uncover policy aims that don't match the results. These findings can assist politicians in figuring out where to intervene and how to make policies work better. NLP gives us a lot of information on climate change policy, but it also has big effects. Ethics and privacy of data are important issues. NLP algorithms work with huge amounts of text data, which may include private information. To protect people's rights, the government must make sure that data is analyzed, gathereed, and stored in accordance with the law and privacy standards. (Greca et al., 2023)

Another effect is that people from different fields work together. NLP alone can't fix all the difficult problems with climate change policy. To turn NLP findings into useful policy actions, climate scientists, policy specialists, and NLP researchers need to work together.

The Future of NLP in Climate Change Analysis: Prospects and Challenges

Natural Language Processing (NLP) could improve climate change research by giving researchers and policymakers fresh information. There are both opportunities and problems with NLP's full potential in this field. One of the most important things that NLP can do is look at a lot of text data about climate change, like case studies, research works, and social media posts along with news updates. Using advanced techniques like text mining and sentiment analysis, NLP systems may detect patterns in this data and make inferences. This helps professionals learn more about climate change.

NLP can also help find bogus news and false information on climate change. It's hard for people to distinguish what's true and what's not because of social media, which spreads false information quickly. NLP algorithms can find false information in online messages and help stop it from spreading. NLP can also help scientists and others who make laws work together. Using NLP algorithms to look over and summarize scientific material helps researchers stay up to date on the latest discoveries. By turning scientific jargon into plain language, NLP can also assist policymakers and the public in understanding and acting on information about climate change. (Guo et al., 2020)

But there are a lot of problems that need to be solved before NLP can be fully used in climate change research. It's a problem when datasets aren't labeled or standardized. NLP models need a lot of labeled data to learn from, yet climate change datasets are hard to find or don't have enough data. To make reliable NLP models, you need climate change datasets that have been well labeled.

Another problem is that the vocabulary of climate change is complicated and particular to the field. Current NLP models may not be able to accurately capture the complex scientific ideas and words related to climate change. To perform a competent job of analyzing something, you need specialist models and understanding of the field. NLP algorithms also need to deal with bias and fairness. When NLP models don't take into account biases in training data, they can give biased or unfair analytical conclusions. We need to come up with and employ evaluation methods and ethical standards to make sure that the analysis is fair and unbiased.

Using NLP to Study Climate Change With Examples

A great number of fascinating case studies have been conducted to demonstrate how Natural Language Processing (NLP) is being utilized in the battle against climate change. To illustrate, I will provide a few examples:

Natural Language Processing (NLP) is making a lot of intriguing progress in the fight against climate change. This section examines people's feelings about climate change. Researchers utilized NLP algorithms to figure out how people on social media felt about climate change. By looking at a lot of Twitter text data, they can figure out how people feel about climate change. This knowledge can help governments and businesses adapt their plans to deal with climate change and how they talk about it. Here are a few examples:

Assessment and Prediction of Climate Risk

NLP is looking at many scientific papers, weather reports, and historical climate data to figure out how climate change might affect the world. Researchers can better comprehend the effects of climate change and come up with ways to lessen them by looking at patterns and predicting what will happen in the future. This helps governments and businesses make better use of their resources and slow down climate change.

Talking About Climate Change

NLP algorithms make it easier to talk about climate change. By looking at text data, these computers uncover important themes, emotions, and ways of framing that work for different groups of people. By using NLP insights to improve how they talk to people, organizations may get people involved in and teach them about climate change action.

Analyzing Environmental Data

NLP looks at unstructured environmental data such as climate reports, research papers, and satellite photographs. Researchers can learn about how climate change works by getting essential data and patterns from a variety of sources. This helps make climate change and environmental policies that are founded on facts. (Hamlet et al., 2020)

USE OF CLUSTERING WITH UNSTRUCTURED CLIMATE DATA

Researchers use clustering and dimensionality reduction to look for patterns and trends in large climate data sets. K-means clustering along with further clustering algorithms can group similar data points, whereas PCA and other dimensionality reduction approaches can cut down on the number of variables in a dataset.

Researchers in climate science utilize methods like clustering and dimensionality reduction to:

a) K-means clustering: This method puts similar data points into groupings. Researchers have utilized this method to identify regions with similar climates and monitor climate change. (Ikegwu et al., 2022)
b) PCA: PCA cuts down on the number of variables in a dataset. It has been used to find important factors that affect climate change and make complicated climate models easier to understand.

The Figure 1 explains how climate research uses big data analytics:

Figure 1. Data Science based Pipeline for Climate Change Mitigation and Adaptation

The process involves PCA and grouping. PCA takes datasets with many dimensions and makes them smaller, uncorrelated dimensions while keeping as much variety (like information) as possible. Cluster analysis puts related observations into groups. Classification or clustering is the process of putting data points into groups that have more in common than not.

Clustering Case Study

We utilize a cluster-based evaluation method to find and keep track of countries' policies to fight climate change. Clustering simplifies the management of policy mixes that include numerous policies from various sectors, types, and levels of strictness. As this article indicates, the strictness of climate laws is a key part of governments' efforts to reduce their impact on the environment. Nascimento and Höhne grouped countries' climate policies based on "policy density," which is the number of climate policies a country has established. They used Eskander and Fankhauser as their source. A better way to score policy instruments and stringency is to use a scale from zero (least strict) to ten (most rigorous). This approach gives a clearer picture of policy packages and how they affect emissions reductions than just looking at the number of active policy instruments in the CAPMF database. D'Orazio42 looks at how different groups have put climate-related finance policies into action and spread the word about them. (Ikegwu et al., 2023)

We show that between 2000 and 2020, all countries added more climate policy measures and made them stricter. Different countries used different policy tools and moved at different rates to implement mitigation policy. These developments led to different methods for mitigation policy. We find four types of ways that countries are trying to slow down climate change and look at how they have changed over time. The most ambitious projects for mitigating climate change today, most of which are from Western Europe, involve a number of tight policy tools. Examples include carbon pricing, technical support subsidies, non-market-focused policies, as well as significant international pledges and climate organizations. This kind of mitigation method works for buildings, power, industry, and transportation, as long as the policies are equally strict. All countries can tighten their regulations, but those with more ambitious mitigation measures can tighten them even further. Other groups have fewer procedures to reduce damage, and the ones they do have are not always strict across all industries.

Second, our results suggest that policy packages that are more thorough can cut emissions more effectively. This discovery helps design policies that work to reduce damage. We show how the four mitigation choices revealed by cluster analysis are related to the emission reductions of a group of countries during the past 20 years. This study shows how important it is for policies to work together. Our panel regression results show that emissions make a country's mitigation policy less strict. This correlation is stronger for mitigation programs that use more tools. This means that countries with broader mitigation policies that cut emissions have more policy synergies.

Figure 2. Mitigation Strategies Across Different Policy Categories

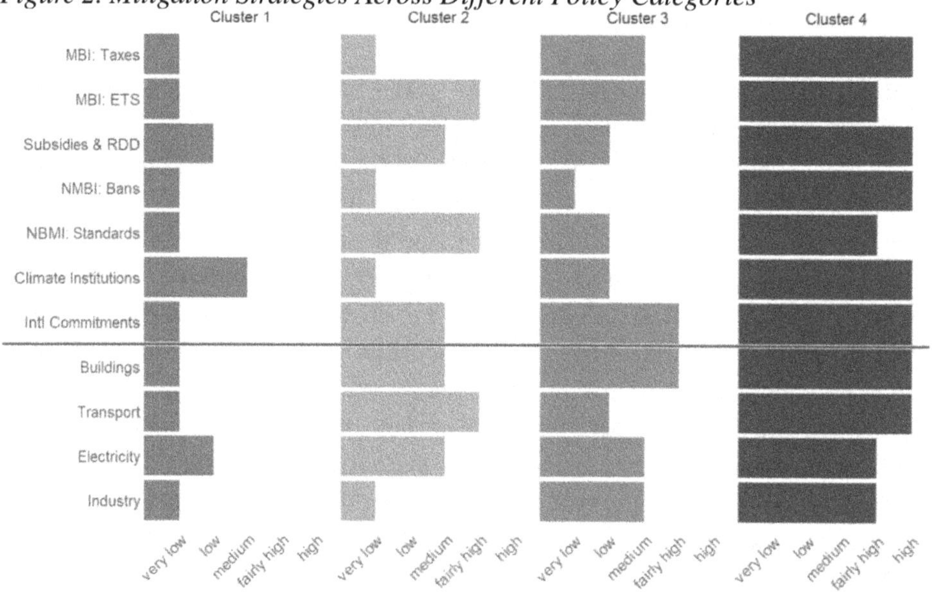

In seven policy areas and industries, Figure 2 demonstrates Cluster 1–4 mitigation actions. We look into mitigation approaches in a lot of detail, both quantitatively and qualitatively, to uncover similarities and differences between clusters. This information helps us figure out the stringency score. Policies that are similarly stringent in all countries in a cluster (low within-cluster variance) but very diverse in stringency between clusters (high between-cluster variance) make the cluster more stable. The horizontal axis illustrates how tight the mitigation strategy is based on our qualitative and quantitative policy analysis.

The policy range and stringency of Cluster 1 mitigation solutions are lower than those of other clusters. They don't give out many subsidies, notably for energy efficiency, and they don't have many market-based policies. Cluster 1 approaches support standards such as minimum performance criteria, labeling regulations, bans, and phase-outs; however, these standards are not as strict as those in other clusters. They also have ways to get money, including preferred loans, to help make energy use more efficient. This group's initiatives have made their climate institutions stronger by clearly defining NDCs, net-zero targets by sector, and international actions. However, most of them still don't record or account for GHG emissions. (Lozo & Onishchenko, 2021)

Cluster 2 mitigation approaches focus on policy groups. For example, there are incentives for energy efficiency and high feed-in tariffs. We have rules about how to utilize energy efficiently and MEPS, but we don't often stop fossil fuel production. Carbon and fuel don't have to pay much or any excise tax. Cluster 2 countries have strict emissions trading programs (ETSs) in areas that aren't controlled, such as transportation and building. Adoption is still unique, though. The fact that countries don't give clear contributions and set net-zero targets shows that climate governance in this group is bad. Cluster 2 initiatives can help with climate governance, carbon prices, and limits and phase-outs.

The most talked-about ways in Cluster 3 to reduce pollution are fuel excise taxes and pollution trading systems. Other ways include getting rid of internal combustion engine vehicles and requiring energy labeling for appliances. Cluster 3 solutions cut down on emissions and MEPS with fewer subsidies. They report GHGs and sign international treaties, even if they don't do a good job of managing the climate. Cluster 3 damage reduction strategies use more resources than Cluster 1, but they don't have the same well-rounded strategy as Cluster 4. Cluster 2 and Cluster 3 have different rules, yet they are both strict.

The ways to lessen the effects of Cluster 4 are more rigid and thorough. This group mostly uses market-based policies like carbon taxes, pollution trading programs, and taxes on gasoline. We also have robust climate governance, rules, limits, phase-outs, and subsidies. These programs emphasize innovative solutions and aim to eliminate coal power plants and vehicles that use internal combustion engines. All EU countries use either Cluster 3 or 4.

CHALLENGES AND LIMITATIONS OF BIG DATA ANALYTICS IN CLIMATOLOGY

Big data analytics has the potential to change climatology, but it has many problems to face. Some of the main problems and limitations are: (Number Analytics, 2025)

Problems With the Quality and Availability of Data

Data availability and quality are crucial in climatology and big data analytics. Weather stations, satellite images, and climate models are used to look at climate changes; however, these methods might be biased, wrong, or leave out important information. To solve these problems, researchers are coming up with new ways to combine data and verify its quality. Some scientists are working on better ways to share and combine data, while others are creating new data standards and methods to detect and rectify mistakes in climate data.

Discusses the Problems Associated With Climate Data and Systems

The complexity and non-linearity of climate systems make climate data analysis and modeling challenging. Because of the way variables interact with each other, high-dimensional climate data is likely to be non-stationary and non-linear. To solve these problems, scientists are coming up with new ways to look at and model complicated climate data. Nonlinear dimensionality reduction approaches, such as t-distributed stochastic neighbor embedding (t-SNE), are being used by one group in order to simplify climate data that is difficult to comprehend. The other group is working on new machine learning algorithms for non-stationarity and non-linearity.

Working Together Across Disciplines and Sharing Data

To do big data analytics in climatology, you need to share data between disciplines and organizations. But differences in data formats, standards, and cultural norms can make it challenging to work together and share information. Researchers are working on innovative ways for people from different fields to work together and share data to solve these problems. Some researchers are establishing metadata protocols and data standards to help with data exchange and integration. Others are making collaborative study platforms.

Table 1. Key Challenges and Limitations in Applying Big Data Analytics to Climate Studies

Challenge / Limitation	Description
Data Quality and availabitity issue	Climate data is often prone to errors, inconsistencies, and biases
Complexity of climate systems and data	Climate systems are inherently complex and non-linear
Need for itherdiscipcynary colmodelsboration and data sharing	Big Data nalytics requires colloboration and data sharing across diccipline and institutions

In climate research, Table 1 shows the major problems with big data analytics. Data quality and availability are crucial because climate data might be wrong, biased, or inconsistent, which can affect how accurate models are. It's challenging to model and predict climate systems because they aren't linear and depend on each other. Finally, initiatives that use data to study climate change need scientists from different fields and institutions to work together and share data. Big data can only work in this situation.

FUTURE DIRECTIONS FOR BIG DATA IN CLIMATOLOGY

Despite its several challenges, big data analytics could become increasingly relevant for climate research. As climate change becomes worse, the amount of data science and climate technologies that work together will grow a lot. More satellite data, IoT devices, and AI analytics will help us make better predictions about the weather and stop disasters from happening. Start a career in climate tech by getting a data science certification and leading the way in Chennai. By mastering data-driven methodologies, professionals can come up with new ways to help the environment.

The following big new data analytics tools and techniques may have an effect on climate research in the future.

New Developments in Big Data Analytics

a) Artificial Intelligence (AI) and its sub division of Machine learning: These technologies will assist climate researchers in identifying complex trends and patterns in enormous data sets.
b) Cloud computing will be more significant for climate research because it helps scientists store and work with enormous amounts of data in a way that is flexible and can grow as needed.

The Internet of Things (IoT) The IoT can link to many sensors and devices, which makes it very useful for climate scientists.

Adapting to and Mitigating Climate Change

There are several ways that big data analytics can help us address and lessen the effects of climate change. Here are a few:

a) Big data analytics can figure out where climate-related hazards are and where they are most likely to happen.
b) Using big data analytics can help cut down on greenhouse gas and other pollutants that change the climate.
c) Big data analytics aid in combating climate change by mitigating the effects of rising sea levels and extreme weather events.

This mind map (Figure 3) shows how big data analytics may help fight climate change and adjust to it:

Figure 3. Role of Big Data Analytics in Combating Climate Change

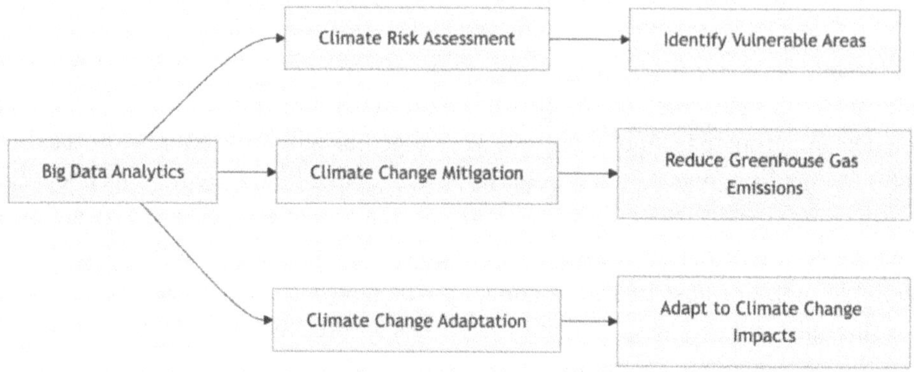

Chances to Work Together and Share Information

For climatology to use big data analytics, institutions need to share their knowledge. There are many chances to work together and share knowledge, such as

a) Online research communities: Researchers can post their work in online research communities.
b) Data repositories: These hubs can hold and share weather data.
c) Conferences and workshops: These are formal occasions where academics meet and share ideas.

CONCLUSION

Big data analytics has brought about a technical revolution in climatology by letting researchers look through massive volumes of data to uncover complicated patterns and trends. There are many problems and rules that need to be fixed before big data analytics can be used for climate research, but the potential is limitless. Researchers can help make the future more sustainable by harnessing new technologies and trends in big data analytics to come up with ways to mitigate and lessen the consequences of climate change. Such work helps make the future better and more sustainable in the long run. When it comes to fighting climate change, information science is a fantastic way to come up with plans. Data science helps with a lot of climate technology tasks, such as predicting extreme weather, building the best renewable energy systems, and lowering carbon emissions. Data science can also help lower carbon emissions. As businesses and governments rely more and more on data-driven methods, the need for qualified workers in this industry is developing quickly. Putting together science-related data We can employ big data analytics to construct long-lasting resilience, and we can use technology to forecast and stop climate disasters. You need both of these skills to stay strong.

REFERENCES

Amato, F., Guignard, F., Robert, S., & Kanevski, M. (2020). A novel framework for spatio-temporal prediction of environmental data using deep learning. *Scientific Reports*, *10*(1), 1–11. DOI: 10.1038/s41598-020-79148-7 PMID: 33335159

Andersson, M., Baccianti, C., & Morgan, J. (2020). Climate change and the macro economy (Econstor No. 243). https://hdl.handle.net/10419/234484

Avand, M., Moradi, H., & lasboyee, M. R. (2021). Using machine learning models, remote sensing, and GIS to investigate the effects of changing climates and land uses on flood probability. *Journal of Hydrology (Amsterdam)*, *595*, 125663. DOI: 10.1016/j.jhydrol.2020.125663

Avand, M., Moradi, H. R., & Lasboyee, M. R. (2021). Spatial prediction of future flood risk: An approach to the effects of climate change. *Geosciences*, *11*(1), 25. DOI: 10.3390/geosciences11010025

Badawy, A., Sultan, M., Abdelmohsen, K., Yan, E., Elhaddad, H., Milewski, A., & Torres-Uribe, H. E. (2024). Floods of Egypt's Nile in the 21st century. *Scientific Reports*, *14*(1), 27031. Advance online publication. DOI: 10.1038/s41598-024-77002-8 PMID: 39505946

Do, S. K., Tran, T.-N.-D., Le, M.-H., Bolten, J., & Lakshmi, V. (2024). A novel validation of satellite soil moisture using SM2RAIN-derived rainfall estimates. *Frontiers in Remote Sensing*, *5*, 1474088. Advance online publication. DOI: 10.3389/frsen.2024.1474088

Greca, S., Shehi, I., & Nuhi, J. (2023). Analyzing climate changes impacts using big data Hadoop. *CEUR Workshop Proceedings*, *3402*, 21–27.

Guo, J., Wu, X., & Wei, G. (2020). A new economic loss assessment system for urban severe rainfall and flooding disasters based on big data fusion. *Environmental Research*, *188*, 109822. Advance online publication. DOI: 10.1016/j.envres.2020.109822 PMID: 32615356

Hamlet, A. F., Byun, K., Robeson, S. M., Widhalm, M., & Baldwin, M. (2020). Impacts of climate change on the state of Indiana: Ensemble future projections based on statistical downscaling. *Climatic Change*, *163*(4), 1881–1895. DOI: 10.1007/s10584-018-2309-9

Ikegwu, A. C., Friday, H., Chioma, N., & Anikwe, V. (2023). Recent trends in computational intelligence for educational big data analysis. *Iran Journal of Computer Science*, *6*(3), 1–28. DOI: 10.1007/s42044-023-00158-5

Ikegwu, A. C., Nweke, H. F., Anikwe, C. V., Alo, U. R., & Okonkwo, O. R. (2022). Big data analytics for data-driven industry: A review of data sources, tools, challenges, solutions and research directions. *Cluster Computing*, *25*(5), 3343–3387. DOI: 10.1007/s10586-022-03568-5

Ikegwu, A. C., Nweke, H. F., Mkpojiogu, E., Anikwe, C. V., Igwe, S. A., & Alo, U. R. (2024). Recently emerging trends in big data analytic methods for modeling and combating climate change effects. *Energy Informatics*, *7*(1), 6. Advance online publication. DOI: 10.1186/s42162-024-00307-5

Lozo, O., & Onishchenko, O. (2021). The potential role of the artificial intelligence in combating climate change and natural resources management: Political, legal and ethical challenges. *Ziran Ziyuan Xuebao*, *4*(3), 111–131.

Number Analytics. (2025). Future climate research: Big data analytics. https://www.numberanalytics.com/blog/future-climate-research-big-data-analytics

Soni, D., Dass, A., Kamble, P., & Kumar, R. (2025). Decoding of long-term spatio-temporal precipitation dynamics over India: Insights from 115 years of meteorological data. *Global and Earth Surface Processes Change*, *3*, 100005. Advance online publication. DOI: 10.1016/j.gespch.2025.100005

Chapter 15 Decolonising Climate Change Education Through Indigenous Knowledge Systems

K. Sunanda
http://orcid.org/0009-0001-3803-9977
SASTRA (Deemed to be) University, India

G. Loganayaki
http://orcid.org/0000-0002-5944-1420
SASTRA (Deemed to be) University, India

S. Sanmathi
SASTRA (Deemed to be) University, India

ABSTRACT

The chapter focuses on India and South Africa for their rich cultural biodiversity and colonial histories. Both these countries share rich ecological diversity and histories of colonialism. In South India, the ritual water tanks, sacred temple groves, community-led conservation tradition, and many showcase how biodiversity is supported by local knowledge and traditional resource management practices. In South Africa, philosophies like Ubuntu and indigenous practices around medicinal plants (Muthi) showcase similar community-oriented approaches to foster environmental protection and climate resilience. Therefore, the chapter interconnects the varied knowledge boards and showcases how the IKS can be imparted into climate change education as a significant step. To make this integration into an effective practice,

DOI: 10.4018/979-8-3373-6975-4.ch015

the following three-dimensional methodology is proposed: 1. Curriculum Reform 2. Pedagogical Change for teacher trainees 3. Experiential Learning that actively imparts local community knowledge.

INTRODUCTION

In the 21st century, climate change is the foremost global and ecological challenge, impacting ecosystems, the economy, social stability, and public health. The global surface temperature has already increased approximately 1.1°C above pre-industrial levels, as stated in the Sixth Assessment Report of the Intergovernmental Panel on Climate Change (IPCC, 2023). This has further intensified heat waves, droughts, erratic rainfall, floods, and biodiversity loss. The impacts are unevenly seen, with a major impact on the Global South, which is rich in natural ecosystems.

To address emerging climate change and achieve Sustainable Development Goal 13, Climate Action, Climate Change Education (CCE) is prioritized by many international institutions, such as UNESCO. Implementation of the Sustainable Development Goals Roadmap also explicitly supports education systems in shifting their technical literacy on climate to incorporate indigenous perspectives and cultural values into teaching and learning processes. Similarly, IPCC Working Group II highlights the effective working of climate strategies when combined with context-specific and indigenous knowledge systems, especially in places vulnerable to climate variability.

Climate education taught in institutions primarily focuses on global climate models, emission statistics, and mitigation frameworks like carbon accounting. Global studies on national curricula in India show that climate education is primarily taught through content-heavy, examination-oriented perspectives, limiting the scope for experiential or community-based learning. For example, an analysis of NCERT textbooks from grades 6 to 9 conducted by Singh and Ahamed (2025) shows that climate change is introduced through scientific explanations in geography and science, with minimal engagement with local ecological practices or indigenous perspectives.

As Paulo Freire (1970) stated, education becomes a tool of domination when it silences local knowledge and lived experience. This reflects the historical and epistemic structure of education in colonial systems. Additionally, Boaventura de Sousa Santos states that the marginalisation of non-Western knowledge systems aids systems aids the systematic Erasure of alternative traditional epistemologies. To understand how Indigenous Knowledge Systems function as alternative knowledge systems and climate education in real-life contexts, the study has examined two Global South countries. Both countries possess ecological richness in community practices and knowledge, yet their education systems primarily focus on Western scientific epistemologies. This comparative view helps in understanding indigenous ecological practices that can influence climate education.

BACKGROUND OF THE STUDY

In this chapter, India and South Africa are analyzed. Both countries have extensive IKS that have been maintained for centuries across diverse ecological zones. In India, ecological practices such as traditional water management systems, including temple tanks in Tamil Nadu and historical Eri systems, support rainwater harvesting, groundwater recharge, and flood mitigation. According to the Tamil Nadu Water Resources Department, improved groundwater levels are due to the restoration of temple tanks, which reduce drought vulnerability. The Ashoka Trust for Research in Ecology and the Environment (ATREE) indigenous study on the sacred groves found in Kerala, Karnataka, and other parts of the Western Ghats are the biodiversity hotspots that preserve endemic species and regulate microclimates.

In South Africa, IKS systems are closely structured by the philosophy of Ubuntu, which emphasizes the relational ethics between communities, humans, and nature. Empirical studies in Limpopo and the Eastern Cape show that indigenous agricultural practices such as crop rotation, intercropping, indigenous weather forecasting, and soil moisture conservation influence food security and climate resilience. Research by Kom et al. (2024) in the Vhembe district shows that households practising indigenous farming methods have greater resilience to rainfall than those relying exclusively on modern inputs. Additionally, as stated by the South African National Biodiversity Institute (SANBI), indigenous medicinal knowledge associated with Muthi practices supports biodiversity use through customary harvesting rules.

Despite these practices, the formal education system in both countries remains dominated by Western-centric ideologies and marginalized indigenous knowledge systems. The National Education Policy (NEP-2020) in India acknowledges the importance of Indian knowledge systems and local traditions. Despite this emphasis, the institutional curriculum prioritises standardised scientific approaches, and teacher education programs rarely integrate indigenous ecological knowledge into classroom practice. In South Africa, the Curriculum and Assessment Policy Statement (CAPS) lacks pedagogical guidance and institutional support for inclusivity in Climate Change Education, as evidenced in studies published in the South African Journal of Higher Education.

Although national policies in both India and South Africa formally recognize IKS, its integration into Climate Change Education remains limited (Table 1).

Table 1. The gap between policy and the classroom implementation of Indigenous Knowledge Systems

Policy	Official Recognition of Indigenous Knowledge	Observed Gap in Educational Practice
National Education Policy (NEP) 2020 – India	Recognizes the importance of Indian Knowledge Systems (IKS) and encourages the integration of local traditions and experiential learning into the curriculum.	Climate Change Education in schools primarily focuses on scientific and textbook-based content. There is very limited incorporation of indigenous ecological practices or community knowledge.
National Curriculum Framework (NCF) (2023). – India	Emphasises place-based learning, experiential pedagogy, and environmental awareness aligned with Education for Sustainable Development.	Limited implementation due to a lack of training and institutional support to integrate indigenous ecological knowledge into classroom teaching.
NCERT School Textbooks (India)	Climate change topics are included within science and social science curricula.	Content is heavy and examination-oriented, focusing on global climate models and environmental awareness rather than local ecological practices or community knowledge systems.
Curriculum and Assessment Policy Statement (CAPS) – South Africa	Acknowledges Indigenous Knowledge Systems and Ubuntu philosophy within educational discourse.	The curriculum provides limited pedagogical guidance for teachers on integrating indigenous ecological knowledge into climate education.
Teacher Training Initiatives (NISHTHA, Samagra Shiksha – India)	It includes environmental education modules designed to improve teachers' awareness of sustainability issues.	Training programmes focus primarily on environmental awareness and conservation messaging rather than on the practical integration of indigenous ecological knowledge into climate pedagogy.
Biodiversity and Environmental Policies (India & South Africa)	The policies recognize community-based ecological knowledge and biodiversity governance.	These knowledge systems are rarely translated into formal classroom learning or curriculum design, leading to a persistent disconnect between policy recognition and educational practice.

The disconnection between pedagogical practices and policy integration shows a critical gap in Climate Change Education. IKS is recognized only at a symbolic level and is rarely integrated into epistemic frameworks across curricula, assessment, and teacher training. Therefore, the learners are often cut off from locally relevant climate knowledge and its significance in enhancing ecological understanding and adaptive capacity. This exclusion is especially notable in postcolonial societies, where historical, rural, and cultural inequalities intersect directly with Climate Change Education.

OBJECTIVES OF THE STUDY

- This chapter aims to critically examine the Western-centric epistemologies and their dominant role in Climate Change Education.
- To analyze climate adaptation knowledge suggested by indigenous knowledge systems based on existing peer-reviewed literature.
- To identify the epistemologies followed in formal climate change education that marginalize indigenous knowledge and bring out comparative empirical studies from India and South Africa, and to understand the contribution of indigenous ecological practices.

CONCEPTUAL MODEL OF THE STUDY

IKS represents knowledge encompassing only locally developed ecological knowledge, cultural practices, lived experiences, etc., transmitted across generations. Integrating IKS into Climate Change Education provides an alternative epistemic perspective on understanding climate and biodiversity stewardship, emphasizing sustainable resource use, community-based resilience practices, and ethical behaviour.

The term Decolonial Pedagogy refers to filling the gap between scientific and indigenous knowledge in pedagogy. By eradicating colonial hierarchies in the knowledge system, IKS serves as a mediator that supports cultural legacy, curricular inclusivity, the integration of alternative epistemic perspectives, and the combination of varied knowledge.

To support this integration, India and South Africa have been taken into a comparative context. The comparative lens also strengthens the conceptual model by highlighting how IKS operates differently in varied geographic, cultural, and historical landscapes. In India, it is temple groves, water tanks, or sacred forests that aid community conservation. In South Africa, Ubuntu ethics, sustainable medicinal plant usage 'Muthi', or communal land management, and bio-cultural knowledge provide a real-life empirical view and eliminate theoretical or abstract ideologies on sustainability.

This chapter suggests a Four-Dimensional Framework to integrate IKS into CCE.

a. Curriculum Reform: Reconstructing the curriculum for learners and teachers can connect the indigenous eco-practices with global climate science.
b. Transformative Pedagogy: Teacher training emphasises the importance of a place that emphasises essential learning, creating a responsive teacher rather than following textbook-only instructions.

c. Experiential learning: knowledge is also enhanced through community participation, field visits, and the reliving of cultural practices—this positions communities as co-educators for the learners and teacher trainees.
d. Ethical Safeguards: ethical awareness in integrating IKS into CCE to address the risks of IKS becoming tokenistic or extractive.

This combined measure is more effective and inclusive for the new CCE model. This enhances the learning plurality that is, combining multiple knowledge systems to achieve a universal perspective. Also, it strengthens the cultural relevance and ecological resilience in climate education adaptation.

RESEARCH METHODOLOGY

Type of Study: This chapter employs a qualitative, descriptive, and conceptual analysis based solely on secondary data. It does not involve any fieldwork or empirical interventions. Instead, it interprets and synthesizes early knowledge to construct a comparative understanding of climate change education when integrated with previous years' knowledge systems.

Inclusion and Exclusion Criteria: The chapter includes only studies written in English for consistent interpretation; peer-reviewed sources; studies on CCE, IKS, and decolonial pedagogy; studies with suitable thematic or comparative analysis; and case studies relevant to India or South Africa's knowledge system contexts. Non-scholarly media content, opinion pieces, studies with unidentifiable methodology or insufficient theoretical grounding and sources that are unrelated to pedagogy, climate discourse education or indigenous studies.

Search method: The secondary data for this study is sourced from multiple categories of documentary materials. Peer-reviewed academic journal articles were taken from international and regional journals, including the *South African Journal of Higher Education*, *Sustainability*, *Discover Education*, *Journal of Education and Practice*, *Agriculture & Food Security*, *Journal of Environmental & Earth Sciences*, *RASSA Journal of Science for Society*, and the *Asian Research Journal of Arts & Social Sciences*. Academic books and edited book chapters were sourced from publications produced by the C.P.R Environmental Education Centre, Bharti Publications, and Springer. Research repository materials were accessed via ResearchGate, including unpublished manuscripts representing emerging scholarship. Government policy guidelines, curriculum frameworks, legal notifications, and official program documents were obtained from the Ministry of Rural Development, Ministry of Water Resources, Ministry of Environment, Forest and Climate Change, Ministry of Education, Ministry of Human Resource Development, Public Works Department

(Tamil Nadu), Hindu Religious and Charitable Endowments Department (Tamil Nadu), the Supreme Court of India, the Department of Basic Education (South Africa), and NCERT. Together, these diverse sources provide a comprehensive foundation for policy analysis, conceptual understanding, thematic synthesis, and contextual comparison across the domains of education, environmental governance, cultural heritage, and sustainability. Additionally, the snowballing approach is used to explore deeper into each selected study.

Analytic method:

a. Comparative Juxtaposition: Analyzing the aspects of South Indian ecological practices alongside South African ecological practices.
b. Conceptual Integration: This conceptual integration is also used to understand the wider view of the stated research objectives in a four-dimensional perspective: 1. Curriculum Reform, which combines indigenous perspectives into global climate science and education. 2. Pedagogical Change for teacher trainees to promote culturally responsive and place-based methods. 3. Experiential learning that actively imparts local community knowledge. 4. Ethical measures that are necessary to ensure that IKS is not simplified or given abstractly in educational contexts.

LITERATURE REVIEW

Recent global policy discussions emphasize IKS in climate adaptation strategies. The Intergovernmental Panel on Climate Change (IPCC) Sixth Assessment Report highlights indigenous ecological knowledge as an important source of local climate adaptation practices. In many regions across the globe, including the African Union, Indigenous Knowledge Systems are being incorporated into environmental policy frameworks to strengthen climate resilience and sustainable development strategies.

The Importance of IKS in Climate Change Education

The study by Acharibasam (2022) reveals that education can be decolonized by integrating indigenous and ecological knowledge into yearly childhood environmental and climate lessons. This integration also counters Western epistemic dominance in the curriculum. Through a community-based participatory research design, the study has found that children's knowledge of the local environment can be improved by re-storing indigenous knowledge, strengthening human-non-human ethical relationships, promoting self-determination, and recognising community-recognised solutions as knowledge rather than as a means of education. Western-centric approaches alone are insufficient for addressing the climate crisis in rural and indigenous communities. Madzivhandila (2024) demonstrates through the theoretical and conceptual methodology, the policies, existing literature and illustrative examples from Africa and other developing regions, to analyze how the indigenous communities cope with climate-related challenges such as water conservation, traditional weather forecasting, agricultural practices, and other coping mechanisms because scientific interventions are often culturally outcast, external and inaccessible. Therefore, integrating indigenous and scientific knowledge systems will make climate change adaptation more effective and contextually relevant.

The study by Lautensach et al. (2025) clearly states that it is important to strengthen the prioritization of climate education content and to address why it is a social necessity. The authors have taken the Climate Education in Teacher Education project in Northern British Columbia as the case study, and they have used an Educational Design-Based Research (EDBR) methodology to show that to promote an effective CCE, it must go beyond technical scientific knowledge and include indigenous knowledge systems, social-emotional learning, and a locally responsive pedagogy. It can be understood that the importance of a strong teacher community and climate education practices is.

In the study by Singh and Ahmad (2025), it is stated how far the curriculum addresses the depth of the content, integration, and pedagogical orientation of related content in accordance with national educational policies and its NCERT textbooks. Through documentary and qualitative content analysis of key policy documents, the curriculum remains content-heavy and only awareness-oriented. The study concluded with suggestions to strengthen teacher capacity and promote action-oriented, experiential pedagogy that helps build a transformative education aligned with the Education for Sustainable Development goals.

The literature review stated above highlights the ecological importance of IKS and their role in climate adaptation. However, limited research has been done on integrating IKS into CCE in a formal education context. The gap in understanding how such knowledge can be translated or integrated into curriculum design, teacher training, and pedagogical practices in school-based climate education contexts persists, as existing studies focus only on documenting ecological practices or community-based adaptation strategies. Also, a comparative analysis of the integration of IKS into CCE across the Global South is relatively limited.

This chapter addresses this gap through a critical study of Indian and South African Indigenous Knowledge Systems. These two regions are rich in indigenous ecological knowledge, yet they coexist with education systems which are largely influenced by Western scientific knowledge. By analyzing the cultural and ecological heritage of these two regions, such as tank hydrology, sacred grove conservation, Ubuntu ethics, and Muthi medicinal knowledge, the chapter demonstrates how IKS can enhance the curriculum design, pedagogical practices, and experiential learning models. Through this approach, the chapter offers a broader discourse on decolonizing Climate Change Education by presenting a four-dimensional framework that bridges indigenous ecological knowledge with formal climate education systems. This also strengthens culturally relevant, context-sensitive climate learning.

The Indian Knowledge System for Climate Change Resilience

Temple Tanks

The temple tanks serve a purpose beyond spirituality and ritual. It acts as the most effective climate-resilience infrastructure. Examples of temple tanks across South India illustrate the ecological and cultural significance of traditional water management systems. In Tamil Nadu, notable examples include the Mahamagam Tank in Kumbakonam, Melukote Kalyani at the Cheluvanarayana Swamy Temple in Melukote, the Virupaksha Kalyani at the Virupaksha Temple in Hampi, and the Padmanabhaswamy Temple in Thiruvananthapuram. These temple tanks represent longstanding indigenous hydrological systems that historically supported water conservation, groundwater recharge, and community-based ecological management across the region.

The government takes many policies to conserve and protect the ecological infrastructure:

i. Wetlands (Conservation and Management) Rules, 2017: Under the Environment (Protection) Act, the Wetlands Rules (2017) list important water bodies that require protection and management. Numerous tanks that are associated with temples are listed as official wetlands in some states and come under these acts:

   - Mandatory zoning protections
   - Pollution control measures
   - Restoration and low-impact use frameworks

Apart from temple tanks, these rules also provide legal protection to many other ecologically sensitive water bodies.

ii. Kudimaramathu Scheme—Tamil Nadu (active since 2017): The Kudimaramathu Scheme was announced in 2017 and is still working; it is one of the strongest and most active initiatives taken by the government. This scheme actively supports the restoration and maintenance of traditional temple tank systems through an integrated community participation.

- Incentives are granted to the village-level community members to desilt, repair, and manage tanks and canals.
- The scheme reflects the ancient Tamil practice of community responsibility (*kudimaramathu*) for water infrastructure.
- Also, funds are granted by the local panchayats for maintenance activities. (tank repairs, bund strengthening and catchment management.)
- The program's outcomes include the restoration of many temple tanks and village irrigation tanks. Therefore, this scheme is still operating and supported by the Tamil Nadu government.

iii. Mission Amrit Sarovar (active since 2024): This mission was launched in recent years by the Government of India.

- Revive and restore water bodies, such as lakes, ponds, and tanks, across India.
- Promote water security and recharge.
- This mission strongly emphasizes relational water bodies (traditional tanks) as a major support for restoration.
- Mapping and census of water bodies are frequently taken under this mission as a part of implementation. Therefore, it will be helpful to strengthen grassroots and planning data. Under this mission, the temple tanks with rich heritage have been given an extra emphasis in the state-level project plans.

iv. National Water Policy (active since 2012 with emphasis in the 2020s): The National Water Policy manages the water body protection activity in the current national framework. The policy includes the following activities:

- Protection and preservation of water bodies. This also includes artificial water systems such as tanks and ponds.
- Many measures are taken to prevent encroachments. The policy also emphasizes the promotion of restoration and rejuvenation of traditional water structures.
- Promoting traditional water-harvesting systems as part of the water security strategy. Therefore, this policy has helped in state-level water tank conservation programs since 2020.

v. State Religious Endowment Department Programs: In states like Tamil Nadu and Karnataka, Hindu Religious and Charitable Endowments (HR&CE) and comparable trusts allocate recurring budgets for:

- Tank desilting and renovation
- Festival-related tank maintenance

- Structural conservation of heritage temple tanks

HR&CE annual plans specifically allocate maintenance funds for temple tanks in their jurisdictions. This fund allocation also continues into the current budgetary cycles.

Sacred Groves

Sacred groves serve as reservoirs of biodiversity and socio-ecological heritage, and as a form of climate resilience. It is culturally supported, and they act as "mini-ecosystems" that support endemic flora, medicinal species, watershed protection, feedbacks, and wildlife. Many researchers report that local communities have historically managed these groves through taboos or clan-based governance. Also, many ritual prohibitions over the groves guarantee protection without any formal governance rules. Due to modernisation, urbanisation, and other religious transformations, these cultural practices eroded, leading to encroachment, resource extraction, and the eventual weakening of traditional belief systems.

Kanojia & Saha (2024), in a case study, state that Mawphlang Sacred Grove in Meghalaya demonstrates how diverse legal systems and community custodianship enable long-term conservation practices. The grove is managed under a traditional law, "Kot Adong Hima Mawphlang". In this case, the Mawphlang sacred grove acts as an excellent model for participatory conservation strategies. This law prohibits the sale of forest products, regulates the collection of fuelwood, and imposes penalties for unauthorized burning or extraction. This also helps in understanding how these sacred groves enhance climate resilience through microclimate cooling, soil conservation, hydrological regulation, aquifer recharge, and carbon sequestration. To create a sustainable future, integrate indigenous knowledge with modern ecological practices, and provide government incentives to increase community participation in conservation.

Various Beneficiary Outcomes of the Grove

a. The natural aid: Mawphlang is home to unique taxa, including Rhododendron formosum, Quercus griffithii, and an endemic blue mushroom. Also, Sural Bhatori in Himachal Pradesh hosts Betula utilis (Bhojpatra) and has been declared a Biodiversity Heritage Site under India's Biological Diversity Act 2002. These findings clearly demonstrate that groves are community-protected genetic reservoirs that can operate independently outside formal, modern networks.

b. Community-based governance: The study confirms that rational governance structures such as clans, Dorbar councils, priests, and chieftains are the central method for resource regulation, taboo enforcement, and responsible ecological management. With this smooth functioning of community-based governance, state governance may interfere with the sacredness or the local community legacy.
c. Weakening belief systems in contemporary governance: There are many alterations made in the belief systems customarily followed by the community, such as Sanskritization, Christianity, and modern secularisation that alter ritual practices. Additionally, youth migration and urban education distort knowledge transfer, and cultural taboos against extraction are declining under economic pressures. This clearly explains the weakening state of cultural conservation mechanisms in the contemporary era.
d. Sacred groves produce many climate resilience benefits: The benefits rendered by sacred groves are atmospheric humidity regulation, microclimate cooling, soil and watershed protection, aquifer recharge, carbon retention and habitat provisioning for pollinators and fauna. The authors also explain that these benefits of the groves reflect the Nature-Based Solutions (NbS) ideology, which strongly supports India's climate and biodiversity agendas.
e. Erasure of the groves: The current destruction of these groves can be prevented through socio-cultural aspects, and not just the ecological aspects. Therefore, a multidimensional view can be taken, including land encroachment, temple expansion and road construction, privatisation of commons, fragmentation and invasive colonisation, erosion of ritual knowledge, and collapse of intergenerational transmission.

Another case study by Tirkey et al. (2022) shows the role of sacred groves, such as Sarna, Mahadani, Buddhadev, and Phool, in the Jashpur district of Chhattisgarh. The authors have taken field surveys and interviews with the members of tribal communities. The functions of groves and how they are maintained by the tribes such as Oraon, Gond, Santhal, Kawar and Pahari Korwa are recorded. These culturally protected ecological sites work as environmental protectors as they are often dedicated to local deities (Sarna Bhuria, Andhari Pat, Mahadania). These groves are integrated with the ritual cycle, festivals, and the community's worship for continued protection and to fulfil provisioning needs such as fuelwood, fodder, food, and medicinal plants. Also, they are maintained by ritual, taboo, and cultural authority rather than state legislation. The Erasure of grove areas due to encroachment and overexploitation underscores the need for ecological responsibility and conservation.

South African Indigenous Knowledge System in Ecological Conservation

Ubuntu is presented as a South African methodology of relationality, communitarianism, sufficiency, and interdependence (Greeff, 2019). While Western climate strategies rely on individual autonomy, Ubuntu situates humans within a network of humans, non-human life, nature, and future generations. It promotes collective labour, shared resources, solidarity, and mutual aid during environmental crises, while also reducing vulnerability during droughts, floods, and food insecurity. It discourages overconsumption and the commodification of nature, and supports the sustainable use of land, water, and ecosystems based on basic needs rather than maximizing profit. It also emphasizes intergenerational responsibility by protecting future generations and aligning practices with climate policy frameworks. However, many Western approaches, including sustainable development models and the commercialization of GM crops, continue to ignore these cultural value systems, thereby limiting their effectiveness in community-based contexts.

This position is further reinforced by Ayayia (2023), who identifies anthropocentrism, individualism, and capitalism as dominant forces contributing to climate change in destructive ways. Ubuntu, in contrast, emphasizes humanness, solidarity, relationality, and care in a cultural sense. It prioritizes the community's collective responsibility through ethical resource use and the maintenance of ecological harmony. Nature is valued alongside human communities, promoting ecological responsibility among community members. It also critiques the exploitative nature of capitalist development by directly opposing consumerism and prioritizing community well-being over profit accumulation. It reframes humans as part of an interconnected web of life, moderating harmful anthropogenic activities, reducing environmental degradation, and supporting climate justice. While this strengthens the ethical argument, the discussion remains largely conceptual and does not fully extend into practical or institutional application.

Similarly, Onuma et al. (2025) argue that dominant Western models contribute to Africa's socio-ecological crisis, including climate destruction, environmental degradation, inequality, and fragmented governance. These models, based on individualism, extractive growth, and market logics, conceal the communal solidarity and ecological interdependence present in African knowledge systems. Ubuntu is identified as a counter-framework capable of guiding sustainable futures by emphasizing communal responsibility, ecological dignity, shared well-being, and interdependence between humans and nature. For African communities, Ubuntu is not only a philosophical concept but also a lived system that supports climate justice, participatory governance, and community-based natural resource management. This ecological dimension considers humans and nature as part of the same web, creating interdependence relevant to Africa's climate risks. However, the emphasis again remains on normative ideals, with limited engagement with institutional or educational integration.

In contrast, Mbelebele et al. (2024) provide an empirical perspective by showing how indigenous knowledge and traditional practices support biodiversity conservation in the Amatol district through Muthi practices. The study highlights community-based production, oral transmission of knowledge, reduced reliance on biomedical systems, and income generation through local sales and healing services. It shows that indigenous knowledge serves as a form of biocultural management that supports food security, livestock health, and economic well-being. At the same time, it identifies structural constraints, including limited knowledge of seed propagation, low awareness of species endangerment, limited financial support, and deforestation. Despite these challenges, in-situ conservation practices support biodiversity retention, and households practising sustainable methods earn higher incomes. This indicates that, while indigenous knowledge is effective in practice, its sustainability is constrained by external factors.

A broader perspective is provided by Sahani et al. (2025), who examine multiple studies to analyze how indigenous knowledge systems support water conservation, biodiversity protection, and climate change adaptation. Practices such as terracing, rainwater harvesting, wetland protection, site governance, and seasonal weather prediction contribute to community-driven environmental management. These practices are low-cost and culturally embedded, relying on collective governance, intergenerational knowledge transfer, and localized ecological understanding. They strengthen responses to climate crises, such as droughts and floods, through planting schedules, crop selection, and local forecasting methods, thereby improving water retention and agricultural productivity. Data also shows measurable outcomes, such as reduced water stress and increased water availability. However, the focus remains largely on environmental management and less on how these practices are translated into formal systems such as education.

These studies establish the ecological and ethical value of Indigenous Knowledge Systems, particularly Ubuntu and Muthi practices, across different contexts. At the same time, most studies remain either conceptual or context-specific, with limited focus on integration into formal structures such as education, curriculum, or policy implementation. This creates a clear gap between recognition and application, underscoring the need for a more structured, integrated approach.

FINDINGS AND DISCUSSION

Curriculum Reform: Integrating Indigenous Perspectives into Global Climate Science

The CCE models remain abstract from indigenous realities, especially in the context of postcolonial societies, regional communities, and historically sustained ecological crises, through community-led conservation and indigenous government systems. Western epistemologies, such as climate models, mission inventories, techno-managerial mitigation, and meteorological datasets, cannot practically teach children how humans and nature work in an interdependent way. The case studies of South India and South Africa clearly show that indigenous knowledge systems are not just folklore but sophisticated ecological management systems. In South India, ritual water tank systems support rainwater harvesting, groundwater research, and flood mitigation, and have stabilized aquifers for centuries. In addition, sacred temple groves also act as bio-cultural refugia that preserve many species, endemic biodiversity, medicinal flora, and pollinators, and regulate hydrological functions. As observed empirically, in the states of Tamil Nadu, Karnataka, Kerala, and Andhra Pradesh, temple tanks were directly linked to watershed cycles, community maintenance practices, and, most importantly, seasonal hydro-ritual calendars.

In South Africa, the Ubuntu policy, multi-medicinal practices, and communal land-use ethics help preserve biodiversity, support water conservation, promote drought-resistant agriculture, and advance climate adaptation strategies. These traditional systems highly reflect the effort of community-driven adaptation. As stated by Madzivhandila (2024), Western approaches remain abstract and insufficient in a practical context without the inclusion of rural and indigenous contexts, because Western frameworks alone fail to account for indigenous water use, weather forecasting, resource sharing, or agricultural adaptation strategies. Acharibasam (2022) clearly demonstrated that engaging community elders, teachers, and learners together can significantly improve students' environmental competencies when they collectively participate in ecological activities such as plant identification, livelihood practices, and medicine preparation. This community participation activity is a form of epistemic co-protection in which knowledge is transferred horizontally through social, ecological, and intergenerational channels, rather than through typical vertical transmission.

Lautensach et al. (2025) similarly showed that climate education must be integrated with indigenous perspectives alongside technical knowledge to build resilient, ethically formed communities. Their work on teacher education showed that integrating these two fields of knowledge can broaden pedagogical understanding of nurturing biodiversity and climate resilience. In South India, Singh and Ahamed

(2025) observed that NCERT climate content remains limited to an exam-oriented perspective. Also, they are content-heavy and descriptive, with minimal engagement with the local ecosystem. Their findings prove that the Indian curriculum structures reflect the colonial legacy, which is a high concentration of Western science as a universal one, while indigenous epistemologies become marginalized. This knowledge is relegated to cultural studies rather than environmental science.

Therefore, the curriculum we form in a postcolonial context requires a four-dimensional shift. Firstly, IKS must be recognized as knowledge, not just as tradition. Secondly, the integration of IKS into climate science. Thirdly, imparting IKS through contemporary educational policy. The National Education Policy 2020 explicitly emphasizes integrating the Indian knowledge system into formal curricula. The fourth measure is the ethical consideration of implementing IKS without exhausting communal rights or ownership. Also, they strongly believe in experiential, local, and interdisciplinary approaches for effective environmental learning for the learner group. Similarly, the National Curriculum Framework (2023) currently emphasizes place-based pedagogy, community learning models, environmental civics, and other experiential learning strategies that focus on policy-level development rather than classroom-level implementation.

Additionally, the Supreme Court's Environmental Education Mandate 2003 focuses on environmental education across school levels by integrating IKS into CCE. Samagra Shiksha and NISHTHA teacher training initiatives include many environmental modules, but their operationalization is oriented toward conservation awareness rather than indigenous ecological governance. The Biodiversity Act (2002) institutionalized People's Biodiversity Registers and Biodiversity Management Committees to establish indigenous ecological documentation at the community level. The Forest Rights Act (2006) guarantees tribal guardianship over forests, thereby maintaining indigenous communities as a form of ecological governance knowledge. All these policies demonstrate that the Indian policy system has already recognized the importance of indigenous ecological knowledge, but the curriculum reform has not yet reflected it. The gap between an abstract idea and a concrete structure is not only overlooked in legislation but also in the curriculum.

In the context of South African ecology, Ubuntu-based ecological ethics create a form of moral obligation between the community and the land. As Greeff (2019) describes, Ubuntu intertwines humans within a web of interdependence, sufficiency, and collective responsibility. This ideology is against the capitalist logic of extractive development. Ayaiya (2023). I have also added to the argument that UBANTU understands environmental responsibility as a community-centred ethics rather than an individual moral choice. It also challenges anthropocentrism. Muthi medicinal knowledge and indigenous weather forecasting are forms of biocultural knowledge that cannot be separated from one another and also from communal practices, seasonal rhythms, and ritual calendars. Mbelebele et al. (2024) demonstrated that medicinal plant knowledge in Eastern Cape communities supports biodiversity conservation, food security, and economic autonomy, while also maintaining conservation ethics through selective harvesting and community regulation. Sahani et al. (2025) report that indigenous water and agricultural practices in South Africa significantly reduced water stress, increased drought resilience, and align clearly with IKS and climate adaptation models, as emphasized by IPCC Working Group II.

Therefore, Ubuntu is not just a heritage; it must be recognised as the greatest climate resilience strategy. Comparing these two regions reveals a shared pattern. The IKS system is a multidimensional ecological knowledge system that aligns with CCE's goals. Despite its recognition, it remains marginalized in the formal education curriculum. The exclusion of IKS from the curriculum continually reinforces the epistemic hierarchies born of colonial education systems. As Singh and Ahmad (2025) noted, the Indian curriculum always prioritizes textbook literacy over experiential ecological learning. The South African curriculum and assessment policy statements have sought to embed pedagogical initiatives integrating Ubuntu and Indigenous practices into the formal curriculum. Therefore, in both contexts, the curriculum reform must be restructured to integrate IKS not only as a cultural annotation but also as a climate science. This integration of Temple Tanks, Sacred Groves, Ubuntu, and Muthi into the curriculum can highly influence climate education beyond the abstract idea by linking hydrology with ritual, biodiversity with community governance, and conservation with the ethical practices of the community members.

It is a form of ecological literacy that includes awareness of resource cycles, interdependence, custodianship, and responsibility at a generational level. For example, Ubuntu practices provide an ethical grammar for climate justice, while temple tanks offer an infrastructural model. The curricular necessity for IKS thus starts with functional use rather than cultural preservation alone. Therefore, integrating them into the curriculum better aligns with ecological realities, helping learners understand the reasons for changes in climate and biodiversity as socio-ecological phenomena. This shift can greatly transform the climate change education from the cognitive domain into a socially responsive field, as stated by Lautensach et al. (2025). Curriculum reform in both these countries requires versatility, a multidimensional perspective, and the construction of a modern climate literacy that reflects cultural histories, current ecological conditions, and community practices.

Pedagogical Change for Teacher Trainee

Curriculum transformation will remain insufficient without a parallel change in teacher education and pedagogy. To integrate IKS into contextual learning, the pedagogical dimension becomes essential, as indigenous ecological knowledge cannot be taught solely through textbooks. This knowledge system is a relational and place-based one that depends on tacit learning, intergenerational transmission, and validation from the respective community. This makes the teacher not just a facilitator but a reflection of practicality. Teacher trainees require competencies in cultural responsiveness and land-based learning to integrate IKS effectively into environmental science and climate education, as Lautensach et al. (2025) have already suggested that Teacher education must expand beyond scientific literacy to include indigenous, socio-emotional, and community-based knowledge domains.

The study also found that teachers trained through an integrated educational model develop broader pedagogical perspectives, enabling them to address climate-related anxieties, ecological ethics, and other adaptation strategies. In South India, temple tanks and sacred groves can be utilized to promote a concrete pedagogical context for land-based climate learning. Teacher trainees interacting with these ecological infrastructures can understand seasonal hydrology, groundwater recharge, and biodiversity through community-driven governance systems. The Temple Tanks, which directly influence every ritual and festival calendar, offer a pedagogical tool that connects seasonal cycles to one's cultural practices. Similarly, groves are a form of biocultural system in which every species and its richness, medicinal plant use, and taboo-based resource regulation can be taught as ecological management, apart from ritualistic and superstitious practices. One example of this community-driven ecological management is seen in the study by Kanojia et al. (2024). In their study, they documented that the Sagar groves of Meghalaya are protected and governed by customary law that restricts extraction, commercialization, and burning. This is one of the clear examples that shows traditional governance enforces much sustainability without state intervention. In addition, Tirkey et al. (2022) note that tribal groves in Chhattisgarh preserve endangered species and provide provisioning services, all of which fall outside formal conservation policies and rules. Therefore, teacher trainees engaging with these case studies can easily understand that indigenous conservation practices are very important for an ethical and ecological understanding of the environment under climate education.

Similarly, taking the South African Traditional Practices into the Teacher Training Curriculum. Ubuntu offers a conceptual framework that is completely antithetical to Western models, which focus primarily on extraction and commercialization. Ubuntu practices always prioritize the well-being of the community over the commodification of resources, as Onuma et al. (2025) state. Therefore, teacher trainees trained under this Ubuntu practice will be able to contextualize climate change by enhancing responsibility, ethical interdependence, and justice by consent. Similarly, the Muthi medicinal knowledge provides pedagogical ideas for teaching biodiversity, ethnobotany, and ecosystem services through community practice, as demonstrated by Mbelebele et al. (2024). This study demonstrates that Muthi knowledge is effectively transmitted through elder mentorship, household healing practices, and local markets rather than through textbooks or schools. Therefore, pedagogical strategies combined with Muthi knowledge will be effective if teachers collaborate with community custodians to involve the community as co-educators with students, rather than approaching an external resource.

This pedagogical intervention in both regional contexts emphasizes teacher-training programs that mediate between scientific and indigenous ecological knowledge. The teacher-training system in India has only recently begun implementing such shifts. The National Education Policy 2020 has suggested that teacher preparation programs emphasize multidisciplinary, experiential, and holistic approaches. NISHTHA teacher training includes environmental awareness modules that help teachers learn and better understand the environment. However, these modules currently focus only on conservation advocacy rather than on learning about the indigenous governance of the ecosystem. Similarly, the National Curriculum Framework (2023) emphasizes place-based learning and the integration of local knowledge.

On the other hand, UGC's IKS Division under the Ministry of Education, and IKS Centres, have established numerous academic infrastructures for indigenous knowledge research. However, these approaches are limited to higher education and to only a very limited downward integration into teacher education programs and institutions. Therefore, teacher trainees lack proper exposure to the IKS context, which highly disconnects them from policy and pedagogical capacity.

This similar problem is also evident in the South African context, where Ubuntu and indigenous land practices are rhetorically recognized in environmental studies. However, teacher education institutions have long struggled to incorporate them. This is due to a lack of curricular frameworks funded by the institution and to the development of alternative pedagogical models. The curriculum and assessment policy statement clearly references indigenous knowledge and is significant, but it lacks proper application of classroom methodologies to teachers; as a result, Ubuntu-informed environmental pedagogy is a symbolic practice rather than a practical application within schools. Therefore, climate pedagogy in both countries remains constrained by epistemic inequality. In other words, indigenous ecological knowledge and governance are widely recognized, but no effort is made to incorporate or acknowledge them in the curriculum.

Experiential Learning

The third dimension of this chapter is Experiential Learning through which IKS is integrated into Biodiversity Conservation and Climate Education. As already stated in the review of this chapter, Indigenous Ecological Practices are naturally embedded in landscapes, collective social processes, and rituals. However, the pedagogical intervention depends on participation and on learning from experience. Therefore, Experiential Learning positions climate knowledge and biodiversity learning within lived experiences in ecological contexts. Thereby, it converts climate change education from a cognitive, abstract idea into socio-ecological engagement. In South India, experiential learning occurs when the learner group collectively visits groves, temple tanks, or other agricultural fields. It is further enhanced through interaction with community custodians and the observation of ecosystems. This interaction helps in understanding climate concepts such as groundwater recharge, evapotranspiration, sea dispersal, and species interactions through a local context rather than through textualized case studies. Ritual calendars linked to monsoon water harvesting and agricultural cycles are also a great example of climate phenomena. This form of experiential learning in the temple tank ecosystem will help understand historical engineering techniques and tactics in hydrological infrastructure, which also help cope with droughts and monsoon variability. All this knowledge is particularly relevant in the current climate change context, given the declining rainfall rates and water scarcity. A small-scale environmental education initiative in Tamil Nadu has connected temple tank restoration projects with school-based learning activities. Students participated in mapping local water bodies, documenting traditional water management practices, and interacting with community elders working on tank maintenance. The initiative demonstrated that experiential learning connected to indigenous ecological knowledge improved students' understanding of water conservation and climate resilience.

Similarly, Sacred groves are natural biodiversity learning environments. The learners can observe endemic plant species, pollinator networks, soil moisture gradients, and canopy stratification in real-life contexts. Because these groups are nurtured by customary governance or community-driven protection within the ecosystem, there are many community-restricted taboos, such as extraction, hunting, or land conversion. All these taboos demonstrate that conservation can be effectively enforced through cultural norms rather than statutory enforcement. The erosion of such belief systems, as stated by Kanojia et al. (2024), critically threatens both ecosystem and cultural sustenance. Therefore, it is again emphasized that experiential learning must be included in a cultural context, alongside the responsibility for the ecosystem.

In the South African context, experiential learning is practised through the traditions of Ubuntu and Muthi. Ubuntu carries ecological responsibility through collective resource management, sustainable land use, and the passing of intergenerational obligation. Also, experiential learning in this context emphasizes resource sharing under ritual direction, decision-making through community acceptance, and seasonal agricultural practices that help mitigate drought, flooding, and crop failure. Muthi knowledge transmits indigenous pharmacology, biodiversity management, and seed preservation through oral folklore, nadavagi, apprenticeships, and household healing. As already stated by Mbelebele et al. (2024). Experiential knowledge transfer improves community adaptation, especially during a climate crisis, by reducing dependence on commercial supply chains. Sahani et al. (2025) also reported that indigenous weather forecasting and agricultural practices help efficiently store water and improve crop resilience. This also demonstrates that community knowledge serves as a functional strategy. For example, in South Africa, community-based agricultural education programmes in the Limpopo region have integrated indigenous weather forecasting and crop rotation practices into school-based environmental education activities. These programmes helped learners understand climate variability through local ecological knowledge and demonstrated how Indigenous Knowledge Systems can complement scientific climate education.

The integration of experiential learning in the curriculum is also supported by the National Curriculum Framework (2023), which strongly emphasizes local project work, field visits, and community engagement to recognize experiential pedagogy. The National Education Policy (2020) similarly emphasizes multidisciplinary, holistic, and hands-on learning. However, there are many implementation barriers due to requirements for standardization in assessment, timetables, and teacher workload. Similarly, in South Africa, management faces many challenges, as schools lack institutional support to integrate community-based learning despite widespread recognition of Ubuntu as an Indigenous Knowledge System.

Therefore, the curriculum authorities, policy makers, school leaders, and educational institutions must work in coordination to delete these epistemic hierarchies and institutional barriers. Therefore, to decolonize the educational system, the corresponding embodiments must acknowledge curriculum design, pedagogy, assessment, and teacher training, rather than merely symbolic recognition of indigenous knowledge. Meaningful integration requires coexistence and coordination between the respective institutions. This approach also aligns with global perspectives, such as those of UNESCO and the Intergovernmental Panel on Climate Change. Both bodies consistently emphasize effective climate adaptation grounded in cultural context and community participation.

GENDER DIMENSIONS AND ASSESSMENT OF PLURIVERSAL LEARNING IN REAL-LIFE CONTEXT

Many recent pilot initiatives showcase how Indigenous knowledge systems (IKs) meaningfully integrate climate change education through community-based learning. In Canadian educational programs, students connect with elders through ecological field trips, seasonal environmental monitoring, and traditional knowledge workshops in land-based environmental education. These initiatives prove that place-based learning can deepen students' understanding of climate education. It also strengthens the relationship between schools and the local intellectuals. Similarly, 18 community-based adaptation programs in Latin America have established local knowledge centres where farmers, students, and researchers can collaborate to discuss the traditional agricultural practices, glacier observation techniques, and the ecosystem management strategies. These initiatives demonstrate how Indigenous ecological knowledge can support scientific climate research and enhance locally relevant climate adaptation strategies. Integration of IKs into CCE requires gendered knowledge transmission. In many communities, women act as the custodians of ecological knowledge. However, these contributors remain underrepresented in formal educational contexts. (UNFCCC, 2013) Recognizing these gendered dimensions in IKS is essential to ensure inclusive and equitable climate mitigation.

Digital technologies also provide new opportunities to document and preserve indigenous ecological knowledge. Digital archives, participatory mapping platforms, and community media initiatives also support intergenerational knowledge transfer and create new educational resources for climate learning. However, these tools must be used with careful attention to maintain ethical consent, cultural ownership, and knowledge sovereignty. These practical strategies also strengthen the real-life application of IKS within CCE while ensuring sustainable knowledge exchange.

CHALLENGES, ETHICS, AND INTERSECTIONALITY IN INTEGRATING IKS INTO CLIMATE EDUCATION

Many practical and ethical challenges complicate the implementation of Indigenous Knowledge Systems (IKS) into climate change education (CCE), which can significantly provide culturally based and contextually relevant learning. Despite the policy-level recognition, such as the National Education Policy NEP 2020 in India and the Curriculum and Assessment Policy Statement CAPS in South Africa, their actual implementation into classroom practice is very limited. One of the primary barriers is the institutional structure of formal education. Many educators are trained in Western scientific traditions and lack the pedagogical preparation to integrate indigenous ecological knowledge into their teaching. Additionally, rigid curriculum frameworks, limited institutional support, and insufficient funding act as barriers to schools' implementation of place-based and experiential learning approaches.

As a result, indigenous knowledge systems are often symbolically acknowledged without meaningful integration into everyday educational practice. Another important dimension is the intersectional nature of indigenous knowledge transmission. In many communities, ecological knowledge is embedded within specific social roles or cultural structures. For example, women frequently act as mediators of environmental knowledge through practices such as medicinal plant use, seed preservation, and traditional healing systems like Muthi in South Africa. Similarly, in India, particularly in the southern parts, women and community groups often participate in maintaining temple tanks and local water management practices.

In contrast, there are also social hierarchies, such as caste-based restrictions or socioeconomic inequalities in rural communities, that influence access to certain ecological spaces or practices. This intersectional analysis shows that indigenous knowledge systems are not socially neutral but are connected to broader social structures such as gender, class, and cultural authority. Integrating IKES into climate education requires careful planning to ensure inclusivity and representation, preventing the reproduction of existing inequalities in educational settings.

A further challenge is the Evaluation of learning outcomes in climate education. Conventional assessment systems based on standardized testing are inadequate for measuring experiential learning, ecological awareness, or community engagement. Alternative assessment approaches, such as reflective portfolios, community-based projects, field observation, and participatory learning activities, may provide a more appropriate measure of student learning. These methods allow educators to evaluate not only the cognitive understanding of climate science but also the development of cultural awareness, environmental ethics, and adaptive resilience.

Ethical considerations must also be taken into account when indigenous knowledge is incorporated into the formal curriculum. Indigenous knowledge systems may be simplified without recognizing the cultural contexts in which they are embedded. Furthermore, translating the oral traditions into written curricular materials may alter the authority or meaning of such knowledge. Meaningful integration requires respectful collaboration with community knowledge holders and recognition of indigenous intellectual property. Similarly, initiatives in other parts of the global South showcase the broader relevance of these approaches. For example, Māori environmental education in New Zealand connects indigenous ecological ethics such as *kaitiakitanga* (guardianship of nature). At the same time, Indian community-based adaptation emphasizes traditional agricultural knowledge and communal environmental accountability. These examples clearly demonstrate that indigenous knowledge systems across diverse cultural contexts provide valuable insights for climate resilience and environmentally responsible education. Addressing these challenges, ethical concerns, and the intersectional dynamics associated with the integration of IKs is essential to developing inclusive and sustainable models in climate change education.

FURTHER RESEARCH

Comparative International Work beyond India & South Africa: Future research can build on comparative studies that include the context of Latin America, Southeast Asia, the Pacific Islands, and Indigenous Australia. These new contextual examinations can be helpful in further investigation of community ecologies. Such expansions would test whether the decolonial CCE model continued in the contemporary education system.

Teacher Training and Professional Identity: Further research extending this core idea can analyze how pre-service and in-service teachers are prepared to engage with IKS, Ubuntu values, sacred grove knowledge, and temple tank practices. This also includes examining teachers' epistemic attitudes, cultural competency, and confidence levels in handling multidimensional knowledge systems.

Policy Implementation Research: A Mixed-Methods study can be conducted as an extension of this study. The research could examine NEP-2020, CAPS, UNESCO–ESD, and local biodiversity policies to understand why the gap in implementation of indigenous knowledge remains.

Digital Mediation of Indigenous Ecological Knowledge: There is scope to understand how the indigenous environmental knowledge can be circulated on digital platforms. With that in mind, the research should be structured to evaluate whether digitization preserves or distorts people's responsibility in indigenous ecology.

IKS Curriculum Design and Prototype Testing: Research must move from an abstract idea to a concrete curriculum approach.

Pilots could test modules based on temple tanks and hydrology, sacred groves and biodiversity, Ubuntu ethics for climate justice and Muthi-based medicinal ecology for resilience. Also, the Evaluation of learning outcomes will be used in effective policy adaptation.

Intergenerational Knowledge Transmission Studies: Future studies can focus on how indigenous knowledge persists amid urbanization, schooling, and other modes of transmission, as the primary modes are through community elders, apprenticeships, and everyday traditional practices.

Classroom-Based Empirical Research: This chapter largely focuses on CCE and IKS at the policy and conceptual level. Future research can examine empirical observations of how teachers actually teach climate concepts and whether indigenous perspectives are included, resisted, or adapted in real classroom environments.

Gender Dimensions of Indigenous Knowledge: The future study can analyze how roles as medicinal knowledge custodians, seed preservers, and sacred grove caretakers help combat the climate crisis. Because gender has been understudied. Even though it is central to Muthi and ecological practices.

MANAGERIAL IMPLICATIONS

This chapter demonstrates how the indigenous knowledge system can be integrated into climate change education, and many requirements from institutional curricula and governance efforts can extend the practicality of this study beyond symbolic representation. Several managerial implications can emerge:

IKS integration in Teacher Training Institutions: Currently, the teacher competencies are taught only at the operational level of CCE. This study suggests that teachers lack exposure to indigenous knowledge on ecology, community conservation, and land-based learning. Therefore, integrating IKS into teacher preparation programs is required across community partnerships, reflective climate pedagogy, cross-cultural pedagogical methods, field-based practicum methods, and related areas. Without these structural interventions, curricular reform remains unimplemented.

Community collaboration models to manage the knowledge: IKS is horizontally transferred to the next generation custodians, ritual calendars, and clan governance. Therefore, the school cannot act as a separate hub that manages the knowledge. Therefore, managers must team up with temple custodians, tribal knowledge holders, grow protectors, local panchayats, healers, mooty custodians, biodiversity committees, etc., and form partnerships. Therefore, a shared knowledge infrastructure will definitely enable the community to act as a co-educator, providing a better understanding of the environment and climate studies.

Coordination in multi-sectoral implementation. This chapter suggests that climate education is not only implemented by the Ministry of Education. Effective implementation of this approach requires coordination among many government agencies, including the Ministry of Tribal Affairs, Water Resource Departments, local governance bodies, Biodiversity Boards, the Ministry of Environment, and the Climate Action Missions. Therefore, managers must develop a coordinated approach to implementing IKS in CCE to improve implementation and policy development.

Government managers should protect IKS by integrating ecological policies. The chapter emphasizes traditional infrastructure to improve the ecosystem. Therefore, governments must also have managerial responsibilities. Such a list of responsibilities requires protecting IKS knowledge holders, legally safeguarding biocultural landscapes, incorporating IKS into climate planning, and encouraging greater protection ofof the biodiversity landscape. The government can incentivize the community or volunteers. This can greatly tie the climate governance to the cultural governance.

Shift the managerial ethics from anthropocentric to a relational ecological model. Ubuntu and Indian practices are currently non-anthropocentric ethical resources. Climate managers should encourage the adoption of relational ethics, such as sufficiency, interdependence, non-human valuation, and intergenerational responsibility. Such ethical attempts can significantly challenge contemporary models of extraction and commercialization. The commercialization aligns with the decolonial climate change education and curriculum.

Educational leaders must prioritize experiential-based learning. School management must restructure school timetables and assessment norms, and foster a link between the school and the community to enable more effective experiential and land-based learning. The cases of temple tanks, sacred groves in India, and Ubuntu-based environmental ethics in South Africa are great ecological responsibilities transmitted through lived participation rather than textual learning. Therefore, redefining the learning environment beyond the classroom can significantly affect the integration of ITS into CCE.

Curriculum designers must adopt a multiversal ideology. Western and scientific epistemologies highly dominate the current curriculum design. The chapter's findings also show that knowledge from temple tanks, Ubuntu ethics, sacred groves, Muthi practices, and community knowledge is valid ecological knowledge and a significant factor in climate literacy, not merely a ritualistic or cultural practice. Therefore, the curriculum managers must build a basic foundation in ministries of education to transition this domination of the knowledge system into a pluriversal one. Therefore, the curriculum includes multiple knowledge systems that are equally credible. This shift can greatly enhance students' cognitive capacity in climate learning. All these measures also align with Inner's course ESD sections.

CONCLUSION

Climate change education in the contemporary context continues to be shaped by colonial and Western knowledge as a dominant epistemic ideology. While scientifically essential, this epistemic does not adequately acknowledge the lived experiences or ecological realities of communities that have long carried the tradition of water management, climate adaptation, biodiversity conservation, etc., through their own respective Indigenous knowledge systems, like yes. Therefore, the comparative analysis taken in this chapter on India, South India, and South Africa clearly demonstrates that temple tank hydrology, sacred grove conservation, Ubuntu ethics, Muthi biocultural practices, and the community's traditional knowledge as a form of ecological knowledge system that enables high responsibility among the community members, climate resilience strategies, and intergenerational ecological continuity.

These environmental values are already recognised; however, the knowledge system that recognises them remains marginalised in schools, teacher training, and assessment cultures. Therefore, policies in both countries, such as India's NEP 2020 and South Africa's CAPS, emphasize indigenous development. Also, implementation of these policies contains gaps due to epistemic hierarchies within the working of curriculum and pedagogy. There is also institutional fragmentation and limited pedagogical capacity. Therefore, the chapter proposes a four-dimensional framework comprising curriculum reform, transformative pedagogy for teacher trainees, and experiential learning. These dimensions address the gaps by prioritizing IK not just as a priority but as a sustainable knowledge system essential to biodiversity conservation and climate change.

REFERENCES

Acharibasam, J. B. (2022). Decolonizing climate cDecolonizingion: Evidence from an empirical study in Ghana. *Journal of Education and Practice, 13*(32), 54–63. DOI: 10.7176/JEP/13-32-06

Ayayia, G. S. (2023). *The concept of "Ubuntu" in African environmental ethics vis-à-vis the problem of climate change* (Unpublished manuscript). ResearchGate. https://doi.org/DOI: 10.13140/RG.2.2.27804.36484

Department of Basic Education. (2011). *Curriculum and Assessment Policy Statement (CAPS).* Government of South Africa.

Freire, P. (1970). Pedagogy of the Oppressed. *Continuum.*

Greeff, A. C. T. (2019). Ubuntu and environmental ethics: The West can learn from Africa in confronting climate change. In Chemhuru, M. (Ed.), *African environmental ethics: A critical reader* (pp. 93–109). Springer. DOI: 10.1007/978-3-030-18807-8_7

Hindu Religious and Charitable Endowments Department. Tamil Nadu. (2024). *Annual temple infrastructure and tank maintenance report 2023–2024* (HR&CE Report No. 18/2024)—Government of Tamil Nadu.

Intergovernmental Panel on Climate Change (IPCC). (2023). *Climate change 2023: Synthesis report. Contribution of Working Groups I, II and III to the Sixth Assessment Report of the Intergovernmental Panel on Climate Change*. IPCC. https://www.ipcc.ch/report/ar6/syr/

Kanojia, K., & Saha, N. C. (2024). Sacred groves: Harmonizing nature, goHarmonizinge. In Das, D. (Ed.), *Sacred groves and biodiversity conservation* (pp. 168–183). Bharti Publications.

Kom, Z., Nicolau, M. D., & Nenwiini, C. S. (2024). The use of Indigenous Knowledge Systems practices to enhance food security in Vhembe District, South Africa. *Environment, Development and Sustainability, 13*(3), 599–612. DOI: 10.1007/s40003-024-00716-8

Lautensach, A., Litz, D., Younghusband, C., Banack, H., Thielmann, G., & Crandall, J. (2025). The what, why, and how of climate change education: Strengthening teacher education for resilience. *Sustainability (Basel), 17*(19), 1–25. DOI: 10.3390/su17198816

Madzivhandila, T. S. (2024). Integrating Indigenous Knowledge Systems (IKS) to climate change adaptation in South Africa: Lessons from the past. *South African Journal of Higher Education, 38*(5), 159–168. DOI: 10.20853/38-5-6417

Mbelebele, Z., Mdoda, L., Ntlanga, S. S., Nontu, Y., & Gidi, L. S. (2024). Harmonizing with environmental preservation: Sustainable strategies for the conservation of indigenous medicinal plants (IMPs) and their implications for economic well-being. *Sustainability, 16*, 5841. Advance online publication. DOI: 10.3390/su16145841

Ministry of Education. (n.d.). *National Initiative for School Heads' and Teachers' Holistic Advancement (NISHTHA)*. Government of India.

Ministry of Education. (n.d.). *Samagra Shiksha*. Government of India.

Ministry of Environment. Forest and Climate Change. (2017). *Wetlands (Conservation and Management) Rules, 2017* (S.O. 425(E)). Government of India.

Ministry of Human Resource Development. (2020). National [Government of India.]. *Educational Policy*, •••, 2020.

Ministry of Rural Development. (2022). *Mission Amrit Sarovar guidelines for restoration of water bodies* (Notification No. 32/2022)—Government of India.

Ministry of Water Resources. (2012). *National Water Policy 2012* (Policy Document No. NWP-2012). Government of India.

National Council of Educational Research and Training. (2023). *National Curriculum Framework for School Education 2023*. NCERT.

Onuma, S., & Innocent, M. H. (2025). Ubuntu and sustainable futures: Reimagining development in contemporary Africa. *Asian Research Journal of Arts & Social Sciences, 23*(11), 102–108. DOI: 10.9734/arjass/2025/v23i11831

Public Works Department, Government of Tamil Nadu. (2017). *Kudimaramathu scheme operational guidelines* (PWD Circular No. 24/2017). Government of Tamil Nadu.

Sahani, A. K., Gupta, G., Anand, S., Sharma, V. R., Singh, R., Kamil, A., Kumar, H., Gagan, A., Mayala, V. K., Mishra, A. P., Jaiswal, S., & Anand, J. (2025). Indigenous knowledge and water conservation practices in South Africa: A systematic literature review. *Journal of Environment and Earth Science, 7*(2), 248–261. DOI: 10.30564/jees.v7i2.7988

Santos, B. de S. (2014). *Epistemologies of the South: Justice against epistemicide*. Routledge., DOI: 10.4324/9781315634876

Singh, M., & Ahmad, A. (2025). Climate change and sustainability: A critical analysis of the Indian curriculum. *Discover Education*, *4*(1), 2–16. DOI: 10.1007/s44217-025-00947-z

Supreme Court of India. (2003). *Order on mandatory environmental education in schools*. Government of India.

Tirkey, J., Toppo, S., Lal, J., Mexudhan, & Singh, L. (2022). Sacred groves: A place of biodiversity conservation centre—A case study from Jaspur District of Chhattisgarh. *RASSA Journal of Science for Society, 4*, 105–108. DOI: 10.5958/2583-3715.2022.00016.8

UNESCO. (2020). *Education for Sustainable Development: A roadmap.*

United Nations Framework Convention on Climate Change (UNFCCC). (2013). *Decision 19/CP.18: Gender and climate change*. https://unfccc.int/resource/docs/2012/cop18/eng/08a03.pdf

About the Contributors

Dumsani Mncube is an associate professor at Unisa's College of Education, Department of Curriculum and Instructional Studies. He holds a PhD in Curriculum Studies from the University of Zululand. In 2003, he was awarded a Ford Foundation Fellowship to study a Master of Science in Environmental Management at the University of the Witwatersrand. His career in Higher Education spans over 19 years. He specialises in Curriculum Studies, Environmental Education, Climate Change Education, Self-Directed Learning and Higher Education. Prof Mncube edited the IGI Global book chapters: Advancing Self-Directed Learning in Higher Education (2022-2023). He also published a book on Climate Change entitled Education and Climate Change: Discovering Strategies for Individual and Collective Action, which was recently published in May 2025. Prof Mncube was a project coordinator for the Keep it Cool Climate (KIC) Change Education project from 2020 to 2022. To date, Prof Mncube has published over 50 research articles in SAPSE-accredited journals and 16 book chapters and is currently an editor of three books (AOSIS, BRILL, and Routledge). He has attended several international and national conferences, allowing him to grow as an academic and researcher. He has supervised over 20 master's students to completion and 4 PhD students to completion. Three of these master's students and two PhD students graduated in 2025.

Anriët van Deventer is a PhD graduate in Humanities Education from the University of Pretoria, South Africa. She currently serves as a lecturer in the Department of Language Education, Arts and Culture, College of Education, School of Teacher Education, at the University of South Africa (UNISA). Her academic work focuses on Creative Arts education, Design Thinking, Open Distance e-Learning (ODeL), curriculum transformation, and the integration of emerging technologies, including artificial intelligence, into teaching and learning. Dr Van Deventer teaches across undergraduate and postgraduate programmes, specialising in Arts Methodology in Creative Arts for the Intermediate, Senior, and Further education and training phase, including Dramatic Arts, Design, and Visual Arts. She is actively involved in module development, digitally mediated pedagogy, and student support within online learning environments. In addition to her teaching responsibilities, she supervises postgraduate research and contributes to academic scholarship through conference presentations and research outputs. Her scholarly interests centre on advancing inclusive, transformative Arts education that integrates Africanism, creativity, critical thinking, climate change education and artificial intelligence. Dr Van Deventer is committed to fostering reflective practitioners and contributing to the development of innovative, contextually responsive curricula in higher education.

***

Joyce Senya Ama Anku is an Educational Linguist and specialist in English Language Teaching at the Department of Teacher Education, University of Ghana. Her research focuses on language (in) education, language teacher identity, gender and curriculum, materials development, school system resilience and competence-based education within multilingual African contexts. She examines how language policy, sociolinguistic dynamics, and gendered power structures shape teaching, learning, and professional practice in schools and universities. Dr. Anku's work equally contributes to ongoing conversations on decolonising English language pedagogy, strengthening literacy development, and aligning academic training with contemporary educational reforms. Her work foregrounds questions of legitimacy, agency, and equity, positioning language education as a critical site for negotiating identity, power, and transformation in African schooling systems.

Tamara Chansa Kabali is a Zambian academic and Associate Professor of Developmental Psychology at the University of Zambia (UNZA), affiliated with the Department of Psychology. Her work focuses on child development, developmental psychology, and inclusive education, with particular emphasis on neurodevelopmental disorders, mental health, and early childhood learning. She holds a PhD in Educational Psychology, where her research examined the acquisition of early reading skills and the influence of the home environment, highlighting the importance of parental support and socio-cultural context in children's literacy development. She has developed a strong research profile with numerous publications, contributing to areas such as autism spectrum disorders, inclusive education, and child development in low-resource settings. Her work explores issues such as awareness of autism, access to services for individuals with disabilities, and caregiver experiences of children with neurodevelopmental conditions in Zambia. In addition to her academic role, she is actively involved in professional and community initiatives, including providing psychological and therapy services. Her work integrates research, teaching, and applied practice, positioning her as a key contributor to advancing developmental psychology, inclusive education, and child mental health in Zambia.

Ntandokamenzi Penelope Dlamini is an academic in the Department of Curriculum and Instructional Studies, College of Education at the University of South Africa. Her research interests are in educational technology, geography education, pedagogy improvement, gender education and mentoring. Holding qualifications in both science and education, she has focused on finding innovative teaching methods to enhance the understanding of geography education among students. Her advocacy for gender equality and technology in education is reflected in her published research in various journals over the past years. Her work is on pedagogy improvement, technology in education and gender education. She is dedicated to improving pedagogy and nurturing the next generation of teachers.

G. Loganayaki is an Assistant Professor in the Department of English at SRC, SASTRA Deemed University. With over 18 years of teaching experience, she has published around 25 research papers and contributed several book chapters. Her areas of specialization include Ecocriticism, Spatial Theory, and Hauntology. She also serves as a research supervisor for Ph.D. scholars and has led projects integrating literature, indigenous knowledge systems, and environmental sustainability.

J. Nandha Gopal received Bachelor Degree in Electronics and Communication Engineering from Odaiyappa College of Engineering and Technology and Master Degree in Power Electronics and Drives from Anna University and Ph D from Anna University, Chennai. Presently he is working as Associate Professor in the Department of Electrical and Electronics Engineering at Velammal Institute of Technology, Chennai. His research interest includes Power Electronics, Renewable Energy Systems and Smart Grid. He has published 12 international journals, 15 national and international conference papers and 1 patent. He received best faculty award from Nature Science Foundation. He is a life member of IE (I).

K. Sunanda is a Research Scholar and Teaching Assistant in the Department of English at SRC, SASTRA Deemed University, Tamil Nadu, India. Her research interests encompass cultural studies, hauntology, trauma theory, and the intersections of literature, identity, and social justice. She has presented and published several research papers in national and international peer-reviewed journals and conferences. Her academic work engages with contemporary literary theories, film studies, and pedagogical innovation in English language teaching. As an emerging scholar, she aims to bridge the gap between critical theory and classroom practice by exploring how literary theories and affect studies can reshape perspectives on narrative, memory, and identity. Her ongoing doctoral research focuses on spectrality, Hauntology, and social justice in Stephen graham Jones' trilogy particularly through the lens of Sadeq Rahimi and Avery Gordon's theoretical frameworks.

Marvin Kapenda is a Zambian development practitioner and Behavioural Science Specialist working with the Food and Agriculture Organization in Zambia. His work is closely associated with climate change and sustainability initiatives, particularly under programmes such as the Facility for Action for Climate Empowerment to Achieve Nationally Determined Contributions (FACE-NDC). He focuses on applying behavioural science approaches to influence attitudes and practices among communities, especially in promoting climate-resilient agriculture and environmental conservation. His role involves designing and implementing interventions that encourage sustainable decision-making among farmers and local communities, using methods such as demonstration plots and peer learning to shift mindsets toward more effective agricultural practices. He contributes to capacity-building initiatives and stakeholder engagement, working with communities, traditional leaders, and development partners to strengthen climate awareness and action. With a background linked to the University of Zambia, his professional profile reflects experience in education, training, and development practice, and a strong focus on integrating behavioural insights into climate education and community-based sustainability efforts.

K. Saravanan, is working as an Associate Professor, Department of Computer Science & Engineering at College of Engineering, Guindy, Anna University, Chennai-25, Tamilnadu. He has done M.E Software Engineering and Ph.D in Computer Science engineering. His research interest includes Cloud Computing, Software Engineering, Internet of Things, Smart cities. He has published papers in 14 international conferences and 27 international journals. He has also written 14 book chapters and edited seven books with international publishers. He has done many consultancy works for Municipal Corporation and Smart City schemes. He is an active researcher and academician. Also, he is reviewer for many reputed journals in elsevier, IEEE etc. He is a member of ISTE, IEI, ISCA, ACM, etc.

Phyllis Bernice Kwarteng Donkor holds a Doctor of Philosophy in Occupational and Technical Education from the Old Dominion University, Norfolk, VA, USA. Dr. Kwarteng Donkor is an ardent advocate for girl child education, especially among traditional rulers and religious leaders, and has organized and co-organized various training programs encouraging girls into STEM education and leadership positions in school and their communities. For several years she was a lecturer at the Languages and General Studies Department of Energy and Natural Resources. She is currently a lecturer at the Teacher Education Department of the University of Ghana, where she also provides counselling and guidance to students, mentoring and finding funds for those in need. Her research interests are in female participation, persistence and success in STEM; Gender and Leadership; Flourishing in Academia; Game-based & Project based pedagogies; and developing requisite 21st century skills among learners; Climate pedagogy; Climate Smart Agriculture; and integrating sustainability education across the curriculum. She is also the minister at the Bethel Methodist Church, in the Taifa Circuit of the Northern Accra Diocese of the Methodist Church Ghana.

Garikai Membele is a Zambian academic and geospatial scientist affiliated with the University of Zambia, where he serves as a Lecturer in the Department of Geography within the School of Natural Sciences. His academic work focuses on Geographic Information Systems (GIS), remote sensing, spatial analysis, and land-use planning, with applications in areas such as urban planning, climate change, and environmental management. He holds a PhD and has developed a strong research profile with over a dozen publications and more than 200 citations, contributing to topics such as indigenous knowledge in climate change adaptation, flood vulnerability mapping, and geospatial analysis for development planning. His work also involves international collaborations, including contributions to projects on population mapping and health planning using geospatial data. Overall, his profile positions him as a specialist in geospatial science and environmental analysis, contributing to both academic research and applied development initiatives within Zambia and beyond.

Inonge K. Milupi (also published as Delphine Inonge Milupi) is a Zambian academic and environmental education specialist affiliated with the University of Zambia (UNZA), where she serves as a Lecturer and Head of Department in the Department of Language and Social Sciences Education within the School of Education. She holds a Doctor of Philosophy (PhD) from the University of Pretoria, a Master of Education in Environmental Education, and a Bachelor of Science with Education from UNZA. Her research is firmly grounded in environmental education, climate change, and environment–society interactions, with a strong emphasis on community-based natural resource management and the role of indigenous knowledge in sustainability. She has an extensive research profile with over 40 publications and more than 250 citations, reflecting contributions to areas such as climate change adaptation, wildlife resource management, environmental conservation, and school-based education. Dr. Milupi has contributed to influential work on environmental education and community-based natural resource management in Zambia, highlighting how education can enhance community participation and the sustainable use of natural resources. Her scholarship integrates empirical research with educational practice, positioning her as a key contributor to advancing environmental and climate change education, as well as sustainability-focused research and policy engagement in Zambia.

Divyanshu Mishra is pursuing a Doctoral degree at the School of Education, Central University of Rajasthan. His research area is Climate Change Education. He completed a Master in Botany in 2018 and Masters in Education [M.Ed.] in 2022 from the Dayalbagh Educational Institute, Agra, India. He qualified for the UGC NET JRF exam. He worked on a research project," Problems Among Adolescence during the post-COVID period." at the Regional Institute of Education, NCERT, Bhopal, India. He also worked as an assistant professor in education at the Regional Institute of Education, NCERT, Ajmer, for a session. He presented research papers at different national and international conferences.

Pauline Namakau Monde is a Zambian academic affiliated with the University of Zambia (UNZA), within the Department of Language and Social Sciences Education in the School of Education. Her work is grounded in environmental education, science education, and teaching and learning, with a strong emphasis on the integration of environmental and climate-related issues into school systems. She holds a Master of Education in Environmental Education from UNZA, where her research examined barriers to the effective implementation of environmental education in Zambian high schools. Her research profile includes over 20 publications and nearly 100 citations, reflecting contributions to areas such as school safety, environmental management, and climate change education. She has co-authored studies on climate change impacts and adaptation among local communities in Zambia, highlighting the importance of environmental awareness and indigenous knowledge in addressing climate challenges. Additionally, her work explores land cover change and ecosystem management in urban contexts, linking geospatial analysis with sustainability education.

Chipo Moonga is a Zambian environmental education specialist and emerging scholar affiliated with the University of Zambia (UNZA), within the School of Education. She holds a Master of Education in Environmental Education, where her research focused on sustainability, including a study examining green building materials and technologies in Lusaka District and their role in promoting environmental sustainability and a green economy . Her academic interests are grounded in environmental education, climate change, and sustainability, with a strong emphasis on how education can contribute to sustainable development.

She is currently a PhD candidate in Climate Change and Sustainability Education, further advancing her research on integrating sustainability and climate change into education systems. Her work reflects a commitment to strengthening climate change pedagogy, promoting sustainable practices, and contributing to knowledge production that supports Zambia's transition toward environmental sustainability. Overall, her profile positions her as an emerging academic working at the intersection of education, climate change, and sustainable development.

Kofi Nkonkonya Mpuangnan is a Postdoctoral Research Fellow and Lecturer in the Department of Educational Foundations at the University of Zululand, South Africa. He holds a PhD in Education from The Maharaja Sayajirao University of Baroda in India, a Master of Education in Educational Administration and Management, and a Bachelor of Education from the University of Education Winneba in Ghana. His research interests focus on curriculum studies, educational management and leadership, climate change education and sustainable development and Teacher Professional Development. Dr Mpuangnan serves as a reviewer for several reputable publishers, including Taylor & Francis, Springer, and SAGE. He has published extensively in accredited peer-reviewed journals and has presented numerous papers at national and international conferences. In addition, he has successfully supervised several master's and doctoral candidates, contributing significantly to research capacity building in education.

Kabwe Harnadih Mubanga is a Zambian academic and Lecturer at the University of Zambia (UNZA), affiliated with the Department of Geography and Environmental Studies. He holds a Doctor of Philosophy (PhD) in Environment and Society from the University of Pretoria in South Africa. His research is grounded in environmental science, climate change, and sustainable agriculture, with a strong focus on the interactions between climate variability and rural livelihoods. His scholarly work includes studies on drought, rainfall variability, land use change, and agricultural adaptation strategies, particularly among smallholder farmers in Zambia. Dr. Mubanga has an established research profile, with over 30 publications and more than 140 citations, reflecting contributions to areas such as food security, conservation agriculture, and climate resilience. His earlier academic work includes research on climate variability and adaptation in Zambia's agro-ecological regions, highlighting coping strategies used by farmers to sustain agricultural productivity under changing climatic conditions.Beyond academia, he is also actively involved in climate policy and practice in Zambia, serving as a climate change specialist on national technical committees and contributing to climate negotiations and adaptation planning processes.

Kaiko L. Mubita (Dr. Kaiko Mubita) is a Zambian academic affiliated with the University of Zambia (UNZA), within the Department of Language and Social Sciences Education in the School of Education. He holds a Doctor of Philosophy (PhD) and is actively involved in teaching, research, and postgraduate supervision. His work is grounded in education, with a strong focus on Geography education, environmental education, and school-based studies, including areas such as school safety and health management, teaching practice, curriculum implementation, and the integration of information and communication technologies in teaching and learning. He has an established scholarly profile with over 70 publications and more than 400 citations, reflecting sustained contributions to education research in Zambia and beyond. His research often addresses practical challenges in schools and proposes context-specific approaches to improving teaching, learning, and school management, positioning him as a significant contributor to education and environmental education research in Zambia.

Manoah Muchanga is a Zambian academic and researcher specializing in Water Science, Climate Change Education, and Research Methodology. He serves as a Senior Lecturer/Researcher/Senator at the University of Zambia (UNZA). His research interests include water science, climate change education, research methodology, geospatial analysis. Dr. Muchanga is a pioneer of the Master's and PhD programmes in Climate Change and Sustainability Education with a behavioural science focus at the University of Zambia, contributing significantly to the development of interdisciplinary climate education in the region. He is the author of a textbook in research methodology and has published extensively in peer-reviewed outlets on topics such as water science, environmental education, research methodology, climate change education, and sustainable ecosystem management in Zambia. Beyond his teaching and research responsibilities, Dr. Muchanga has contributed to several environmental and climate-related initiatives, including projects focused on river restoration and community adaptation to environmental change. His work integrates geographical science, environmental education, and applied research, with the aim of supporting sustainable development and climate resilience in Zambia and across the Southern African region. He currently serves as Principal Investigator for the Facility for Climate Empowerment to Achieve Nationally Determined Contributions (FACENDC) Project, Work Package 4, and the Sub-Saharan Africa Teacher Leadership for Education for Sustainable Development (SSATL-ESD) Project, Work Package 2. Dr. Muchanga has also extensively supervised and examined Master's and PhD research, contributing to the development of the next generation of scholars in climate change and sustainability education.

Tiwine Zimba Muchipa is a Zambian communications and development professional associated with the Food and Agriculture Organization (FAO), where she works as a Communications and Digital Content Specialist. Her work focuses on strategic communication, digital storytelling, and public engagement around climate change, agriculture, and sustainable development initiatives. She is actively involved in initiatives linked to climate action and national development processes, including programmes that support climate education, awareness, and stakeholder engagement. Her professional interests include youth engagement, behaviour change communication, and the use of media to promote sustainability and community participation. With a background in peace, leadership, and conflict resolution, her work integrates communication, governance, and social transformation. Overall, her profile reflects a practitioner operating at the intersection of communication, climate action, and sustainable development, contributing to efforts that translate technical knowledge into accessible and impactful messages for diverse audiences in Zambia.

Shirley Mukhari is an experienced teacher and academic with a strong background in technology integration and language education and curriculum innovation. She holds a Doctorate in Education with a focus on ICTs in teaching and learning. Shirley has served as a Senior Lecturer at the University of South Africa and has contributed to research in schools and higher education, particularly on the use of modern technologies in education, blended learning, and AI integration.

Ecloss Munsaka is a Zambian academic and Associate Professor of Psychology at the University of Zambia (UNZA), affiliated with the Department of Educational Psychology, Sociology, and Special Education. He holds a PhD in Psychology and has extensive experience in teaching educational psychology at university level. His work is grounded in educational and developmental psychology, with a strong focus on child and adolescent development, school psychology, and learning processes. His research spans areas such as school dropout, adolescent development, comprehensive sexuality education, and psychosocial support in education systems, examining the social and behavioural factors that influence learners' outcomes. In addition to his research, he is an established author of psychology textbooks used in universities and teacher training institutions, and his work integrates teaching, research, and applied practice, positioning him as a significant contributor to educational psychology and learner development in Zambia.

Liberty Mweemba is a Zambian academic and Dean of the School of Education at the University of Zambia, based in the Department of Language and Social Sciences Education. He holds a PhD (EEng) obtained in China, an MSc in Geography, and a Bachelor of Arts with Education from the University of Zambia, along with additional training in Education for Sustainable Development. His research is interdisciplinary, focusing on environmental education, climate change and sustainability, geography, rural development, natural resource management, and conservation. With over 30 publications and more than 200 citations, his work has addressed issues such as climate variability and pastoral livelihoods in Zambia, environmental degradation, civic education pedagogy, and human-wildlife conflict. His scholarship combines empirical, Zambia-focused research with educational theory and policy relevance. In addition to teaching undergraduate and postgraduate students, he has supervised graduate research and previously served as School Experience Coordinator. As Dean, he contributes to teacher education, curriculum development, and research leadership, positioning him as a key figure in advancing climate change education and sustainable development within Zambia's education system.

Matilda Kanyampa Nakazwe is a Zambian academic and teacher educator affiliated with the University of Zambia, specifically within the Department of Language and Social Sciences Education in the School of Education. Her scholarly work focuses on Geography Education, teacher education, and environmental education, reflecting an interest in how teaching and learning processes intersect with environmental and sustainability issues. Her research contributions include studies on school safety, transport risks affecting pupils, and broader educational and environmental challenges in Zambia. She has co-authored work examining the safety of transport modes used by pupils, highlighting risks such as inadequate infrastructure and the need for improved safety education and policy interventions.

Progress H. Nyanga is a Zambian academic and Senior Lecturer at the University of Zambia (UNZA), affiliated with the Department of Geography. He holds a PhD in Environmental and Development Studies and has established expertise in rural development, agriculture, and sustainability. His research focuses on food security, conservation agriculture, rural livelihoods, and sustainable development, with particular attention to smallholder farming systems in Zambia. His work examines issues such as agricultural adaptation, climate variability, and the socio-economic factors influencing farming practices. He adopts interdisciplinary approaches, including participatory methods and emerging analytical techniques, to better understand and address livelihood challenges among farming communities. Overall, his profile positions him as a key contributor to research on climate change, agriculture, and sustainable rural development in Zambia.

Jo-Anna Reed-Johnson is a Associate Professor in the Institute of Education at the University of Reading, UK. She obtained her PhD from Manchester

Metropolitan University, UK. Her research interests and expertise include STEM education, Education for Sustainable Development, Global Education, Climate and Sustainability Education. She is a Senior Fellow of the Higher Education Academy (SFHEA) and a Member/Chartered Physicist of the Institute of Physics, UK

S. Sanmathi completed her undergraduate studies at AVC College, Mannampandal, Mayiladuthurai, and is currently pursuing her postgraduate degree at SRC, SASTRA Deemed University, Kumbakonam, with a specialization in Microbiology. Her academic project focuses on decolonising climate change education through Indigenous Knowledge Systems.

Debbie Anne Sanders, holds a Doctorate in Education. Debbie serves as a Senior Lecturer at the University of South Africa. Her research focus interests are in special needs and language education. Her research informs inclusive education, curriculum transformation, engaged scholarship and the integration of ICTs to address unique challenges, foster innovation, and promote equity.

John Sedofia is a Senior Lecturer at the Department of Teacher Education, University of Ghana, Accra, Ghana. John's research interests include academic counselling, teacher education and development, and pedagogy.

Kirsty Shakespeare is an Engagement Project Manager at the University of Reading specialising in Climate and Environmental Sustainability, with a particular focus on advancing climate and nature education for local schools and the community. She is also a PhD student at the university, studying part time researching climate education policy and practice in England. Her role involves the design and delivery of programmes that provide opportunities for nature connection, strengthens climate literacy, builds confidence in sustainability action, and empowers people at all levels to understand and respond to environmental and climate challenges, by transforming complex climate science and policy into accessible, engaging learning experiences. Driven by a commitment to education and positive change, she works to ensure that climate understanding becomes a shared foundation for decision making and collective action.

Anjali Sharma (Dr) is a Professor at the School of Education, Central University of Rajasthan. She has twenty-four years of Teaching /research experience. She has published around fifty research papers in Web of Science and Scopus journals and fifteen chapters in edited National /International books. Seven Research Scholars have been awarded their degree under her supervision. She has presented around 50 papers at national and international conferences/seminars. She has completed three sponsored research projects in diverse areas. She designed an educational game and presented it at Toycathon 2021. She is a UGC nominated Master Trainer for the Community-based Participatory Research Program under Unnat Bharat Abhiyan2.0, Government of India. She is a UGC-nominated committee member constituted to establish guidelines for the Department of Education to implement the National Education Policy 2020. Dr. Anjali Sharma has organized many national seminars/open-house discussions and conducted capacity-building and personality development programs, awareness programs, and workshops for homemakers in collaboration with various government autonomous agencies in India. She received the "Best Paper Award" at an international conference in Jakarta, Indonesia, and the 26th Edition of the International Young Scientist Awards, the "Best Researcher Award" for Contribution and Honourable Achievement in Innovative Research. Certificate No: 13312, Date: 24th Dec 2024, youngscientistawards.com. She is the Executive nominee of Global Inclusive Education Network. She was an associate editor for the Teacher Support Journal published by NCTE, New Delhi, India. She was also appointed as a Reviewer for Sage Publication. She is the author of the book "Creating a New Horizon in Pedagogy through Growth Mindset," published by Cambridge Scholar Publishing, UK. She has a patent under the Literary Category: Growth Mindset Pedagogical Framework for Elevate Learning Paradigm, IPR, Copy Right Office, Government of India.

Amos Sikayile is a Zambian development and climate change professional primarily affiliated with UNESCO, where he contributes to initiatives related to education, sustainability, and knowledge dissemination. His work aligns with global frameworks such as Education for Sustainable Development and Action for Climate Empowerment, with a focus on strengthening climate education, public awareness, and stakeholder engagement. He is also associated with the Food and Agriculture Organization (FAO), where he has been involved in programmes supporting climate action, agriculture, and community-based sustainability, including initiatives linked to Nationally Determined Contributions, and his role across these organizations involves capacity-building, project coordination, and collaboration with government institutions, civil society, and community-based organisations to promote climate-resilient practices.

Steria Monica Simooya is a Zambian academic and Lecturer at the University of Zambia (UNZA), affiliated with the Department of Language and Social Sciences Education in the School of Education. She has an established research profile in environmental education and communication. Her scholarly work focuses on environmental education, climate change, urbanisation, and land use change, with particular attention to how these processes affect livelihoods and learning. She has contributed to research on environmental learning, community participation in natural resource management, and climate change impacts and adaptation in Zambia.

Everisto Simweene is a Zambian education and environmental studies scholar affiliated with the University of Zambia (UNZA), where he is currently pursuing doctoral studies with scholarship support from the FACENDC Project. His academic work engages with environmental issues, including research on sedimentation and sustainable environmental management, as reflected in contributions to the UNZA repository. His scholarly interests are situated at the intersection of environmental education and sustainability, with a focus on addressing local environmental challenges through research and informed practice. In addition to his academic pursuits, he is connected to the Ministry of Education, where he contributes to teaching within the social sciences, thereby linking research with classroom practice.

Georina Westraadt holds a D.Ed. in Visual Art Education and an M.Ed. (cum laude) from the Cape Peninsula University of Technology (CPUT). She has extensive experience in teacher education across pre- and postgraduate B.Ed. programmes and previously taught art at primary, secondary, and tertiary levels. After retiring as a senior lecturer at CPUT, she served as Head of Department for Intermediate Phase Education at the Two Oceans Graduate Institute. She is the author of Die onderrig van Visuele Kuns in die Laerskool (2015), later translated as Visual Arts Education in the Primary School, and has published widely in accredited journals and conference proceedings. Her expertise includes art education, visual literacy, curriculum design, assessment, mentoring, and staff development.

Hong Yang is a Professor in the Department of Geography and Environmental Science at the University of Reading, UK. He obtained his PhD from University College London (UCL), UK. His research interests and expertise include STEM education, Education for Sustainable Development, transnational education, environmental pollution, and climate change. He is a Senior Fellow of the Higher Education Academy (SFHEA) and a Fellow of the Royal Geographical Society (FRGS), UK.

Vincent Ziba is a Zambian development and environmental professional associated with the Food and Agriculture Organization (FAO), where he works as a Project Coordinator under initiatives such as the Facility for Action for Climate Empowerment to Achieve Nationally Determined Contributions (FACE-NDC). His work focuses on climate change education, community engagement, and capacity-building, particularly in supporting NGOs, youth groups, and community-based organisations to integrate behavioural science into climate action and sustainability practices. He has extensive experience in community-based natural resource management and environmental governance, having worked with organisations such as Oxfam, WWF, and regional civil society networks. His professional work emphasizes climate justice, sustainable livelihoods, and the strengthening of local participation in environmental decision-making. He has also been involved in advocacy efforts related to environmental protection, including engagement in issues concerning conservation and responsible resource use in Zambia.

Index

A

B

C

D

E

G

I

L

M

O

P

R

S

T

U

V

www.ingramcontent.com/pod-product-compliance
Lightning Source LLC
LaVergne TN
LVHW062339300926
849161LV00007B/1756
* 9 7 9 8 3 3 7 3 6 9 7 5 4 *